Advances in Planetary Science — Vol. 7

NEUTRAL-ATOM ASTRONOMY
PLASMA DIAGNOSTICS FROM THE AURORA TO THE INTERSTELLAR MEDIUM

Advances in Planetary Science

Print ISSN: 2529-8054
Online ISSN: 2529-8062

Series Editor: Wing-Huen Ip *(National Central University, Taiwan)*

The series on Advances in Planetary Science aims to provide readers with overviews on many exciting developments in planetary research and related studies of exoplanets and their habitability. Besides a running account of the most up-to-date research results, coverage will also be given to descriptions of milestones in space exploration in the recent past by leading experts in the field.

Published

Advances in Planetary Science – Vol. 7

NEUTRAL-ATOM ASTRONOMY
PLASMA DIAGNOSTICS FROM THE AURORA TO THE INTERSTELLAR MEDIUM

Ke Chiang Hsieh
University of Arizona, USA

Eberhard Möbius
University of New Hampshire, USA

World Scientific

EW JERSEY · LONDON · SINGAPORE · BEIJING · SHANGHAI · HONG KONG · TAIPEI · CHENNAI · TOKYO

Published by

World Scientific Publishing Co. Pte. Ltd.

5 Toh Tuck Link, Singapore 596224

USA office: 27 Warren Street, Suite 401-402, Hackensack, NJ 07601

UK office: 57 Shelton Street, Covent Garden, London WC2H 9HE

Library of Congress Control Number: 2022932734

British Library Cataloguing-in-Publication Data
A catalogue record for this book is available from the British Library.

Advances in Planetary Science — Vol. 7
NEUTRAL-ATOM ASTRONOMY
Plasma Diagnostics from the Aurora to the Interstellar Medium

ISBN 978-981-3279-19-3 (hardcover)
ISBN 978-981-3279-20-9 (ebook for institutions)
ISBN 978-981-3279-21-6 (ebook for individuals)

For any available supplementary material, please visit
https://www.worldscientific.com/worldscibooks/10.1142/11241#t=suppl

Desk Editor: Joseph Ang

Typeset by Stallion Press
Email: enquiries@stallionpress.com

To the community of past, present, and future investigators in space-plasma research beyond the reach of spacecraft and those who wish to venture even farther.

Preface

We want this book to be *a primer for those who wish to learn about the diagnostics of space plasmas beyond the reach of spacecraft by detecting and analyzing energetic neutral atoms (ENAs) emanating from afar. We hope it will spawn novel ideas and approaches that could expand the investigations into new domains.*[1]

Like ENAs, both of us crossed boundaries, albeit artificial ones, to discover, learn, and teach. Eventually, our journeys converged and led us to writing this book to convey a message to a receptive audience.

K. C. Hsieh learned from C. Y. Fan and J. A. Simpson at the University of Chicago to detect ^{2}H and ^{3}He nuclei of galactic and solar origins in interplanetary space. At the University of Arizona, Fan inspired him to develop neutral-atom analyzers aiming at interplanetary neutral atoms. G. Gloeckler and D. Hovestadt, who won his respect in Simpson's group, led him to the time-of-flight (TOF) technique and to partake in *SOHO* HSTOF, which turned into the first heliospheric ENA sensor. D. M. Hunten, E. Keppler, H. Lauche, H. Rosenbauer, S. Grzedzielski, and W. I. Axford gave him advice and support in his early ENA research. Working with C. C. Curtis, D. S. Evans, J. L'Heureux, W.-H. Ip, Arne Richter, A. L. Broadfoot, B. R. Sandel, J. Giacalone, J. R. Jokipii, J. Kóta, K.-L. Shih,

---

[1]For a refresher on space-plasma physics, we refer to George J. Parks' *Physics of Space Plasma: An Introduction*, 2nd Ed. (2003, Westview Press, Boulder, USA) and a more advanced text by Thomas E. Cravens' *Physics of Solar System Plasmas* (1997, Cambridge University Press, UK).

M. A. Gruntman, M. Hilchenbach, A. Czechowski, D. Mitchell, S. Barabash, S. Orsini, and A. Milillo was most gratifying. The technical support from UA's Physics Department greatly eased his writing of this book after retirement.

E. Möbius started with laboratory plasma physics, studying H. Alfvén's Critical Ionization Velocity together with A. Piel under the tutelage of H. Schlüter and G. Himmel at the Ruhr Universität Bochum. He then took a turn into space plasma physics at the Max-Planck-Institut für Extraterrestrische Physik. In D. Hovestadt's group, he built one of the first space-borne TOF spectrographs, the SULEICA sensor for the *AMPTE* mission, led by G. Haerendel and S. M. Krimigis. This sensor enabled the detection of interstellar He^+ pickup ions, whose interpretation succeeded in collaboration with G. Gloeckler, D. Hovestadt, F. M. Ipavich, B. Klecker, M. A. Lee, and M. Scholer. This pivotal find moved EM's interest to the heliosphere-interstellar medium interaction. Now at the University of New Hampshire, he collaborated with P. Bochsler, M. Bzowski, S. Fuselier, A. B. Galvin, J. Geiss, G. Gloeckler, P. A. Isenberg, L. M. Kistler, B. Klecker, H. Kucharek, M. A. Lee, D. McComas, D. Rucinski, N. A. Schwadron, M. Wieser, and P. Wurz, among many others. After being invited by J. Geiss to co-lead with him the first Workshop at the International Space Science Institute (ISSI) on the heliosphere and the interstellar medium he led two ISSI Science Teams on the interstellar gas in the heliosphere. This effort has now come full circle with the ENA and interstellar flow sensors on IBEX and their implementation on IMAP.

Together, we express our gratitude to the dedicated engineering teams at our own and collaborating institutions for their tireless efforts in developing and finalizing the novel instrumentation that enabled the pioneering observations.

After our paths met, we deepened our appreciation of the significance of ENA diagnostics of space plasmas beyond the reach of spacecraft. We gladly accepted W.-H. Ip's invitation to share our enthusiasm with future investigators and students of space plasmas through this book. We hope the readers will benefit, as we have, from the forerunners and collaborators in this field.

We begin with defining "ENA", where ENA-diagnostics are valuable, and end with forecasting where ENA-diagnostics might lead. Between these two bookends, we start with the discovery of ENAs in space and their early detection in near-Earth space, followed by the fundamental relations between the observed ENAs and the properties of the remote plasma whence they come. Then, we introduce the requirements for and the implementation of ENA instruments, followed by sample results from the Earth's magnetosphere, other planet atmospheres and magnetospheres, the outer heliosphere, and near interstellar space. With lower energy thresholds for detection, *ambient neutral atoms* (ANAs) crucial for the birth of ENAs, enter the purview of detection and analysis under special conditions.

To write this book, we choose *sufficiency over completeness* as our guide. We are incredibly grateful to the UNH graduate Plasma Physics students in spring 2016 for providing a constructive review of the emerging Chapter 3 on the physics of ENAs from the perspective of future users of the book, as a bonus homework assignment. We thank M.A. Lee, N.A. Schwadron, and R. Winslow for their feedback on the critical aspects of ENA analysis and the flow of the text. E. Möbius appreciates greatly the support and inspiring atmosphere of ISSI during extended visits in 2018 and 2019, while preparing the draft of four book chapters.

The history of heliospheric ENA research would be wanting without noting a little-known effort that benefitted both of us and led to significant international collaborations as we write. From 1984 to 1988, annual workshops on heliospheric-ENA physics were held, thrice in Poland, once in Günzburg and in Moscow. They facilitated meetings of researchers from Eastern Europe, the Soviet Union, and non-communist countries with minimal financial and political risks. Under the shadow of the Cold War, S. Grzędzielski of Poland, joined by H. Rosenbauer and H.-J. Fahr of West Germany, V. B. Baranov and M. A. Gruntman of the USSR, led this visionary undertaking. With the support from W. I. Axford, then President of COSPAR, the 6th workshop became a COSPAR Colloquium held in Warsaw, 19–22 September 1989, launching the *COSPAR Colloquia*

Series with *Physics of the Outer Heliosphere* (Pergamon, 1990). The international ENA-research community was born.

During the proofreading stage, a book by M. Gruntman was published.[2] It provides an enlightening account of the pioneering research and development efforts toward ENA imaging and on the physics of the heliospheric boundary layers, especially in Moscow and Warsaw in the 1970s and 1980s. This lively narrative perfectly complements the description of the early ENA history in our Chap. 2. We appreciate this valuable contribution.

We gratefully acknowledge Daniela Möbius' help with the design for the book cover. Finally, we could not have completed this incredibly long writing adventure without the untiring support and patience of our wives, Shigeko Hsieh and Hannelore Möbius. We are indebted and grateful for providing us a sense of place and sanity, especially when we seemed lost in our thoughts and despaired on repeated obstacles.

Ke Chiang Hsieh & Eberhard Möbius

---

[2]Gruntman, M., *My Fifteen Years at IKI, the Space Research Institute: Position-Sensitive Detectors and Energetic Neutral Atoms Behind the Iron Curtain* (Interstellar Trail Press, California, 2022), ISBN 979-8-9856687-0-4.

Contents

List of Tables

List of Figures

List of Acronyms

"Misnomers yield incoherent discourse; incoherent discourse fails the intent."

(From *Zi Lu* in *The Analects of Confucius, circa* 200 BCE)

Acronym	Full Name	First Used in Chapter
1D, 2D, 3D	1-, 2- and 3-dimensional (*adj.*)	Here
ANA	Ambient neutral atom	1
BS	Bow shock	1
CAM	Cluster Analysis Method	7
CEM	Channel electron multiplier	5
CELIAS	Charge ELement Isotope Analysis System	4
CLM	Confidence Limit Method	7
CMA	Crustal magnetic anomaly	6
CME	Coronal mass ejection	1
CS	Conversion Surface	4
CX	Charge exchange	1
DOY	Day Of Year	4

Acronym	Full Name	First Used in Chapter
EM	Electron Multiplier	4
ENA	Energetic neutral atom	1
ESA	Electrostatic analyzer	4
ESA	European Space Agency	4
EUV	Extreme ultraviolet: $10 < \lambda$ (nm) <124	1
FOV	Field-of-view	1
FWHM	Full Width at Half Maximum	4
GDF	Globally distributed flux	1
GIC	Geocentric Inertial Coordinates	4
GSE	Geocentric Solar Ecliptic coordinates	4
HENA	High Energy Neutral Atom imager	4
HESA	Hemispheric electrostatic analyzer	5
HP	Heliopause	1
HSTOF	High Suprathermal energy Time-Of-Flight sensor	4
IBEX	Interstellar Boundary Explorer	1
IHS	Inner heliosheath	1
IMAGE	Imager for Magnetopause-to-Aurora Global Explorer	1
IMAP	Interstellar Mapping and Acceleration Probe	7
IMB	Induced magnetic boundary	6
IMF	Interplanetary magnetic field	1
IP	Interplanetary	1
IS	Interstellar	1
ISM	Interstellar medium	1
ISMF	Interstellar magnetic field	1
ISN	Interstellar neutral	1
ISP	Interstellar Probe	7

Acronym	Full Name	First Used in Chapter
LAE	Low-altitude emission	6
LENA	Low Energy Neutral Atom imager	4
LHS	Left-hand side (of an equation)	Here
LIC	Local interstellar cloud	7
LOS	Line-of-sight	1
LT	Local Time	4
MCP	Microchannel plate	4
MENA	Medium Energy Neutral Atom imager	4
MLT	Magnetic local time	6
MP	Magnetopause	1
MS	Magnetosheath	1
MT	Magnetotail	1
NASA	National Aeronautical & Space Administration (USA)	4
OHS	Outer heliosheath	1
PHA	Pulse-height analysis	4
PSD	Phase space density	3
PUI	Pickup ion	1
RC	Ring current	1
RHS	Right-hand side (of an equation)	Here
SE	Secondary Electron	4
S/C	Spacecraft	1
SI	Secondary ions	4
S/N	Signal-to-Noise ratio	4
SSD	Solid state detector	2
SW	Solar wind	1
TOF	Time-of-Flight	4
TS	Termination shock	1

Acronym	Full Name	First Used in Chapter
TWINS	Twin Wide-angle Imaging Neutral-atom Spectrometers (also for S/C, but in *italics*)	1
UV	Ultraviolet: $10 < \lambda$ (nm) < 400	4

Symbols and Units for Useful Quantities

Quantity	Symbol	Units	Remarks
Coordinates			
Rectangular	(x, y, z)		
Spherical	(r, θ, ϕ)		
Cylindrical	(r, ϕ)		
Ecliptic	(r, λ, β)	$(AU, °, °)$	
Dimensions			
Length	d, l, r	Varies from cm to AU	Units according to the SI System
Area	a	cm^2	
Volume	V	cm^3	
Particle parameters			
ENA or ion mass	m	$m = A \times Da$	Da = unit atomic mass or 1.66×10^{-27}kg for $\pm 1\%$ precision
Atomic Mass number	A	Integer	

Quantity	Symbol	Units	Remarks
Particle parameters			
Electron mass	m_e	kg	9.11×10^{-31}kg
Atomic number	Z	Integer	See below.
Charge	q	$q = Q \times e$	e = electronic
Charge state	Q	Integer	unit charge
			or
			1.60×10^{-19}C
Pitch angle	α	$\circ$	
cos α	μ		
Velocity	$\vec{v}$	Varies from cm/s	$\hat{v}$ is the unit
	Magnitude: $|\vec{v}|$	to AU/yr	vector
	or v		
Kinetic energy	E	keV	
(K.E.)	E/m	$keV/A \cdot Da$	
	$E/nucleon$	keV/nucleon	
	E/q	keV/e	
Total energy	$W = K.E. + P.E.$	keV	
Linear	$\vec{P}$	keV s m^{-1}	$\hat{P}$ is the unit
momentum	Magnitude: $|\vec{P}|$		vector
Angular	$\vec{L}$	keV s	$\hat{L}$ is the unit
momentum	Magnitude: $|\vec{L}|$		vector
Ensemble of particles			
Temperature	T	K	
Number density	n	particle/cm^3	
Phase-space	$f(\vec{r}, \vec{v})$	$(cm^6 s^{-3})^{-1}$	
distribution			
Differential flux	$J(E, \phi, \theta)$	$(cm^2 sr\ s\,keV)^{-1}$	Energy space
	$j(v, \phi, \theta)$	$(cm^2\ sr\,s\,(\frac{cm}{s}))^{-1}$	Velocity space

Quantity	Symbol	Units	Remarks
Ensemble of particles			
Arrival direction	$\hat{l}$		Unit vector along LOS
CX cross-section	$\sigma_{aiCX}(E)$	cm^2	Atom a and ion i
Frequency	ν	s^{-1}	Collisional Interaction
Circular Frequency	ω	rad/s	Collective Interaction
Mean free path (MFP)	λ	cm, km or AU	λ [°] also for ecliptic long. & invariant lat.
Survival Probability	S	Fraction, %	Measure of extinction
Electromagnetics			
Electric potential	Φ	V	
Potential differ.	$U = \Delta\Phi$	V	
Electric field	$\vec{E}$ Magnitude: $|\vec{E}|$	V/m or kV/cm	Up to usage
Magnetic field	$\vec{B}$ Magnitude: $|\vec{B}|$	T in nature G in instruments	$\hat{B}$ is the unit vector
Magnetic dipole moment	$\vec{M}$	J/T	$\hat{M}$ is the unit vector
Point on a dipole field line	(L, λ, ϕ)	$(R_E, °, MLT^*)$	For L-shell model

Quantity	Symbol	Units	Remarks
Instrument parameters			
Solid angle	Ω	sr	
Angular resolution	$\Delta\xi$	$^\circ$	
Angular response	$T(\xi)$	Fraction, %	Resolution in one plane
Geometrical factor	G	cm^2 sr	
Energy interval	ΔE	keV	Sometimes E resolution
Efficiency	$\eta(E)$	Fraction, %	
E-dependent G	$G \cdot \eta(E)$	cm^2 sr	
Collection Power	$G \cdot \eta(E) \cdot \Delta E/E$	cm^2 sr eV/eV	
Counting rate	R	counts/s	Never Hz

Chapter 1

Introduction

"The understanding can intuit nothing; the senses can think nothing. Only through their unison can knowledge arise."

Immanuel Kant, in *Transcendental Logic*
in *The Critique of Pure Reason* (1781)

Imagine a foggy evening, with just one streetlight and the glimpse of another visible farther away in either direction up and down the street. You conclude that the street extends equally far in both directions because the surroundings look identical in either direction. Then, as the fog lifts, you see an almost endless row of lights up the street, but the street ends after the third light down the street. Clearly, the fog misled you.

The plight of a *spacecraft* (S/C) carrying a comprehensive suite of *in-situ* charged-particles and fields sensors echoes this dilemma. Charged particles follow the local *magnetic field* ($\vec{B}$), and collectively they shield the *electric field* ($\vec{E}$), so such a S/C senses only its immediate vicinity, reminiscent of the fog but in a plasma. Such a "fog" is caused by the fields and does not vanish. Thus, *in-situ* measurements lack the global view of the large-scale structure, while they provide potentially comprehensive information on the local plasma conditions and particle-fields interactions.

For sufficiently dense plasmas, some information is available on the structure and ion kinetics from the emission or absorption lines of excited ions. Analogous to optical spectroscopy, we introduce *energetic neutral atom* (ENA) diagnostics or *neutral-atom astronomy*

1

to study space plasmas in regions beyond the reach of the S/C. Where appropriate, we also include the *ambient neutral atoms* (ANAs) of those regions. For convenience, often used terms will appear in acronyms or symbols as our story unfolds (see List of Acronyms & Symbols and Units for Useful Quantities).

1.1. Members of the Cast

Let us begin with *space plasmas*, or ensembles of electrons and ions with no net charge, that are ubiquitous in space above $\approx 10^2$ km or $\approx 0.02\,R_E$ (Earth radius) altitude and beyond. Because moving charged particles cling to $\vec{B}$, ions only provide local information along the S/C paths, unlike ENAs, which are unimpeded by $\vec{E}$ and $\vec{B}$ and escape along ballistic trajectories.

ENAs have *kinetic energies* $E > 10\,\mathrm{eV}$, well above the typical thermal energies of neutral atoms in gases of *temperatures* $T < 10^5\,\mathrm{K}$, which we refer to as the ANAs of that region. ENAs are neutralized *energetic ions* at $E > 10\,\mathrm{eV}$, which gained energy in electric and time-varying magnetic fields or fast-moving plasma irregularities, such as shocks or reconnection regions. An energetic ion turns into an ENA through *charge exchange* (CX) with an ANA, which may serve as a stationary target for the more energetic ion according to

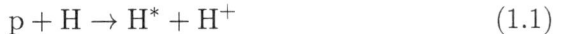

$$\mathrm{p + H \rightarrow H^* + H^+} \tag{1.1}$$

The plasma proton (p) picks up an electron from a H ANA and becomes H*, a H ENA, often in an excited state. In turn, the H ANA becomes a thermal $\mathrm{H^+}$ ion trapped by the local $\vec{B}$. The H ENA escapes with the ion's momentum, enabling detection by an ENA sensor far away (Fig. 1.1).

CX entails minimal energy transfer ($<10\,\mathrm{eV}$), and $\vec{B}$ does not deflect neutral particles. Thus, ENAs are authentic samples of the source ions in *energy* (E), *mass* (m), and *pitch angle* (α), whose spectra are modified only by the ANA *number densities* n_k and the E-dependent *CX cross-sections* σ_{ik} [1, 2]. Figure 1.2 shows σ_{ik} for CX of $\mathrm{H^+}$ with O as a function of energy (left) and with H and He as a function of speed (right). These three ANA species are

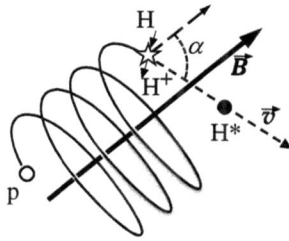

Fig. 1.1.   Birth of an ENA. An energetic ion (open circle) on a spiral path about $\vec{B}$ at pitch angle α becomes an ENA (solid circle) after a CX (star) and escapes with the ion's velocity $\vec{v}$.

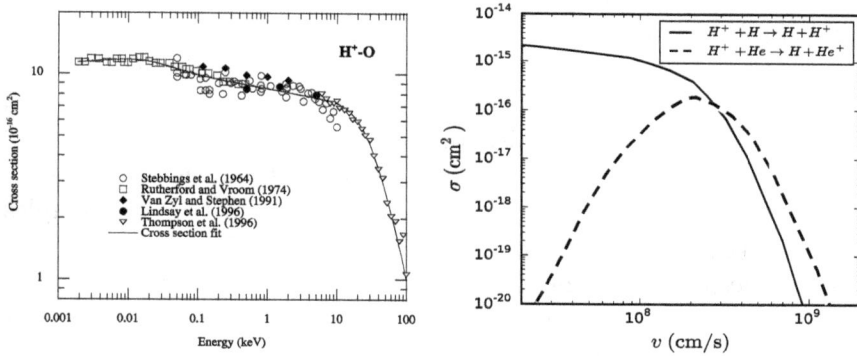

Fig. 1.2.   CX cross sections between p and three common ANA species H, He and O. Left: $p + O \rightarrow H^* + O^+$ as function of the proton energy [3]. (© AGU. Reproduced with permission) Right: $p + H \rightarrow H^* + H^+$ (solid line) and $p + He \rightarrow H^* + He^+$ (dashed line) as functions of the proton speed [4] (© IOP. Reproduced with permission).

relevant for producing H ENAs. Both E and v pertain to the rest frame of one of the collision partners. This distinction is important if both collision partners have substantial kinetic energy due to relative motion and/or temperature. If either the ions or the ANAs are at rest or have a low temperature, E or v of the other partner can be used to obtain σ_{ik}.

ENAs are produced in plasmas co-located with gases of large enough density n_k. Detected at a remote vantage point and analyzed by m, E, and *arrival direction* along the *line-of-sight* (LOS) $\hat{l}$, ENAs provide insight into the spatial, velocity, and mass distribution of the ions. Sometimes, *ENA diagnostics* even reveals the orientation of $\vec{B}$ in the remote source regions. Collating ENAs based on a set $\{\hat{l}\}$ into a 2D map constitutes *ENA imaging*, a unique visualization tool for space-plasma research.

ANAs, the essential partners in the birth of ENAs in the remote plasma, are not accessible to observation unless they happen to move toward the observer at a substantial bulk velocity like a wind. *Neutral-wind atoms* can indeed be detected and analyzed by an ENA sensor with a suitably low E threshold, making ENA diagnostics applicable to them as well.

Depending on the radiation and particle environment along the ENA paths, some may be ionized by photons or particles before reaching the ENA sensor for detection. Chap. 3 will discuss how the ENA production at the source, their extinction along the way, and their detection are analytically connected. Chaps. 4 and 5 will present the guiding principles behind ENA instruments and their implementation. Starting at the Earth, Chap. 6 presents illustrative examples of diagnosing planetary magnetospheres, as well as atmospheres and surfaces of small bodies in the solar system with ENAs. We turn to the heliospheric boundary regions and the local *interstellar medium* (ISM) in Chap. 7. Throughout this book, the term ISM stands solely for the environment just outside the heliosphere. We emphasize that an ENA image contains the LOS integral of ENA fluxes across the source region for each map pixel. Thus, obtaining the ion source distribution from ENA images requires deconvolution, facilitated by modeling. We will introduce such techniques in Chap. 6 and use them again in Chap. 7. We will highlight cases where *in-situ* ion and remote ENA observations complement each other.

Before we move on, let us touch upon the ion that shares the birth of an ENA, *i.e.*, the H^+ in Eq. 1.1. This new-born ion joins the local plasma at the much lower E of the ANA. However, suppose the gas and plasma move at a substantial speed relative to each other, as for the *solar wind* (SW) and the *interstellar neutral* (ISN) gas. In that case, this ion is injected with the bulk velocity $\vec{v}$ and then gyrates about $\vec{B}$ with its perpendicular speed $v_\perp$, forming

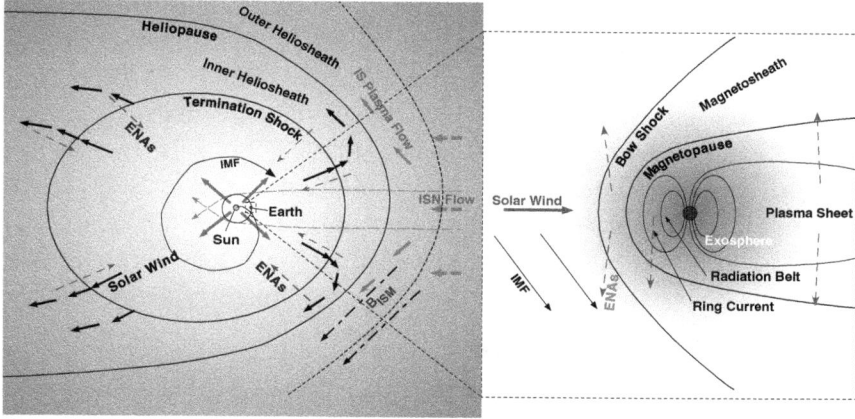

Fig. 1.3.   Schematic representation of the heliosphere (left) and Earth's magnetosphere (right), including the most relevant ENA and ANA observation regions. Dashed lines in different thicknesses indicate various ENA and ANA populations and their origin. The magnetospheric sketch does not include the daily and seasonal tilt of the Earth's dipole relative to the SW nor the resulting seasonal and diurnal north-south asymmetry.

a *pickup ion* (PUI). The first PUIs detected at Earth's orbit [5] were produced mainly by photoionization of the ISN, which will be discussed further in Sec. 1.3.1 and Chap. 7. Collectively, PUIs can affect the bulk momentum of the plasma through "mass-loading" [6] while conveying information on the ANAs. Therefore, PUIs are closely related to ENA diagnostics, as we will see in Chaps. 3 and 7.

Figure 1.3 shows schematic sketches of two regions of space first explored by ENA diagnostics, *i.e.*, Earth's magnetosphere and the heliosphere, which are the foci of Chaps. 6 and 7. After serving as a guide for this introduction, we will refer to Fig. 1.3 again in these chapters.

1.2.   Earth's Magnetosphere

The space around Earth controlled by its intrinsic $\vec{B}$ is our *magnetosphere* (Fig. 1.3, right). The interactions between Earth's dipolar $\vec{B}$ with its trapped plasma and the SW with its embedded *interplanetary magnetic field* (IMF) define the magnetosphere's size, shape, and outer layers, occasionally altered by solar transients, such as shocks and *coronal mass ejections* (CMEs). Ionospheric currents modify its inner regions. In preparation for Chap. 6, we outline our *magnetosphere's* main features while referring to reviews and textbooks for the underlying physics [7–9].

Earth's $\vec{B}$ presents an obstacle to the SW flow and IMF. Dynamic pressure balance controls the *magnetopause* (MP, Fig. 1.3, right) that separates the two plasma domains. An estimate for the geocentric distance r_{MP} of this boundary results from pressure balance at the subsolar point

$$(m_p n_p + m_\alpha n_\alpha)v_{SW}^2 = \frac{\mu_0}{2}\left(\frac{\mu_0 M_E}{4\pi r_{MP}^3}\right)^2 \tag{1.2}$$

ignoring the trapped plasma and the IMF. n_p and n_α are SW H$^+$ and He^{2+} number densities, v_{SW} the SW speed, and M_E Earth's dipole strength. Even though the relative abundance of α-particles is typically only $\approx 6\%$, their contribution to the dynamic pressure is significant because $m_\alpha \approx 4m_p$. Depending on the SW pressure, $r_{MP} \approx 9$–$12\,\mathrm{R_E}$.

The supersonic SW forms a *bow shock* (BS) at a well-defined stand-off distance from the MP. Between the MP and BS is the *magnetosheath* (MS), where the shocked, heated, and, thus, subsonic SW flows around the MP. Further downwind, some 50–100 R$_E$ (not shown in Fig. 1.3), the MS envelopes the *magnetotail* (MT) on the nightside. On the dayside, compression of the Earth's dipolar $\vec{B}$ forms a dimple or cusp in the MP above each pole, through which the shocked *interplanetary* (IP) *plasma* can reach lower altitudes at high magnetic latitudes, forming the dayside auroral oval. This particle transport is most effective when a south-pointing IMF meets Earth's north-pointing $\vec{B}$, leading to reconnection [10].

In the inner magnetosphere, $\vec{B}$ can be represented by dipolar $\vec{B}$-lines shaped as $r(L, \lambda) = L\cos^2\lambda$, with geocentric radius r and *magnetic latitude* λ. $L = r/R_E$ at $\lambda = 0°$ (magnetic-equator crossing) defines the so-called *L*-shells, which intersect Earth ($r = 1\,\mathrm{R_E}$) at $\lambda_L = \pm\cos^{-1}\sqrt{L^{-1}}$ [11]. Charged particles spiral along $\vec{B}$, conserving their *1st adiabatic invariant* and thus bounce between respective mirror points depending on their equatorial pitch angles (α_0). Small α_s with mirror points below ≈ 100 km altitude result in loss of the ion to the atmosphere, forming a *loss cone* for a range in α void of trapped ions. Between bounces, trapped electrons and ions drift in opposite directions along fixed L-shells around the dipole axis due to the gradient and curvature of $\vec{B}$, forming a westward (opposite to Earth's rotation) current flow. These trapped particles form the *radiation belt* and *ring current* (RC) (Fig. 1.3, right) centered on $\lambda = 0°$. The former, which extends to $\lambda \approx \pm 60°$, is located between L-shells ≈ 1.2–3. It consists primarily of MeV protons from β^+ decay of neutrons generated by cosmic rays in the upper atmosphere (Fig. 2.6). The RC in the equatorial plane extends to $L \approx 3$–8 and consists of energetic H^+, He^+, and O^+ ions from the ionosphere or the processed SW in the MT.

At $L > 10$, the dipolar $\vec{B}$ is stretched by the SW flow on the nightside and forms the MT. Closed field lines beyond $L > 10$ contain the plasma sheet, which sustains a dawn-to-dusk cross-tail current. *Field-aligned currents* and the partial RC form current loops that connect the ionosphere at high latitudes to the equatorial plasma sheet. On the nightside, they include SW particles coming in from the tail along the stretched field lines to reach the nightside auroral oval.

The westward flowing RC reduces B_z, the north-pointing component of the Earth's dipolar $\vec{B}$. During the passage of CMEs, the enhanced SW compresses $\vec{B}$ on the dayside, suggesting a B_z-increase. On the contrary, ground-based magnetometers show a net drop in B_z, due to the injection of the processed SW plasma from the MT, thus enhancing the RC. Global averages of magnetometer data form the basis of the *Dst, SYM-H,* and *Kp* indices [12], which serve as magnetic storm indicators. Their comparison with *in-situ* ion and remote ENA observations provides a better understanding of the structure and dynamics of the current systems.

How these current systems [13] and the boundaries react to solar activity forms the core of magnetospheric and space-weather studies. To diagnose the magnetosphere with ENAs, the *exosphere* (shaded area in Fig. 1.3, right), whose ANAs produce the ENAs, elicits our immediate attention.

1.2.1. *Earth's exosphere*

Earth's *exosphere* [14] has been studied *via* its H Ly-α (121.6 nm, 10.2 eV) emission, the *geocorona* glow [15, 16]. Similar to deducing the parent ion distribution from ENA images, obtaining the spatial distribution of exospheric H atoms from the *extreme ultraviolet* (EUV) data requires deconvolution of LOS integrals, which usually does not yield unique solutions. Thus, comparing the EUV emission from exospheric H atom distribution models and observations constrains the model parameters.

The exobase caps Earth's thermosphere at $\approx$ 500 km altitude, beyond which H atoms (largely photo-dissociated from H_2O and CH_4) dominate but rarely collide. Depending on their velocities, they may return to the exobase, follow satellite orbits, escape on hyperbolic trajectories, or are lost to photo or electron-impact ionization. Exospheric H atoms scatter solar Ly-α photons resonantly, *i.e.*, absorb and re-emit at the same energy, generating the Ly-α geocorona. Each absorption imparts an anti-sunward momentum on the atom, but the re-emission is nearly isotropic [17], resulting in a net anti-sunward *radiation pressure* on the exosphere.

The first comprehensive exospheric H-density model, $n^M(\boldsymbol{r})$, was spherically symmetric, parametrized at the exobase in geocentric radius r_c, temperature T_c, and density n_c, with or without satellite orbits, neglecting radiation pressure and ionization loss [18]. For $T_c \approx 1400$ K, $n(r) \propto r^{-3}$ (Fig. 1.4, right). Low Earth orbiters at various geocentric locations $\vec{r'}$ have surveyed the geocorona in the radial LOS integrated Ly-α intensity $I(\vec{r'}, t)$ (Fig. 1.4, left). Observations at lower r' include multiple scattering in the optically thick region between r_c and $r = 3.5 R_E$. When augmented with the radiative transfer [19], the Chamberlain model matched EUV data between 1.09 and 4.65 R_E, obtained with *Dynamics Explorer 1* [20]. The resulting $n^M(r)$ was once widely used for ENA studies.

The need for accurate exospheric H density distributions $n(\vec{r}, t)$ for ENA diagnostics, especially during active times, prompted the inclusion of Ly-α imagers alongside ENA cameras on the *Imager for Magnetopause-to-Aurora Global Exploration* (*IMAGE*) and the

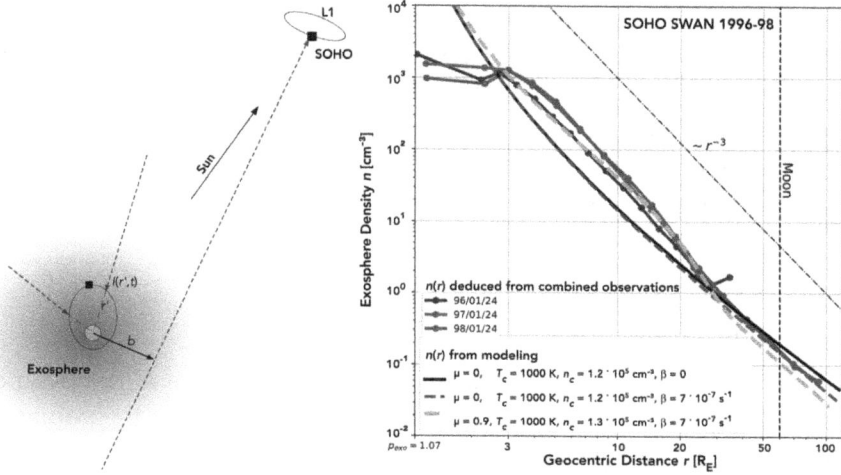

Fig. 1.4. Probing the exosphere *via* resonantly scattered Ly-α. Left: Schematic of various LOS lines (dashed lines). For low Earth orbits, zenith LOS emissions at distances r' are indicated and, for *SOHO*, the LOS with impact parameter b. Right: Radial distribution of the exospheric H-atom number density n as a function of geocentric distance r derived from same-day *SOHO* SWAN data over three successive years. The black curve represents the Chamberlain model with exobase temperature T_c and density n_c. The dashed magenta line includes the ionization rate β, and the dashed yellow line also radiation pressure μ [21]. (© AGU. Reproduced with permission).

Twin Wide-angle Imaging Neutral-atom Spectrometers (TWINS) (Sec. 4.1.2). Observations with the Geocorona Imager on *IMAGE* led to an improved empirical formula of $n^M(r, t)$ with a sum of two exponentials for $r > 3.5 R_E$ [22]. *TWINS'* stereo capability in Ly-α enables the approximation of $n^M(\vec{r}, t)$ in 3D with spherical harmonics at 3–8 R_E, where multiple scattering is rare [23, 24]. The *TWINS* observations revealed asymmetries and variations of the exosphere due to changes in solar radiation but no correlations with *Dst* or *Kp*. For the lack of understanding, concurrent measurement of ENAs and exospheric Ly-α emission $I(r', t)$, like on *IMAGE* and *TWINS*, is crucial.

At $r > 8 R_E$, the IP Ly-α background dominates, first noticed in 1962 with a H-absorption cell on a rocket-borne spectrometer to separate the Lyα distributions by their Doppler shift [25]. We now know this background is due to the ISN gas (Sec. 7.2) moving relative to the Sun [26], [27]. By toggling a similar H-absorption cell [28], *SOHO* SWAN at L1 ($r \approx 230\,R_E$), with its LOS crossing the entire *exosphere* (Fig. 1.4, left), successfully measured the geocorona to $r = 100\,R_E$. Figure 1.4 (right) shows the resulting density profiles based on models that include radiation pressure compared to the Chamberlain model [18]. The ISN flow leads us to the heliosphere.

1.3. The Heliosphere

Like the pressure balance between the SW and Earth's $\vec{B}$ that shapes the magnetosphere, the interaction between the ISM, including its *plasma, gas*, and *magnetic field* (ISMF), and the SW with the embedded IMF, shapes the heliosphere. Here, the Sun's motion relative to its surrounding ISM at ≈ 25.5 km/s [29, 30] replaces the ≈ 400 km/s SW flow. In response (Fig. 1.3, left), the *interstellar* (IS) plasma flows around the *heliopause* (HP), the outer boundary of the heliosphere. The radially expanding supersonic SW forms a *termination shock* (TS), where the SW plasma is thermalized, fills the *inner heliosheath* (IHS), and flows tailward. We will return to the heliosphere [31], ISM [32], and their interactions in Chap. 7, but leave the general description of increasingly sophisticated simulations [33] to reviews.

Unlike in the magnetosphere, *in-situ* sampling of the outer heliosphere has been sparse. *Pioneers 10* and *11*, launched in 1972 and 1973, provided scientific data up to 67 and 44 AU, respectively, about halfway to the TS. *New Horizons*, launched in 2006 on its way to the Kuiper Belt and beyond, has passed 45 AU and may not reach the TS. Only *Voyager 1* and *2*, launched in 1977, have traversed the TS and HP and reached the *outer heliosheath* (OHS), the boundary layer of the ISM, and significantly enhanced our understanding of the ISM and heliosphere. Complementary ENA diagnostics of the heliosphere from 1 AU was deemed plausible [34, 35] and eventually implemented on *SOHO* [36, 37], followed by the *Interstellar Boundary Explorer* (*IBEX*) and *Cassini* INCA. Chap. 7 will feature imaging the outer regions of the heliosphere in ENAs by the three missions and combined studies with *in-situ* ion measurements on the *Voyagers*. In preparation for Chap. 7, the following sub-sections will outline the significance of the two heliospheric *foci*: the ISN flow through the inner heliosphere (ANAs that create the heliospheric ENAs) and the wealth of information on the heliospheric boundary regions revealed by ENA imaging.

1.3.1. The interstellar neutral-atom flow in the heliosphere

While the plasma, including the ISMF, is diverted around the HP and separated from the solar domain, the ISN gas penetrates the heliosphere almost unimpeded as a *neutral interstellar wind* relative to the Sun at 25.5 km/s (dashed lines in Fig. 1.3, left). As it approaches the Sun, the wind is attracted by solar gravitation (modified by radiation pressure for H, like exospheric H) and progressively depleted by ionization. The resulting spatial distribution and flow pattern of the ISN gas in the inner heliosphere became accessible to observation through 1) resonant backscattering of solar EUV for H [21, 26–28] and He [38], 2) He [5] and H [39] PUIs, and 3) direct detection of He [40], O, and H [41]. Together with advanced modeling, the latter observations enable in-depth probing of the physical parameters of the ISN gas beyond the HP for several key species, including Ne [42].

Apart from the diagnostic opportunities that these ISN atoms present for the ISM (Sec. 7.2), they are the ANAs that give birth to ENAs from the parent ion populations in the heliospheric boundary regions; hence of eminent interest here (Secs. 7.3–7.5). The simultaneous observation of the ANAs and the resulting ENAs with the same instrumentation dramatically enhances the value of ENA diagnostics, the very topic of this book.

1.3.2. Global view of the heliospheric boundary with ENAs

The closest sign of the boundary to the ISM is the SW TS, crossed by *Voyager 1* on Dec 16, 2004, at 94 AU after decades of anticipation [43]. *Voyager 2* followed in 2007 at 85 AU [44], thus suggesting an asymmetric shape of the TS and the entire heliosphere [45]. *In-situ* measurements at the heliospheric boundary are very localized, sparse, and a once or twice in a lifetime opportunity, thus calling for global viewing from 1 AU. This call led to implementing *NASA*'s Small Explorer Mission *IBEX* at precisely the right time [46]. It delivers an ongoing (at this writing) series of all-sky ENA maps at a 6-month cadence. Because images do not contain depth or distance information, the *Voyager in-situ* observations are invaluable and, combined with global heliospheric modeling, provide true synergism.

Models anticipated the ENA emission from CX between the heated subsonic SW plasma in the IHS and the ISN gas [34, 35]. It appeared in the *IBEX* ENA maps [46] as the *globally distributed flux* (GDF) (Sec. 7.5), LOS integrated from the TS to the HP. *IBEX*-Lo [47], *IBEX*-Hi [48], *Cassini* INCA [49], and *SOHO* HSTOF [36, 37] measured the GDF from 0.1 to 88 keV.

LOS integrals do not distinguish sources at different distances or layers in the same *field-of-view* (FOV) unless invoking different energy spectra or spatial patterns. As seen in Fig. 1.3 (left), the OHS plasma flow is diverted. CX between its ions and the ISN gas forms another population of ENAs predicted as *secondary ISN* atoms [50]. The so-called H-wall [51], a related enhancement of the IS H density in the OHS, has been confirmed in the H absorption line [52] for the heliosphere and even for astrospheres of nearby stars [53]. Secondary He [54, 55] and O [41, 56] have been identified and analyzed (Sec. 7.4). They are observed at energies lower than the ENAs from the IHS and may be partially responsible for placing a lower energy limit on the IHS ENA observations thus far.

The biggest surprise in the *IBEX* ENA maps has been the very bright "Ribbon" with intensities 2–3 times that of the GDF [57]. Its alignment perpendicular to the ISMF [58] is consistent with the ISMF orientation deduced from the heliospheric asymmetry based on the distances of the *Voyager* TS crossings [43] and the IS H and He flow directions [59, 60]. Sec. 7.3 will discuss its importance and diagnostic opportunities. Obviously, the GDF, discussed in Sec. 7.5, had to be disentangled from the "Ribbon" in the same ENA maps [61]. ENA diagnostics have opened the heliosphere and the local ISM to close inspection from 1 AU. At this writing, opportunities for yet more profound insight into this active area of research are under implementation or in planning.

We have introduced the cast, their relations, and the stage on which they play out the emerging story of ENA diagnostics. Chap. 2 presents a brief history, starting with the discovery of ENAs in space in the mid-1900s, followed by pioneering efforts up to the end of the 20th century. They paved the way for neutral-atom astronomy — ENA diagnostics of space plasmas that have now earned their place among the space-physics tools for current and future heliophysics investigations. We encourage the reader to augment the contents of this book with relevant review articles [62–68] and those yet to come.

References

1. Stier, P. M., & Barnett, C. F. (1956). Charge exchange cross sections for hydrogen ions in gases, *Phys. Rev.*, **103**(4), 896–907.
2. Barnett, C. F., & Stier, P. M. (1958). Charge exchange cross sections for helium ions in gases, *Phys. Rev.*, **109**(2), 385–390.
3. Lindsay, B. G., & Stebbings, R. F. (2005). Charge transfer cross sections for energetic neutral atoms data analysis, *J. Geophys. Res.*, **110**, A12, 213.
4. Friedman, B., & DuCharme, G. (2017). Semi-empirical scaling for ion-atom single charge exchange cross sections in the intermediate velocity regime, *J. Phys. B: At. Mol. Opt. Phys.*, **50**, 115202 (11pp).
5. Möbius, E., Hovestadt, D., Klecker, B. *et al.* (1985). Direct observation of He^+ pick-up ions of interstellar origin in the solar wind, *Nature*, **318**(6045), 426–429. Doi: 10.1038/318426a0
6. Szegö, K., Glassmeier, K.-H., Brinca, A. *et al.* (2000). Physics of Mass Loaded Plasmas, *Space Sci. Rev.*, **94**, 429.
7. Lyons, L. R., & Williams, D. J. (1984). *Quantitative Aspects of Magnetospheric Physics* (Springer, Dordrecht, Germany).
8. Cravens, T. E. (1997). *Physics of Solar System Plasmas* (Cambridge University Press, UK).
9. Parks, G. J. (2003). *Physics of Space Plasma: An Introduction*, 2nd Ed. (Westview Press, Boulder, USA).
10. Dungey, J. W. (1961). Interplanetary magnetic field and the auroral zones, *Phys. Rev. Lett.*, **6**(2), 47–48.
11. McIlwain, C. E. (1961). Coordinates for mapping the distribution of magnetically trapped particles, *J. Geophys. Res.*, **66**(11), 3681–3691.

12. Wanliss, J. A., & Showalter, K. M. (2006). High-resolution global storm index: *Dst* versus SYM-H, *J. Geophys. Res.*, **111**, A02202, doi:10.1029/2005JA011034.
13. Ohtani, S.-I., Fujii, R., Hesse, M., & Lysak, R. L. (2000). Eds. *The Magnetospheric Current Systems* (American Geophysical Union, Washington, DC, USA).
14. Spitzer, L. (1952). *The Atmospheres of the Earth and Planets*; Ed. G. P. Kuiper, 2nd Ed., 211–247 (University of Chicago Press, Chicago).
15. Kupperian, J. E., Jr., Byram, E. T., Chubb, T. A., and Friedman, H. (1959). Far ultraviolet radiation in the night sky, *Planet. Space Sci.*, **1**(1), 3–6.
16. Shklovsky, I. S. (1959). On hydrogen emission in the night glow, *Planet. and Space Sci.*, **1**(1), 63–65.
17. Brandt, J. C., & Chamberlain, J. W. (1959). Hydrogen radiation in the night sky, *Astrophys. J.*, **130**, 670–682.
18. Chamberlain, J. W. (1963). Planetary coronae and atmospheric evaporation, *Planet. Space. Sci.*, **11**, 901–960.
19. Chandrasekhar, S. (1950). *Radiative Transfer* (Oxford University Press, UK).
20. Rairden, R. L., Frank, L. A., & Craven, J. D. (1986). Geocorona imaging with Dynamic Explorer, *J. Geophys. Res.*, **91**(A12), 13,613–13,630.
21. Baliukin, I. I., Bertaux, J.-L., Qumerais, E. *et al.* (2019). SWAN/SOHO Lyman-α mapping: the hydrogen geocorona extends well beyond the moon, *J. Geophys. Res.: Space Phys.*, **124**, 861–885.
22. Østgaard, N., Mende, S. B., Frey, H. U., Gladstone, G. R., & Lauche, H. (2003). Neutral hydrogen density profiles derived from geocoronal imaging, *J. Geophys. Res.*, **108**(A7), 1,300–1,311.
23. Zoennchen, J. H., Nass, U., Lay, G., & Fahr, H. J. (2010). 3-D geocoronal hydrogen density derived from TWINS Lya data, *Ann. Geophys.*, **28**, 1221–1228.
24. Bailey, J., & Gruntman, M. (2011). Experimental study of exospheric hydrogen atom distribution by Lyman-alpha detectors on the TWINS mission, *J. Geophys. Res.*, **116**, A09302 1–9, doi:10.1029/2011JA016531.
25. Morton, D. C., & Purcell, J. D. (1962). Observations of the extreme ultraviolet radiation in the night sky using an atomic hydrogen filter, *Planet. Space Sci.*, **9**(8), 455–458.
26. Bertaux, J.-L., & Blamont, J. E. (1971). Evidence for a source of an extraterrestrial hydrogen Lyman-a emission, *Astron. Astrophys.*, **11**, 200–217.

27. Thomas, G. E., & Krassa, R. F. (1971). OGO-5 measurements of the Lyman Alpha sky background, *Astron. & Astrophys.*, **11**, 218–233.
28. Quèmerais, E., Bertaux, J.-L., Lallement, R. *et al.* (2000). SWAN/SOHO H cell measurements: the first year, *Adv. Space Res.*, **26**(5), 815–818.
29. Möbius, E., Bzowski, M., Chalov, S. *et al.* (2004). Synopsis of the interstellar He parameters from combined neutral gas, pickup ion and UV scattering observations and related consequences, *Astron. & Astrophys.*, **426**, 897–907.
30. McComas, D. J., M. Bzowski, A. Galli, O.A. *et al.* (2015). Six Years of Directly Sampling of the Local Interstellar Medium by the Interstellar Boundary Explorer, *Astrophys. J. Supp.*, **220**:22.
31. Axford, W. I. (1972). The interaction of the solar wind and the interstellar medium, *Solar Wind*, eds. Sonett, C. P., Coleman, P. J., & Wilcox, J. P., 609–660 (NASA, Washington, DC).
32. Frisch, P. C., Bzowski, M., Grn, E. *et al.* (2009). The galactic environment of the Sun: interstellar material inside and outside of the heliosphere, *Space Sci. Rev.*, **146**, 235–273.
33. Zank, G. P., Pogorelov, N. V., Heerikhuisen, J. *et al.* (2009). Physics of the solar wind-local interstellar medium interaction, *Space Sci. Rev.*, **146**(1–4), 295–327.
34. Hsieh, K. C., Shih, K. L., Jokipii, J. R., & Grzedzielski, S. (1992). Probing the heliosphere with energetic hydrogen atoms, *Astrophys. J.*, **393**(7), 756–763.
35. Hsieh, K. C., & Gruntman, M. A. (1993). Viewing the outer heliosphere in energetic neutral atoms, *Adv. Space Sci.*, **13**(6), 131–139.
36. Hovestadt, D., Hilchenbach M, Brgi A. *et al.* (1995). CELIAS — Charge, Element and Isotope Analysis System for SOHO, *Sol. Phys.*, **162**(1–2), 441–481.
37. Hilchenbach, M., Hsieh, K. C., Hovestadt, D. *et al.* (1998). Detection of 55–80 keV hydrogen atoms of heliospheric origin by CELIAS/HSTOF on SOHO, *Astrophys. J.*, **503**(2), 916–922.
38. Weller, C. S., & Meier, R. R. (1974). Observations of helium in the interplanetary /interstellar wind — the solar wake effect, *Astrophys. J.*, **193**, 471–476.
39. Gloeckler, G., Geiss, J., Balsiger, H. *et al.* (1992). The solar wind ion composition spectrometer, *Astron. Astrophys. Supp.*, **92**(2), 267–289.
40. Witte, M., Banaszkiewicz, M., & Rosenbauer, M. (1996). Recent results on the parameters of the interstellar helium from the Ulysses/GAS experiment, *Space Sci. Rev.*, **78**(1–2), 289–296.

41. Möbius, E., Bochsler, P., Bzowski, M. *et al.* (2009). Direct observations of interstellar H, He, and O by the Interstellar Boundary Explorer, *Science*, **326**, 969–971.

42. Bochsler, P., Petersen, L., Möbius, E. *et al.* (2012). Estimation of the neon/oxygen abundance ratio at the heliospheric termination shock and in the local interstellar medium from IBEX observations, *Astrophys. J. Supp.*, **198**(2), 13–17.

43. Stone, E. C., Cummings, A. C., McDonald, F. B. *et al.* (2005) Voyager 1 explores the termination shock region and the heliosheath beyond, *Science*, **309**, 2017–2020.

44. Stone, E. C., Cummings, A. C., McDonald, F. B. *et al.* (2008). An asymmetric solar wind termination shock, *Nature*, **454**, 71–74.

45. Opher, M., Stone, E.C., & Liewer, P. C. (2006). The effects of a local interstellar magnetic field on Voyager 1 and 2 observations, *Astrophys. J.*, **640**, L71–L74.

46. McComas, D. J., Allegrini, F., Bochsler, P. *et al.* (2009a). IBEX — Interstellar Boundary Explorer, *Space Sci. Rev.*, 146(1-2), 11–33.

47. Fuselier, S. A., Bochsler, P., Chornay, D. *et al.* (2009). IBEX-Lo Sensor, *Space Sci. Rev.*, **146**(1), 117–147.

48. Funsten, H.O., F. Allegrini, F., Bochsler, P. *et al.* (2009). The interstellar boundary explorer high energy (IBEX-Hi) neutral atom imager, *Space Sci. Rev.*, **146**(1), 75–103.

49. Mitchell, D. G., Krimigis, S. M., Cheng, A. F. *et al.* (1996). Imaging-neutral camera (INCA) for the NASA Cassini mission to Saturn and Titan, *Proc. SPIE*, **2803**, 154–161.

50. Ripken, H. W., & Fahr, H.-J. (1983). Modification of the local interstellar gas properties in the heliospheric interface, *Astron. Astrophys.*, **122**, 181–192.

51. Baranov, V. B., & Malama, Y. G. (1993). Effect of local interstellar medium hydrogen fractional ionization on the distant solar wind and interface region, *J. Geophys. Res.*, **100**(A8) 14,755–14,761.

52. Linsky, J. L., & Wood, B. E. (1996). The α-Centauri line of sight: D/H ratio, physical properties of local interstellar gas, and measurement of heated hydrogen (the "hydrogen wall") hear the heliopause, *Astrophys. J.*, **463**, 254–270.

53. Wood, B. E., Redfield, S., Linsky, J. L., Müller, H.-R., & Zank, G. P. (2005). Stellar Lya emission lines in the Hubble Space Telescope archive: intrinsic line fluxes and absorption from the heliosphere and astrospheres, *Astrophys. J. Supp.*, **159**, 118–140.

54. Kubiak, M. A., Bzowski, M., Sokol, J. M. *et al.* (2014). Warm Breeze from the starboard bow: a new population of neutral helium in the heliosphere, *Astrophys. J. Supp.*, **213**:29.

55. Kubiak, M. A., Swaczyna, P., Bzowski, M. *et al.* (2016). Interstellar neutral helium in the heliosphere from IBEX observations. IV. Flow vector, Mach number, and abundance of the Warm Breeze, *Astrophys. J. Supp.*, **223**:25.

56. Park, J., Kucharek, H., Möbius, E. *et al.* (2015). Statistical analysis of the heavy neutral atoms measured by IBEX, *Astrophys. J. Supp.*, **220**:34.

57. McComas, D. J., Allegrini, F., Bochsler, P. *et al.* (2009). Global observations of the interstellar interaction from the Interstellar Boundary Explorer (IBEX), *Science*, **326**(5955), 959–962.

58. Schwadron, N. A., Bzowski, M., Crew, G. B. *et al.* (2009). Comparison of Interstellar Boundary Explorer observations with 3D global heliospheric models, *Science*, **326**(5955), 966–968.

59. Lallement, R., Quemerais, E., Bertaux, J. L. *et al.* (2005). Deflection of the interstellar neutral hydrogen flow across the heliospheric interface, *Science*, **307**(5714), 1,447–1,449.

60. Izmodenov V. V., Alexashov D. B., & Myasnikov A. V. (2005). Direction of the interstellar H atom inflow in the heliosphere: role of the interstellar magnetic field, *Astron. Astrophys.*, **437**, L35–L38.

61. Schwadron, N. A., Allegrini, F., Bzowski, M. *et al.* (2011). Separation of the IBEX ribbon from globally distributed energetic neutral atom flux, *Astrophys. J.*, **731**:56.

62. Gruntman, M. A. (1997). Energetic neutral atom imaging of space plasmas, *Rev. Sci. Instrum.*, **68**(10), 3,617–3,656.

63. Hilchenbach, M., Hsieh, K. C., & Czechowski, A. (2000). Energetic neutral atoms, Chap. 10 in *The Outer Heliosphere: Beyond the Planets*, Ed. Scherer, K., Fichtner, H., & Marsch, E. (Copernicus Gesellschaft e. V., Katlenburg-Lindau, Germany).

64. Wurz, P. (2000). Detection of energetic neutral atoms, Chap. 11 in *The Outer Heliosphere: Beyond the Planets*, Eds. Scherer, K., Fichtner, H., & Marsch, E. (Copernicus Gesellschaft e. V., Katlenburg-Lindau, Germany).

65. Goldstein, J., & McComas, D. J. (2013). Five years of stereo magnetospheric imaging by TWINS, *Space Sci. Rev.*, **180**, 39–70.

66. Hsieh, K. C. (2015). Detecting energetic neutral atoms in and out of the heliosphere, *Chin. J. Space Sci.*, **35**(3), 253–2292 (in Chinese).

67. Möbius, E., Galvin, A. B., Kistler, L. M., Kucharek, H., & Popeck, M. A. (2016). Time-of-flight mass spectrographs — from ions to neutral atoms, *J. Geophys. Res.*, **121**(12), 11,647–11,666, 10.1002/2016JA022553.

68. Brandt, P. C., Hsieh, S. Y., DeMajistre, R., & Mitchell, D. G. (2018). ENA imaging of planetary ring currents, Chap. 6 in *Electric Currents in Geospace and Beyond, Geophysical Monograph*, **235**, First Edition, Ed. Keiling, A., Marghitu, O., and Wheatland, M., American Geophysical Union (John Wiley & Sons).

Chapter 2

First Encounters with ENAs

"No great discovery was ever made without a bold guess."

Sir Isaac Newton

"One who perceives the new from reviewing the old may teach others."

from Wei Zheng in *The Analects of Confucius*
circa 200 BCE

The first signs of ENAs in space appeared indirectly through optical spectroscopy, validated in laboratory experiments. Their direct detection had to await progress in understanding the interaction between neutral atoms and plasmas and substantial development efforts in space-based particle instrumentation. After the first carefully planned ENA measurement, several serendipitous observations conspired to kick-start ENA imaging as a novel remote sensing technique for studying space plasmas. In the following, we will present the critical steps along the way.

2.1. The Discovery: From the Telescope to the Accelerator

Pointing the telescope toward the magnetic zenith at Yerkes Observatory (magnetic 52.6° N) in 1950, A. B. Meinel found that the Hα (656.3 nm) line in the auroral spectrum had a peculiar feature. It was blue-shifted relative to the one taken toward the magnetic horizon and showed a tail indicating H atoms of E up to $\approx$90 keV (Fig. 2.1).

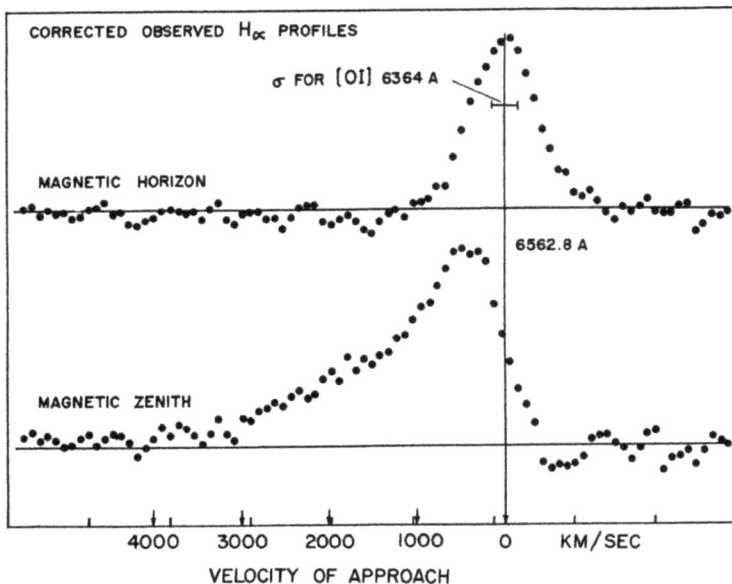

Fig. 2.1. The first evidence of extraterrestrial ENAs: Auroral Hα line profiles after correction for the first positive bands of N_2. The lower panel, taken from the magnetic zenith, shows the blue-shifted 656.3-nm line with a tail indicating H atoms of E up to $\approx$90 keV [1]. (© AAS. Reproduced with permission).

He suggested that the Hα emitters are energetic H atoms formed through CX with atmospheric molecules from high-speed precipitating protons that approach along $\vec{B}$. These precipitating protons also excite atmospheric N_2 and O_2 molecules producing the aurorae [1].

To test his hypothesis, Meinel recruited C.Y. Fan, a fresh Ph.D. in nuclear physics from the University of Chicago, to perform the necessary experiment. They used the 230-keV proton beam from the Kevatron at the university's Institute for Nuclear Studies to excite the residual air molecules in a $\approx$26-cm long absorption tube to simulate the proposed mechanism. The spectrum between 375 and 480 nm matched that of the aurorae at 100-km altitude (Fig. 2.2) [2].

They built an absorption chamber 91.4 cm long and 10.2 cm in diameter with a window at the end facing the proton beam and windows on the side perpendicular to the proton beam to simulate

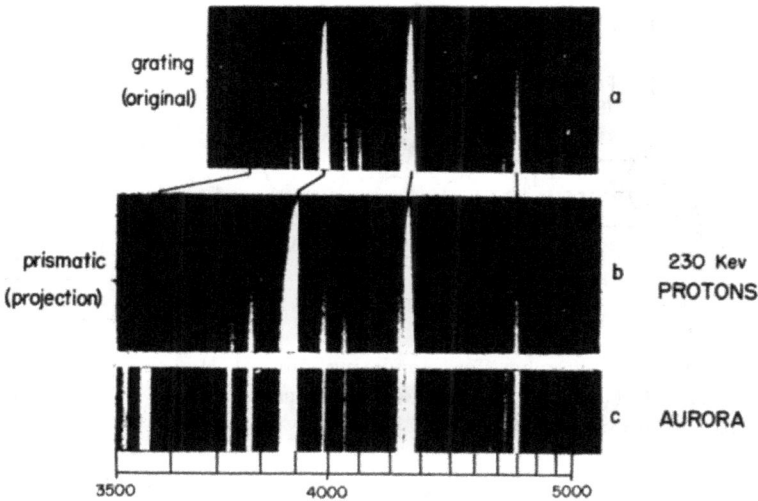

Fig. 2.2. Proton-induced and auroral spectra compared. The absence of lines <370.0 nm in panels a and b is due to absorption by the glass lens of the laboratory camera [2]. (© AAS. Reproduced with permission).

the auroral Hα emissions from the magnetic zenith and horizon, respectively (Fig. 2.3) [3].

Using this chamber, they reproduced the blue-shifted Hα line through CX of the injected H^+ with the residual air molecules. They found that CX produced the elevated state for Hα emission ($n = 3$) and thus eliminated the need for intermediate excitation steps. They also obtained N_2 emission spectra at different pressure (as a *proxy* for altitude) with 40–230 keV H^+, 75–320 keV D^+, 150–450 keV He^+, and 400 keV Ne^+ beams, further supporting the energetic H^+ origin of aurorae [3].

The discovery of ENAs in space through spectroscopy also gave birth to a theory of the aurora 12 years after the original suggestion of the presence of H in aurorae [4]. Besides suggesting an energetic-particle origin of aurorae, Meinel deduced that the aurora-generating solar protons had to travel on a curved path through IP space (Sec. 2.2) based on the time delay between the optical onset of a solar flare and that of an aurora [1]. Subsequent aurora observations and ion-beam excitation measurements provided a first estimate of the

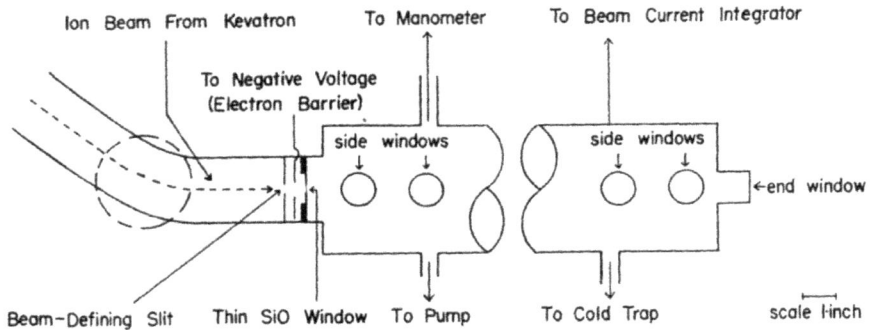

Fig. 2.3. Vacuum chamber for simulating auroral Hα emissions from the magnetic zenith (*via* the end window) and magnetic horizon (*via* the side windows) with a proton beam [3]. (© AAS. Reproduced with permission).

average proton flux at $E > 100\,\mathrm{keV}$ from the associated solar event at the upper atmosphere to $\approx 10^7\,(\mathrm{cm}^2\,\mathrm{s})^{-1}$ [5]. Hence, the discovery of ENAs in aurorae also paved the way for solar-terrestrial physics and space-weather studies ahead of the Space Age.

2.2. First Diagnostic Use of ENAs in Space

Unlike the accidental discovery of auroral ENAs, detecting ENAs at high altitudes 15 years later was intentional and *in-situ*. Four significant developments in the early 1960s paved the way for the new messengers:

• *New findings in space plasma physics*: The Space Age and the interest in solar particles — from their release on the Sun *via* travel through IP space to their interaction with Earth's magnetic field and atmosphere — ushered in a new era. Parker [6] predicted the supersonic SW and the spiral structure of the IMF with the "curved trajectories" deduced by Meinel [1]. Dessler and Parker demonstrated the link of the magnetospheric storm recovery to CX between trapped energetic protons and exospheric H atoms [7]. Chamberlain published the classic *Physics of the Aurora and Airglow* [8]. Parker's *Interplanetary Dynamical Processes* laid a firm theoretical foundation for *in-situ* observations of IP ions and

IMF dynamics [9]. The *in-situ* study of the IP plasma by *Mariner 2* confirmed Parker's predictions [10, 11].

In analogy to ENAs from aurorae, these observations led to the question: How much of the SW is neutral? An early estimate [12], refuted by the absence of the expected optical emissions at twilight, was later reduced significantly by considering UV and electron impact ionization in the solar corona [13, 14]. The *neutral SW* detected 36 years later showed a H/H^+ ratio of 10^{-3}–10^{-4}, 10^{-2}–10^{-3} lower than the first estimate [15] (Sec. 6.2.1.1). Studies of the interactions between SW and IS plasmas suggested the ISN gas as a new ANA population at 1 AU [16–19].

- *Understanding energetic ion passage through matter*: CX is a process that is central to ENA diagnostics of plasmas. In that connection, the study of the interactions of $E < 1\,\mathrm{MeV}$ ions with matter, notably with gases, proved crucial to understanding precipitating ions in the upper atmosphere [20, 21]. At these energies, CX between the projectile and the target plays an important role, leading to more detailed investigations [22]. Two reviews of ion passage through solids helped with radiation shielding and the use of *solid-state detectors* (SSDs), which also led to improved ENA sensors [23, 24].

- *New particle-detectors for ENA sensors*: For use in ENA sensors, the detectors must be sensitive to particles at $E < 100\,\mathrm{keV}$. However, early space-borne particle sensors relied on gas-filled Geiger tubes [25, 26] and proportional counters [27, 28]. Their casing for gas containment and radiation shielding raise their detection thresholds to $>1\,\mathrm{MeV}$, thus too high for ENAs. As discussed in Sec. 4.2.2, not even the early SSDs with >200-keV thresholds or the bulky dynode-chain *electron multipliers* (EM) were usable for the low energies of ENAs. Finally, *channeltrons*, much smaller and lighter EMs capable of detecting single particles at $E \geq 0.3\,\mathrm{keV}$, arrived and enabled effective ENA sensor technology [29].

- *New sounding rockets for research*: Particle detectors must reach altitudes higher than 200 km to detect the precipitating particles

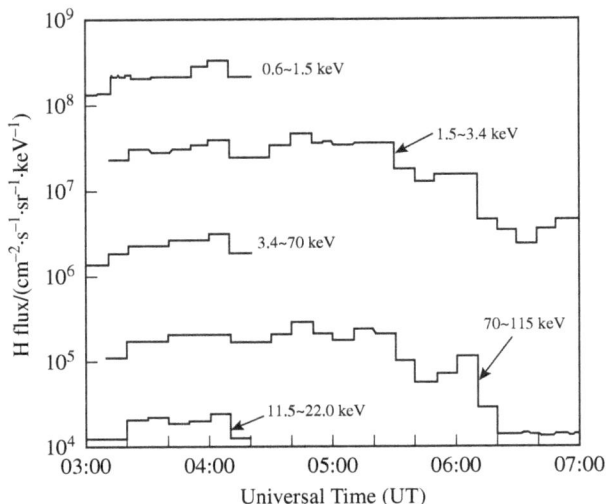

Fig. 2.4.   Direct detection of auroral H^0 generated by precipitating protons at the top of the atmosphere, 200 km apart. Time profile of H^0 fluxes in five energy channels, three of which dropped out due to telemetry failure [35]. (© AGU. Reproduced with permission).

in-situ, including auroral ENAs. The Nike-Tomahawk sounding-rocket, which could reach 215 km with a 115-kg payload, became available for scientific research in 1963.

These developments enabled Bernstein and Wax at TRW [30] to launch the first ENA spectrometer (Sec. 5.1.2) from Fort Churchill, Canada, into an auroral breakup at 06:01:40 UT, on Apr 25, 1968. They measured the H ENA (H^0) flux at 250 km altitude and deduced the total precipitating proton flux over 1–20 keV at the top of the atmosphere. Their indirect approach relied on the knowledge of particle CX and energy loss in gases [20] to obtain the charge state abundances of H^-, H^0, and H^+ after penetration of the protons to 250 km. At the observed energies, charge equilibrium makes H^0 the dominant *charge state* $(Q = 0)$. From the observed H^0 fluxes, they deduced the total H^+ flux at 450 km altitude consistent with the electron precipitation and optical data from the same auroral breakup [31]. Hence, they have conceived the first diagnostic use of

ENAs in space! They also measured H ENA fluxes during another auroral breakup [32] and in a solar-eclipse rocket campaign [33].

2.3. Invoking ENAs to Explain New Satellite Observations

After detecting H ENAs at rocket altitudes, satellites in low Earth orbits observed proton fluxes near the geomagnetic equator that could be traced to ENAs from the radiation belts. Heikkila reported 0.01–10 keV protons at $90° \pm 8°$ relative to B, using an ion spectrometer on the international satellite *ISIS-1* [34]. Hovestadt, Häusler & Scholer [35], and Moritz [36] surveyed the proton population in the radiation belt, using two SSD instruments for 0.5–1.5 MeV and 0.25–1.65 MeV on the West-German satellite *Azur*, followed by Mizera & Blake [37] with SSDs (0.2–1.5 MeV) and an ion spectrometer (12–185 keV) on the *USAF OV1-17*.

To obtain the altitude-independent proton flux shown in Fig. 2.5, Moritz reasoned [36] that a source whose strength increases with a similar altitude dependence must compensate the higher proton loss rate to the denser atmosphere at lower altitudes. He proposed that CX of the resulting H ENAs with O atoms, the dominant constituent at altitudes 400–1,000 km, replenished the original protons according to

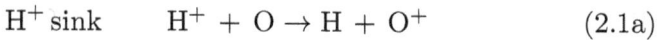

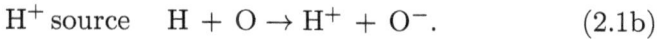

$$\text{H}^+ \text{ sink} \qquad \text{H}^+ + \text{O} \rightarrow \text{H} + \text{O}^+ \qquad\qquad (2.1a)$$

$$\text{H}^+ \text{ source} \qquad \text{H} + \text{O} \rightarrow \text{H}^+ + \text{O}^-. \qquad\qquad (2.1b)$$

The high proton concentration near the equator suggested an origin of the H ENAs in the inner radiation belt at or near the geomagnetic equator, where protons are trapped at $\alpha \approx 90°$. After CX with exospheric H as in Eq. (1.1), their charge neutrality allowed them to escape the magnetic trap, enabling their detection at lower altitudes centered around the magnetic equator. He deduced the approximate inner radiation-belt proton flux at the magnetic equator at altitudes $1.5 \leq L \leq 4.5$ from the 0.25–1.65 MeV proton measurements at low altitudes. His simple model invoked the production of ENAs and their ability to travel large distances. Hence,

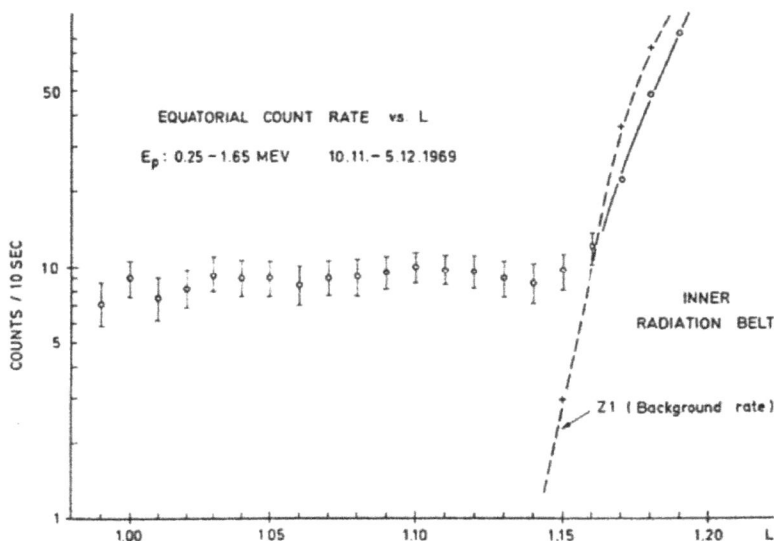

Fig. 2.5.   The ion spectrometer EI-92 aboard the West German satellite *Azur* detected a constant proton flux at 0.25–1.65 MeV as a function of altitude (or *L*-shells, Sec. 1.2) at the geomagnetic equator between Nov 10 and Dec 5, 1969 [36]. (© DGG. Reproduced with permission).

he has conceived the very principle of remote sensing space plasmas *via* ENAs!

Based on the constant H^+ flux shown in Fig. 2.5, Moritz assumed equilibrium between proton sink (Eq. (2.1a)) and source (Eq. (2.1b)) at each given altitude

$$v \cdot n_O \cdot \sigma_{H+HCX}^{O}(E) \cdot J_p(E) = v \cdot n_O \cdot \sigma_{HH+CX}^{O}(E) \cdot J_H(E).$$

$$(2.2)$$

$J_p(E)$ and $J_H(E)$ are the differential H^+ and H fluxes, respectively. The superscript in the CX cross-section denotes the enabling collision partner in the H CX processes. Both CX processes require O, the ANA with altitude-dependent density $n_O(L)$, thus reducing Eq. (2.2) to

$$J_H(E) = \frac{\sigma_{H+HCX}^{O}(E)}{\sigma_{HH+CX}^{O}(E)} J_p(E).$$

$$(2.3)$$

Using the known CX cross-sections and the observed H$^+$ flux, he deduced the H ENA flux $J_H(E) = 3.18 \cdot 10^{-3} \cdot E^{-5.77}$ 1/cm^2 s sr MeV. Based on CX between the protons in the inner radiation belt and exospheric H with density $n_H(r)$ [38], he derived the maximum radiation belt proton flux at the equator by inversion of the ENA production (Eq. (3.11)) according to a method outlined in Sec. 6.1.2.2. Invoking CX and ENAs as the transport mechanism across $\vec{B}$ to explain the H$^+$ flux measured *in-situ*, Moritz performed the first quantitative ENA study of the inner radiation belt and demonstrated the *principle of remote sensing space plasmas via ENAs*.

We note that neutrons mimic ENAs in the formation of the inner radiation belt. Cosmic rays produce neutrons in the interaction with the upper atmosphere, some of which escape to higher altitudes and decay into protons in the radiation belt region. They become trapped in the dipolar magnetic field like PUIs (Sec. 3.4.4), through a process coined *Cosmic Ray Albedo Neutron Decay* (CRAND) [39, 40]. Together with the multi-step process to the *in-situ* detection of H$^+$ fluxes described above and illustrated in Fig. 2.6, this radiation belt diagnostics is a harbinger of the *IBEX* ENA Ribbon (Sec. 7.3) [41] detected 40 years later. In both cases, neutral particles (neutrons

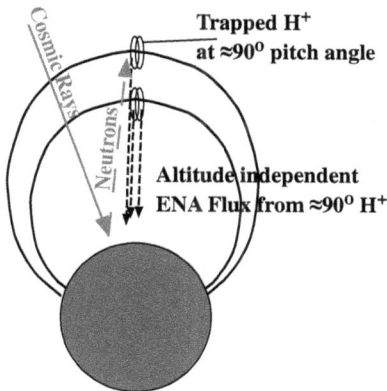

Fig. 2.6. Cosmic rays produce neutrons in the Earth's atmosphere, which then decay on their way out. Some of them get trapped as H$^+$, preferentially at $\approx 90°$ pitch angle in the Earth's $\vec{B}$-field [39, 40]. CX with exospheric atoms turns some H$^+$ into ENAs, as observed on S/C in low Earth orbit [36].

or neutral SW) get trapped after becoming charged (into protons or PUIs), and then ENAs, resulting from another CX, enable the detection from afar. In a way, they provide the appropriate bookends for our story of neutral-atom astronomy.

A lack of recognition for CX and its importance for magnetospheric dynamics persisted into the late 1970s. In 1979, a joint proposal by W.H. Ip, E. Keppler and A. K. Richter of the Max-Planck Institute for Aeronomy, Lindau-Katlenburg, Germany, and C. Y. Fan, C. C. Curtis, and K. C. Hsieh of the University of Arizona in Tucson to study the magnetospheric dynamics with satellite-borne ENA instrumentation was rejected on the grounds that CX had no significant role in magnetospheric dynamics. However, Tinsley's comprehensive and timely review of CX between energetic H^+, $He^{+,}$ and O^+ ions and ambient thermal H atoms in the geocorona [42] kept alive the dream of future ENA studies.

2.4. First ENA Images from Space

In 1985, Roelof, Mitchell, and Williams reported the detection of ENAs produced by CX between magnetospheric H^+ and geocoronal H atoms on Nov 22–23, 1973, and Dec 17, 1977, with their satellite-borne ion detectors, EPE on *IMP-7* and *IMP-8*, and MEPI on *ISEE-1* (Fig. 2.7). During the observations, the particle sources and the ion detectors were not magnetically connected, which assured them that the sensors could detect only ENAs [43].

Roelof used the sectoring of MEPI to identify the arrival directions of the particles and organize the ENA fluxes into a pixelated map of the magnetic storm on Nov 29, 1978 (Fig. 2.7b) [44]. Simulation of the ENA flux distribution emitted from the field-aligned Birkland currents into the *ISEE-1* FOV and variation of the underlying current distribution to match the observed ENA fluxes led to the model ENA distribution shown in Fig. 2.7a. The use of ion detectors to image substorms in ENAs during times with no magnetic connection to the source plasma continued with *POLAR* [45]. These

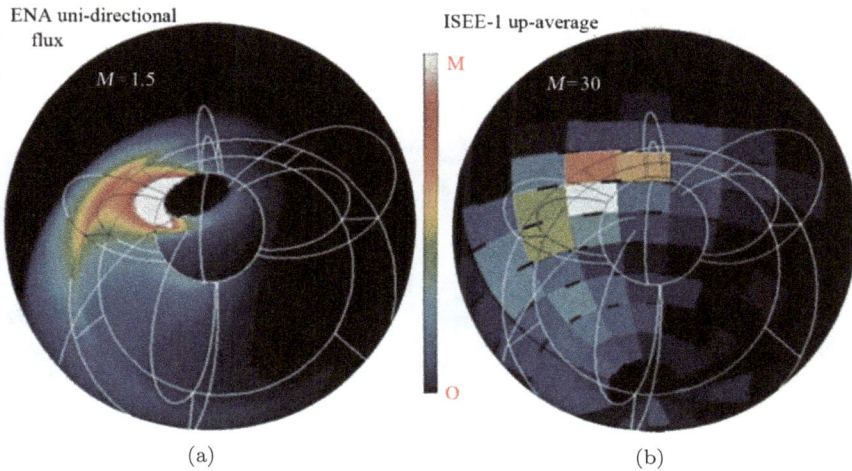

Fig. 2.7.   First ENA image of a space plasma showing the field-aligned currents during the magnetic storm on Nov 29, 1978. Computer simulated (a) and observed (b) H ENA flux organized by the arrival directions at the ion detector MEPI aboard the *ISEE-1* S/C [44]. (© AGU. Reproduced with permission).

results provided the proof-of-concept for ENA imaging and thus revitalized the quest for space-based ENA diagnostics.

The first dedicated magnetospheric ENA imager was Prelude In Planetary Particle Imaging (PIPPI) on the Swedish microsatellite *Astrid-1* (Sec. 5.2.2). It was launched in Jan 1995 from Plesetsk, Russia, into a low-altitude (10^3 km) circular polar orbit at 83° inclination [46, 47]. Figure 2.8 shows two ENA images taken by *Astrid-1* PIPPI [46], whose analysis suggested that the ENAs are produced by CX between exospheric neutrals at 300–400 km altitude and precipitating or mirroring ions [47].

In July 1991, a *NASA* Science Definition Team submitted a mission plan, the *Inner Magnetosphere Imager* (*IMI*), to diagnose the inner magnetosphere with ENAs, which led to the launch of *IMAGE* in 2000 (Sec. 4.1.2). Thus, the end of the 20th century marked the dawn of wide-spread ENA diagnostics of space plasmas, including the first detection of heliospheric ENAs [48].

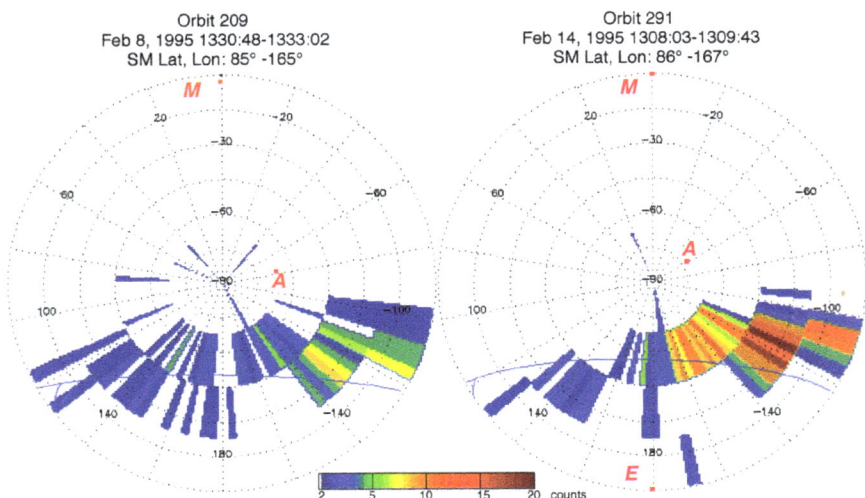

Fig. 2.8. ENA images from the first orbiting ENA imager [46]. Sampled are 26–37 keV ENAs in the anti-sunward hemisphere taken during two substorms. Azimuth and radius of the polar coordinates represent the PIPPI sensor's spin angle and polar angle, respectively. The spin axis lies in the plane of PIPPI's 2π-FOV (Fig. 5.7) and points roughly at the Sun. The thin blue arc indicates the Earth's limb and the small "hook" the terminator. The red dot labeled A is the anti-sunward direction, E the nadir, and M the magnetic pole. ENAs originating from trapped and precipitating ions appear in the bright sectors along the Earth's limb. (© Elsevier. Reproduced with permission).

References

1. Meinel, A. B. (1951). Doppler shifted auroral hydrogen emission, *Astrophys. J.*, **113**(1), 50–54.
2. Meinel, A. B., & Fan, C. Y. (1952). Laboratory reproduction of auroral emission by proton bombardment, *Astrophys. J.*, **115**, 330–331.
3. Fan, C. Y., & Meinel, A. B. (1953). Laboratory ionic-impact emission spectra, *Astrophys. J.*, **118**(2), 205–213.
4. Vegard, L. (1939). Hydrogen showers in the auroral region, *Nature*, **144**, 1,089–1,090.
5. Fan, C. Y., & Schulte, D. H. (1954). Variations in the auroral spectrum, *Astrophys. J.*, **120**(3), 563–565.
6. Parker, E. N. (1958). Dynamics of the interplanetary gas and magnetic field, *Astrophys. J.* **128**, 644–676.
7. Dessler, A. J. and Parker, E. N. (1959). Hydromagnetic theory of geomagnetic storms, *J. Geophys. Res.*, **64**(12), 2,230–2,252.

8. Chamberlain, J. W. (1961). *Physics of Aurora and Airglow* (Academic Press, New York).

9. Parker, E. N. (1963). *Interplanetary Dynamical Processes.* (Interscience Publishers, John Wiley & Sons, New York).

10. Neugebauer, M., & Snyder, C. W. (1966). Mariner 2 observations of the solar wind: 1, Average Properties, *J. Geophys. Res.*, **71**(19), 4,469–4,484.

11. Neugebauer, M., & Snyder, C. W. (1967). Mariner 2 observations of the solar wind: 2, Relation of the plasma properties to the magnetic field, *J. Geophys. Res.*, **72**(7), 1,823–1,828.

12. Akasofu, S.-I. (1964). The neutral hydrogen flux in the solar plasma flow-I, *Planet. Space Sci.*, **12**(10), 905–913.

13. Brandt, J. C., & Hunten, D. M. (1966). On ejection of neutral hydrogen from the Sun and the terrestrial consequences, *Planet. Space Sci.*, **14**, 95–105.

14. Cloutier, P. A. (1966). A comment on "The neutral hydrogen flux in the solar plasma flow" by S.-I. Akasofu, *Planet. Space Sci.*, **14**, 809–812.

15. Collier, M. R., Moore, T. E., Ogilvie, K. W. *et al.* (2001a). Observations of neutral atoms from the solar wind, *J. Geophys. Res.*, **106**(A11), 24,893–24,906.

16. Davis, L. Jr. (1955). Interplanetary magnetic fields and cosmic rays, *Phys. Rev. Lett.*, **100**(5), 1,440–1,444.

17. Dessler, A. J. (1967). Solar wind and interplanetary magnetic field, *J. Geophys. Res.*, **5**(1), 1–41.

18. Morton, D. C., & Purcell, J. D. (1962). Observations of the extreme untralviolet radiation in the night sky using an atomic hydrogen filter, *Planet. Space Sci.*, **9**(8), 455–458.

19. Patterson, T. N. L., Johnson, F. S., & Hansen, W. B. (1963). The distribution of interplanetary hydrogen, *Planet. Space Sci.*, **11**(7), 767–778.

20. Allison, S. K. (1958). Experimental results on charge-changing collisions of hydrogen atoms and ions at kinetic energies above 0.2 keV, *Rev. Mod. Phys.*, **30**(4), 1,137–1,168.

21. Allison, S. K., Cuevas, J., & Garcia-Muñoz, M. (1962). Partial atomic stopping power of gaseous hydrogen for hydrogen beams, *Phys. Rev.*, **127**(3), 792–798.

22. Stebbings, R. F., Smith, A. C. H., & Ehrhardt, H. (1964). Charge transfer between oxygen atoms and O+ and H+ ions, *J. Geophys. Res.*, **69**(11), 2,349–2,355.

23. Lindhard, J., Scharff, M., & Schiøtt, H. E. (1963). Range concepts and heavy ion ranges, *Mat. Fys. Medd. Dan. Vid. Selsk.*, **33**, 1–42.

24. Committee on Nuclear Science chaired by S. K. Allison (1964). *Nuclear Science Series Report Number 39*: Studies in Penetration of Charged Particles in Matter (National Academy of Sciences – National Research Council Publication 1133).

25. Van Allen, J. A., Ludwig, G. H., Ray, E. C., & McIlwain, C. E. (1958). Observation of high intensity radiation by Satellites1958Alpha and Gamma, *Jet Propulsion*, **28**, 588–592.

26. Van Allen, J. A., McIlwain, C. E., & Ludwig, G. H. (1959). Radiation observations withj satellite 1958 e. *J. Geophys. Res.*, **64**(3), 271–286.

27. Fan, C. Y., Meyer, P., & Simpson, J. A. (1960a). Cosmic radiation intensity decreases observed at the earth and in nearby planetary medium, *Phys. Rev. Lett.*, **4**(8), 421–423.

28. Fan, C. Y., Meyer, P., & Simpson, J. A. (1960b). Preliminary results from the Space Probe Pioneer V, *J. Geophys. Res.*, **65**(6), 1862–1863.

29. Evans, D. S. (1965). Low energy charged particle detection using the continuous channel electron multiplier, *Rev. Sci. Instr.*, **36**(3), 375–382.

30. Wax. R. L., & Bernstein, W. (1967). Energy-independent detector for total hydrogen fluxes in the range 1–10 keV for space and laboratory applications, *Rev. Sci. Instr.*, **38**(11), 1,612–1,615.

31. Bernstein, W., Inouye, G. T., Sanders, N. L., & Wax, R. L. (1969). Measurements of precipitated 1–20 keV protons and electrons during a breakup aurora, *J. Geophys. Res.*, **74**(14), 3,601–3,608.

32. Wax. R. L., & Bernstein, W. (1970). Rocket-borne measurements of emissions and energetic hydrogen fluxes during an auroral breakup, *J. Geophys. Res.*, **75**(4), 783–787.

33. Wax, R. L., Simpson, W. R., & Bernstein, W. (1970). Large fluxes of 1-keV atomic hydrogen at 800 km, *J. Geophys. Res.*, **75**(31), 6390–6393.

34. Heikkila, W. J. (1971). Soft particle fluxes near the equator, *J. Geophys. Res.*, **76**(4), 1,076–1,078.

35. Hovestadt, D., Häusler, B., & Scholer, M. (1972). Observation of energetic particles at very low altitudes near the geomagnetic equator. *Phys. Rev. Lett.*, **28**(20), 1,340–1,344.

36. Moritz, J. (1972). Energetic protons at low equatorial altitude: a newly discovered radiation belt phenomenon and its explanation, *Z. Geophys.*, **38**, 701–717.

37. Mizera, P. F., & Blake, J. B. (1973). Observations of ring current protons at low altitudes, *J. Geophys. Res.*, **78**(7), 1058–1062.

38. Meier, R. R. (1970). Observations of Lyman-α and the atomic hydrogen distribution in the thermosphere and exosphere, *Space Res.*, **10**, 572–581.

39. Singer, S. F. (1958). Trapped albedo theory of the Radiation belt, *Phys. Rev. Lett.*, 1, 181–183.
40. Hess, W. N. (1959). Van Allen Belt protons from cosmic-ray neutron leakage, *Phys. Rev. Lett.*, **3**, 11–13.
41. McComas, D. J., Allegrini, F., Bochsler, P. *et al.* (2009). Global observations of the interstellar interaction from the Interstellar Boundary Explorer (IBEX), *Science*, **326**(5955), 959–962.
42. Tinsley, B. A. (1981). Neutral atom precipitation — a review, *J. Atmo. Terres. Phys.*, **43**(A5), 617–632.
43. Roelof, E. C., Mitchell, D. G., & Williams, D. J. (1985). Energetic neutral atoms (E ∼ 50 keV) from the ring current: IMP 7/8 and ISEE 1, *J. Geophys. Res.*, **90**(A11), 10,991–11,008.
44. Roelof, E. C. (1987). Energetic neutral atom image of a storm-time ring current, *Geophys. Res. Lett.*, **14**(6), 652–655.
45. Henderson, M. G., Reeves, G. D., Spence, H. E., Sheldon, R. B. *et al.* (1997). First energetic neutral atoms images from Polar, *Geophys. Res. Lett.*, **24**(10), 1,167–1,170.
46. Barabash, S., C:son Brandt, P., Norberg, O., Lundin, R., Roelof, E. C., *et al.*, (1997). Energetic atom imaging by the Astrid micro satellite, *Adv. Space Res.*, 20 (4/5), 1,055–1,060.
47. Brandt, P., Barabash, S., Norberg, O., Lundin, R., Roelof, E. C., *et al.*, (1997). ENA imaging from the Swedish micro satellite Astrid during the magnetic storm of 8 February 1995, *Adv. Space Res.*, 20 (4/5), 1,061–1,066.
48. Hilchenbach, M., Hsieh, K. C., Hovestadt, D. *et al.* (1998). Detection of 55–80 keV hydrogen atoms of heliospheric origin by CELIAS/HSTOF on SOHO, *Astrophys. J.*, **503**(2), 916–922.

Chapter 3

Remote Sensing of Space Plasma Through ENA Observations

"But although all our knowledge begins with experience, it does not follow that it arises from experience."

Immanuel Kant, 1724–1804

ENA imaging means taking pictures of plasmas beyond the reach of the S/C with a camera that responds to ENAs instead of photons. In other words, this diagnostic tool uses neutral atoms over a wide range of E and m as messengers of the state and structure of the parent plasma. This method is possible because the production of ENAs maintains the ions' identity $(m, \boldsymbol{v})$, and their propagation is unaffected by intervening $\vec{E}$ and $\vec{B}$ fields. They follow straight trajectories unless deflected by gravitational fields, radiation pressure, or collisions. These attributes make ENAs suitable for imaging, just like photons across the electro-magnetic spectrum. Thus, the use of ENAs as messengers in space and astrophysics settings can be coined *"neutral-atom astronomy"*.

This chapter will establish the analytic relationship between the observed ENA flux and their parent space-plasma properties, including potential flux modifications in transit. We will specify the information that we can extract from the remote plasma and the limitations of this method.

We focus on analyzing ENAs from ions in remote plasmas (after CX with ANAs) to gain insight into their ionic ensembles.

The interacting plasmas may be partially ionized, containing both charged particles and ANAs. Suppose the ANAs of the remote population have a substantial bulk velocity and reach the ENA imager, as is the case for the ISN flow through the heliosphere. Then ENA imaging can simultaneously investigate the ANA population. Related diagnostic methods, instrumental techniques, and their applications have been reviewed [1–3] and will be discussed in Chaps. 4 and 5.

3.1. ENA "Imaging": Determination of the Spatial, Energy & Elemental Distributions of Remote Ion Populations

The space plasmas under consideration are "collisionless". In other words, the frequencies ν_{Coll} for collisions among pairs of electrons or ions, between electrons and ions, or between neutral atoms and electrons or ions are rare compared with the frequencies that characterize the collective behavior of plasmas. The latter are the cyclotron frequencies (ω_{ce}, ω_{ci}) and the plasma frequency (ω_{pe}), or $2\pi\nu_{\text{Coll}} \ll \omega_{\text{ce}}, \omega_{\text{ci}}, \omega_{\text{pe}}$. Alternatively, the collisional *mean free path* λ_{Coll} is large compared with the Debye length λ_{D}, the electron and ion gyroradii r_{ce} and r_{ci}, or the typical scale length for gradients in these plasmas. For ENA imaging, it is essential that only a few of the "messengers" encounter any type of collision between birth and detection. Generally, collisions in transit from the source to the observer limit the usefulness of ENA imaging because they disturb the ENA trajectory. We will return to this limitation to the method in Sec. 3.4. However, ENA imaging relies on CX between the energetic ions and ANAs that can often be taken as stationary to create the very messenger ENAs that provide information on the plasma component.

CX between an ion and a neutral atom is distinct from elastic collisions (Sec. 3.4.3). In elastic collisions, the colliding partners keep their identity, while CX collisions alter the partners' *charge states* (Q) through exchanging an electron (Fig. 3.1). The parent ion i_1 leaves the encounter as an ENA a_1, and the plasma incorporates the ANA a_2 of the background gas as a positive ion i_2:

$$i_1 + a_2 \rightarrow a_1 + i_2 \tag{3.1}$$

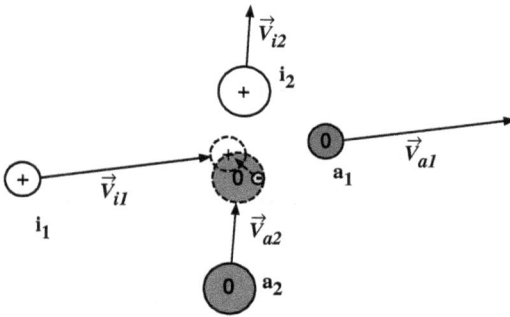

Fig. 3.1. This schematic of a CX collision between a positive plasma ion i_1 and an ANA a_2 provides a blown-up view of the star in Fig. 1.1.

We refer the reader to textbooks that describe the CX processes [4] and the theory of the related interactions [5]. Most importantly, the electron mass is only $1/1836$ of the nucleon mass. Also, in most cases relevant to our discussions, the difference in the electron binding energy before and after the encounter is negligible compared to the kinetic energy of the parent ion and the daughter ENA. Thus, the ENA carries the ion's momentum. However, even small momentum changes and the resulting angular scattering gain significance for the lowest energies considered, *e.g.*, for *secondary interstellar neutrals* (Sec. 7.4) [6, 7]. The most important CX interactions in space plasmas are resonant interactions, featuring the largest cross-sections. No net energy change occurs in a resonant process, which is generally the case for ions and atoms of the same species, *e.g.*, for the CX pairs of the most abundant elements (H, H^+) and (He, He^+). Also, CX for (O^+, H) and (H^+, O) occurs almost resonantly because their ionization potentials (O: 13.618 eV and H: 13.598 eV) differ only by 0.02 eV. In essence, due to the minute energy change and mass transfer, the change in momentum is negligible for CX, and the daughter ENA leaves the interaction with the momentum of the parent ion. In addition, the number of charge carriers in the plasma remains the same.

The above-mentioned features make CX in plasmas a fascinating interaction in two distinct ways:

- Firstly, CX provides an effective remote diagnostic tool for the ion distribution in plasmas constrained by a $\vec{B}$ field, such as fusion plasmas in their containment fields, planetary ionospheres and magnetospheres, or the heliosphere. An image of their ion distribution with spectral information can be obtained from a distant vantage point, using ENAs that leave these plasmas. This book will concentrate on the diagnostics of space plasmas and go beyond earlier reviews [1–3]. Many techniques and physical processes are also applicable to laboratory and fusion plasmas, with adjustments to the parameter regimes [8, 9].

- Secondly, the interaction between ions and neutral atoms of distinctly different energy distributions modifies the original ion population through the injection of PUIs into the plasma. Over time, such interaction either heats or cools the original plasma, depending on the energy content of the injected neutrals. A prime example is the injection of ENA beams into fusion plasmas as one of the standard heating methods installed in all major fusion research facilities [10, 11]. The ionization of the neutral beam injects substantial directed energy into the plasma, which is converted very effectively into heat. In addition, the possibility of removing impurities from fusion plasmas through neutral beam injection has been explored [12]. Similar interactions occur in magnetospheres of planets, *e.g.*, through the injection of Io's volcanic ejecta into the Jovian magnetosphere (Sec. 6.2.1) [13, 14]. At the boundary of the heliosphere, the unimpeded *interstellar gas* flow injects faster ions into the *interstellar plasma* that slows down, piles up in front of the HP. The IS plasma then flows around the heliosphere (Sec. 7.4) [15–18]. This injection of newly generated ions leads to distinct secondary ion populations that affect the momentum and energy balance in these regions. ENA diagnostics is the method of choice to determine the resulting ion populations [19–21]. We will encounter several related applications in this book, starting with the description of plasma parameters and processes that can be diagnosed with ENAs.

3.2. Ion Distribution Functions and Plasma Processes

The ENA distribution reflects the ion velocity distribution in the region of interest. Thus, a measurement of the ENA velocity distribution characterizes the ion population of a remote plasma. Among others, we can deduce the bulk quantities of the parent plasma as the moments of the measured velocity distribution. In the following, we provide a brief review of particle velocity distributions and their relation to plasma bulk properties, such as density, flow velocity, and temperature, while leaving a comprehensive treatment to plasma physics textbooks [22, 23].

3.2.1. *Velocity distributions in thermal equilibrium*

While the motion of charged particles in electromagnetic fields can be described entirely deterministically, it is impossible to track each particle in a plasma. Instead, we can use a statistical description of the particle ensemble as a distribution in velocity and location $f(\vec{v}, \vec{r})$. In thermal equilibrium, the particle distributions are Maxwellian. Let us start with a 1D Maxwellian distribution f_M, which represents, for example, the distribution of the velocity component v_l along the LOS $\vec{l}$, which is equivalent to $\vec{r}$ from the Sun for most heliospheric targets observed at 1 AU.

$$f_M{}^1(v_l) = a \exp\left(-\frac{mv_l^2}{2k_BT}\right) \tag{3.2}$$

T is the temperature of the ensemble, m the particle mass, k_B the Boltzmann constant, and a a normalization constant. The expression $f_M(v_l)dv_l$ describes the number of particles with velocity component v_l in the infinitesimal 1D velocity volume element dv_l. Integrating f_M over the entire velocity space yields the total particle density n:

$$n = \int_{-\infty}^{\infty} f_M{}^1(v_l)dv_l = a \int_{-\infty}^{\infty} \exp\left(-\frac{mv_l^2}{2k_BT}\right) dv_l \tag{3.3}$$

and provides the normalization constant $a = n(m/(2\pi k_BT))^{1/2}$. The Maxwellian distribution has the functional form of a Gaussian whose width $\sigma_M = (k_BT/m)^{1/2}$ determines the temperature T. Often, a characteristic speed of the Maxwellian is defined as $v_T = (2k_BT/m)^{1/2}$, which is different from the mean thermal speed v_{Th}. To obtain the 3D distribution, we multiply three 1D Maxwellians and renormalize them to match the particle density. This distribution describes the particle distribution in 6D phase space (3D for $\vec{r}$ and 3D for $\vec{v}$)

$$f_M^3(\vec{v}) = a_3 f_{Mx}^1(v_x) f_{My}^1(v_y) f_{Mz}^1(v_z) = n \left(\frac{m}{2\pi k_B T}\right)^{\frac{3}{2}} e^{\left(-\frac{m(v_x^2+v_y^2+v_z^2)}{2k_B T}\right)}$$

(3.4)

Any symmetries in the observation geometry or the velocity distribution substantially simplify the analysis. For example, thermal equilibrium requires isotropic particle distributions. Then, it is useful to consider a Maxwellian speed distribution, which arises from integrating the 3D distribution function in spherical coordinates over the two angular directions to obtain:

$$f_M(v) = 4\pi v^2 n \left(\frac{m}{2\pi k_B T}\right)^{3/2} \exp\left(-\frac{m(v^2)}{2k_B T}\right)$$

(3.5)

The appropriate volume element for an isotropic distribution in spherical coordinates $4\pi v^2 dv$ appears in the new normalization constant. The spherical symmetry of the Maxwellian in the source plasma rest frame is invoked to obtain the mean thermal speed v_{Th} as the first moment of the Maxwellian speed distribution (Eq. (3.5)).

$$v_{Th} = \frac{1}{n} \int_0^\infty f_M(v) v \, dv = \left(\frac{8k_B T}{\pi m}\right)^{\frac{1}{2}}$$

(3.6)

Similarly, the second moment of the velocity distribution can provide the pressure tensor. As discussed below, a tensor description is necessary for distributions that are not in thermal equilibrium and contain anisotropies. When in or close to thermal equilibrium, a scalar ideal gas pressure $p = nk_B T$ suffices, which relates to the mean thermal energy E_{Th} from the second moment of the Maxwellian speed distribution:

$$E_{Th} = \frac{1}{n} \int_0^\infty f_M(v) \left(\frac{mv^2}{2}\right) dv = \frac{2k_B T}{2}$$

(3.7)

3.2.2. *Non-thermal velocity distributions*

Particle populations that are not in thermal equilibrium may require descriptions without these convenient symmetries. In particular, magnetic fields imprint anisotropies on the velocity distributions, requiring different distributions perpendicular ($\perp$) or parallel ($\parallel$) to $\vec{B}$. With ENA observations from different vantage points, such anisotropies of the distributions reveal themselves. Also, we can assume that the distributions are gyrotropic. Hence, 2D descriptions in $\perp$ and $\parallel$ suffice for all applications relevant to ENA observations.

However, in typical space plasmas, also deviations in shape from a Maxwellian become important. A non-thermal distribution that appears to be common in space plasmas is the so-called κ-distribution. It was first formulated empirically based on observations [24]. Meanwhile, its application has been successful for space plasmas in many settings, such as magnetospheres [25–28], the inner heliosphere [29–32], the outer heliosphere, and its boundary region [21, 33–39]. More recently, it has also been identified as a more fundamental distribution that can be deduced from non-extensive statistical mechanics [40] as discussed in a recent review [41]. In a nutshell, many space plasmas experience a continuous input of directed energy, while re-distribution into thermal energy is slow due to the lack of collisions. Therefore, the plasma never reaches thermal equilibrium. The particles in the high-speed (or suprathermal) tail of the distribution on average retain a higher energy content than in an equivalent Maxwell distribution. Most appropriate to the applications covered here may be the κ-distribution for speeds f_κ, which is the non-thermal extension of the Maxwellian speed distribution in Eq. (3.8):

$$f_\kappa(v) = \frac{a}{v_T^3} \cdot \left[1 + \frac{(v - v_B \cos \vartheta)^2}{\left(\kappa - \frac{3}{2}\right) v_T^2 + v_B^2 \sin \vartheta^2} \right]^{-\kappa - 1} \cdot v^2 \qquad (3.8)$$

where ϑ is the angle between the directions of v and the bulk flow velocity $\vec{v}_B$. It has been shown rigorously that as $\kappa \to \infty$, the distribution approaches a Maxwellian. For smaller values of κ, the distribution exhibits an elevated suprathermal tail. Figure 3.2

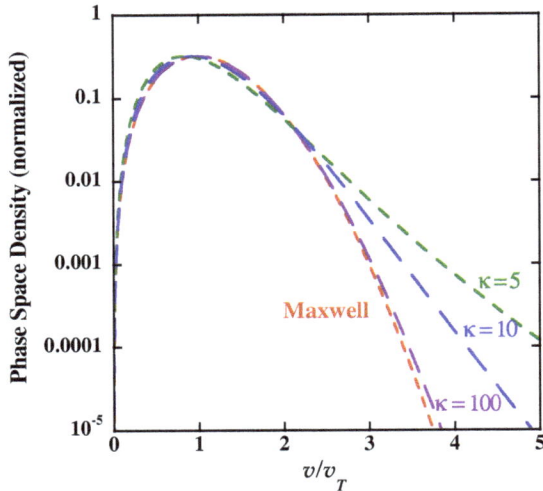

Fig. 3.2.   Maxwell's speed distribution and κ-distribution for $T = 8000$ K and $v_B = 0$ m/s with $\kappa = 5, 10$, and 100, normalized at their respective maximum values.

illustrates this behavior, which compares a Maxwellian and a κ speed distribution with several κ-values for the case $v_B = 0$ and $T = 8000$ K.

Numerous recent ENA spectra and images can be described by κ-distributions. Yet, when exploring non-thermal effects in the parent space plasmas of the observed ENAs, be aware that the energy spectra in ENA images of the heliospheric boundary taken over the 0.01–6 keV range (Sec. 7.5) can also indicate the presence of multiple source populations [19], which may be co-located. PUIs may be implanted into a thermal plasma when the plasma moves relative to an ANA population at the bulk speed $v_B \perp$ to $\vec{B}$, as briefly introduced below in Sec. 3.4.4. After injection, they fill a sphere in $\vec{v}$-space with radius v_B in the plasma rest frame, as observed in the SW for IS PUIs [42–44], and expected in the OHS due to the deceleration and deflection of the plasma flow around the HP (Sec. 7.4). In these cases, the observed ENA fluxes may contain information on an overlay of different sources staggered at various distances along the LOS, as we will see in Sec. 3.3. Disentangling these populations in the ENA spectra and images requires some independent insight into some of them through complementary *in-situ* observations and/or modeling of the relevant physical processes. Hence, ENA analysis is often an iterative exercise.

3.2.3. *Moving distributions and frame transformations*

In non-equilibrium situations, including relative motion, the plasma particle distribution function contains a broader range of information on the physical processes in the plasma. The moments of the distribution function $f_s(\vec{v})$ reflect the plasma bulk parameters, such as density n_s (s indicating atomic species), bulk velocity $\vec{v}_B$, and temperature T. They are accessible through the integration of $f_s v^k$, with $k = 0, 1, \ldots$, over the entire velocity space. In this sense, the total density n_s of species s is the zeroth moment of f_s, introduced in Eq. (3.3). To carry out these integrations requires defining the volume element in velocity space $v^2 dv d\Omega$ for the differential *phase space density* (PSD) f, shown in a Cartesian and spherical coordinate system in Fig. 3.3 for location (v, θ, ϕ).

If the remote plasma under consideration is moving toward the observer or the observing S/C is moving toward the plasma, the bulk speed v_B along a specific LOS, in direction (θ_0, ϕ_0), can be

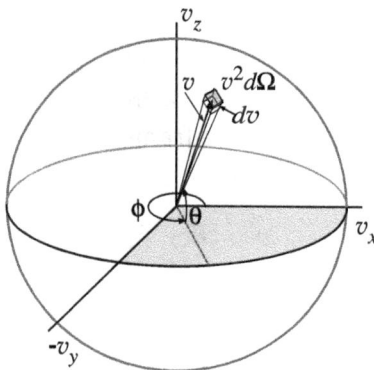

Fig. 3.3. The volume element $v^2 dv d\Omega$ in velocity space that applies to the differential PSD f is shown in spherical coordinates (v, θ, ϕ).

obtained as the first moment of the PSD $f(\theta, \phi, v)$. We introduce the distribution in spherical coordinates here because this is convenient for the connection to ENA imaging. Direction (two angles) and particle speed (or energy) provide the natural reference. For the first moment, we must now consider a vector quantity, the particle velocity $\vec{v}$, and the solid angle element becomes $d\Omega = \sin\theta d\theta d\phi$. It may be convenient to start with the two angle integrations to obtain the flow direction:

$$\theta_0 = \frac{\int_0^\infty \iint_{4\pi} f(\theta, \phi, v)\theta \sin\theta d\theta v^2 dv}{\int_0^\infty \iint_{4\pi} f(\theta, \phi, v) \sin\theta d\theta v^2 dv} \tag{3.9a}$$

$$\phi_0 = \frac{\int_0^\infty \iint_{4\pi} f(\theta, \phi, v)\phi \sin\theta d\theta v^2 dv}{\int_0^\infty \iint_{4\pi} f(\theta, \phi, v) \sin\theta d\theta v^2 dv} \tag{3.9b}$$

For the direction (θ_0, ϕ_0), we can now obtain the bulk flow speed v_B:

$$v_B = \frac{\int_0^\infty \iint_{4\pi} f(\theta, \phi, v)v \sin\theta d\theta v^2 dv}{\int_0^\infty \iint_{4\pi} f(\theta, \phi, v) \sin\theta d\theta v^2 dv} \tag{3.9c}$$

Equations (3.9a) and (3.9b) are equivalent to finding the maximum signal in a 4π image, while Eq. (3.9c) provides the bulk flow speed. For a Maxwellian distribution that drifts with velocity (θ_0, ϕ_0, v_B), it suffices to consider an integration over a 1D distribution in direction (θ_0, ϕ_0) in the third step. This method is related to the appropriate analysis of ENA images presented in Sec. 3.3.

Depending on symmetries in the PSD, it may be useful to perform the integrations either in the plasma rest frame, as in Eq. (3.9), or the observer frame. Generally, the frame transformation is a Galilean transformation in 3D according to:

$$\vec{v'} = \vec{v} + \vec{v}_B \tag{3.10}$$

We will use primed quantities for the observer frame to distinguish it from the plasma frame, which is particularly important when discussing the ENA fluxes in Sec. 3.3 and the ISN flow in Sec. 7.2. If the ENA distribution follows approximately a power law, a simplification by a Compton-Getting transformation may be applicable [45, 46].

3.3. Relation Between the Observed ENA Flux and the Remote Ion Distribution

The quantity of diagnostic interest is the ENA flux $j_{ENA}(v', \theta', \phi')$, which is directly related to the ion velocity distribution f_i:

$$j'_{ENA}(v', \theta', \phi') = v' f_{ENA}(v', l) = v'[n_a(l)\sigma_{aiCX}(v)\Delta l f_i(v, l)] \tag{3.11}$$

Ions of the parent plasma interact with co-located ANAs of density n_a via CX with cross-section σ_{aiCX} over a finite segment Δl along the LOS around distance l from the observer. Note that the ENA flux is evaluated in the observer frame, $i.e.$, at speed v'. In contrast, the CX occurs in the plasma frame at speed v. For simplicity, we take a single ANA species ($\sum_a n_a \sigma_{aiCX}$ replaces $n_a \sigma_{aiCX}$ in Eq. (3.11) for multiple species). Also, the plasma ions are much faster than the ANAs. Thus, $f_i(v, \theta, \phi)$ and n_a are sufficient. In Eq. (3.11), the dimensionless quantity

$$P_{\Delta l}(v) = n_a \sigma_{aiCX}(v)\Delta l = \Delta l / \lambda_{aiCX}(v) \tag{3.12}$$

represents the probability for a typical ion of the distribution f_i to convert into an ENA over the distance Δl. λ_{aiCX} is the mean free path for CX. Within Δl, both n_a and f_i are considered constant.

The observable that represents the ion distribution function is the ion flux $j_i(v, \theta, \phi)$, which we will evaluate with a sensor in the plasma frame for simplicity. To obtain the flux element, dj_i, in $\boldsymbol{v}$-space, we multiply $v f_i(v)$ by $v^2 dv d\Omega$, as illustrated in Fig. 3.3.

$$dj_i(v, \theta, \phi) = v f_i(v, \theta, \phi)v^2 dv d\Omega \tag{3.13}$$

A typical sensor measures differential particle flux $dJ/(dEd\Omega)$ as a function of E rather than v. Therefore, transforming $\boldsymbol{v}$-space into E-space is necessary. Let us trace a few critical steps between the distribution function and the observable. The count rates R recorded by the sensor are related to the differential flux through the sensor's geometric factor G and its finite energy window ΔE:

$$R(\theta, \phi) = G \cdot \frac{dJ(\theta, \phi)}{d\Omega dE} \cdot \Delta E \qquad (3.14)$$

G is composed of the collection area a, the finite solid opening angle $\Delta\Omega$ that defines the FOV, and the detection efficiency $\eta_s(E)$ at the center of ΔE for species s, making G energy- and species-dependent.

$$G_s(E) = a\eta_s(E)\Delta\Omega \qquad (3.15)$$

The product $G \cdot \Delta E$ refers to the sensor collection power for particles, which naturally decreases when the sensor's angular and energy resolution ($\Delta\Omega$ and ΔE) are improved. Obviously, the finite angular and energy widths broaden the observed distributions. We will discuss the geometric factor and resolution of ENA sensors with their dependencies in more detail in Chaps. 4, 5, and Appx. A. We will ignore finite resolution in the following derivation and use only differential quantities until we return to the sensor. Using

$$\frac{dE}{dv} = \frac{d}{dv}\left(\frac{m}{2}v^2\right) = mv \qquad (3.16)$$

Equation (3.13) turns into:

$$\frac{dJ(E, \theta, \phi)}{dEd\Omega} = f(v, \theta, \phi) \cdot \frac{v^2}{m} \qquad (3.17a)$$

where $dJ/dEd\Omega$ now replaces the flux element $dj/dvd\Omega$ in v-space. Expressed in the typical units, E in keV and n in cm^{-3}, we obtain

$$\frac{dJ(E, \theta, \phi)}{dEd\Omega}\left[\frac{counts}{s\,cm^2\,sr\,keV}\right] = f(v, \theta, \phi)\left[\frac{s^3}{cm^6}\right]\frac{1.835 \cdot 10^{30}E[keV]}{A^2}$$

$$(3.17b)$$

Equation (3.17b) is universally applicable to any particle species for ions and ENAs, with A denoting the particle mass in atomic units.

Combining Eq. (3.17a) with Eq. (3.12), we obtain the differential ENA flux (LHS) in the observer frame from a small segment Δl along the LOS, while the CX interaction occurs in the plasma frame (RHS).

$$\frac{J(E', \theta', \phi')}{dE'd\Omega} = n_a\sigma_{aiCX}(E)\Delta l \cdot f_i(v, \theta, \phi) \cdot \frac{v^2}{m} \qquad (3.18a)$$

The observed ENA flux results from the LOS integration along the source plasma from the near l_1 to the far edge l_2.

$$\frac{J(E', \theta', \phi')}{dE' d\Omega} = \sigma_{aiCX}(E) \cdot \frac{v^2}{m} \cdot \int_{l_1}^{l_2} S(E, l) \cdot n_a(l) \cdot f_i(l, v, \theta', \phi') dl$$

(3.18b)

The integral in Eq. (3.18b) allows both n_a (depicted as constant in Fig. 3.4 for simplicity) and $f_i(v)$ to vary along l. For magnetospheric ENAs where the source thickness $(l_2 - l_1)$ is comparable to its distance l from the observer, *i.e.*, $l_1 \approx 0$, the integral must include the ENA extinction. It is included *via* the survival probability $S(E, l)$ for the ENAs from their source distance l to the observer at $l = 0$;

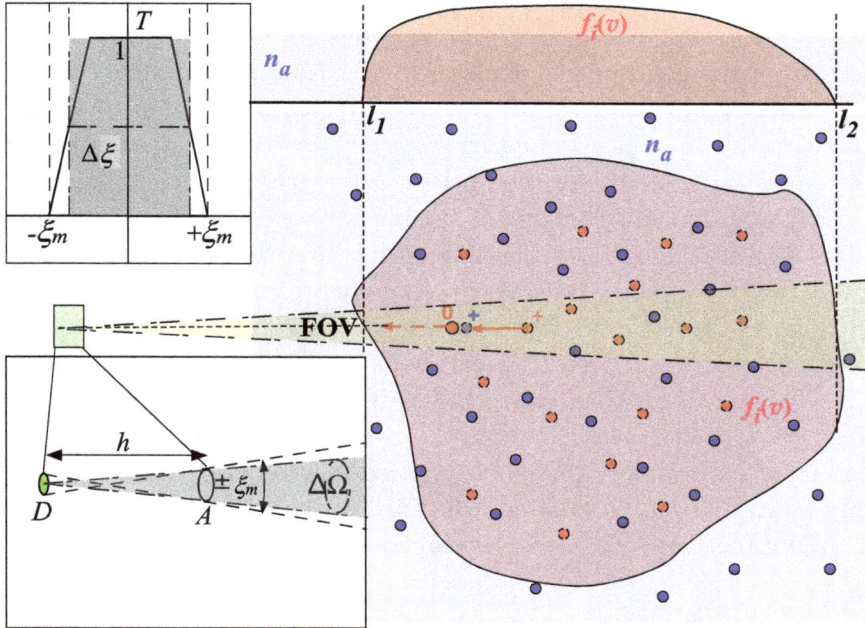

Fig. 3.4. Schematic view of ENA observations of a distant plasma cloud, along with the sensor FOV (lower left) and resulting angular response $T(\xi)$ (upper left). ENAs arise from CX of the ion distribution $f_i(v)$ (red circles) with the ANAs of number density n_a (blue circles) inside the cloud. They are integrated along the LOS from the near (l_1) to the far (l_2) edge and over the entire sensor FOV using $T(\xi)$.

hence a line integral. S_{So} inside the source region and S_{Tr} in transit to the observer may differ, which we distinguish by subscripts in the following. For ENAs from the heliospheric boundary region, most losses scale as $1/l^2$ and thus occur mostly from l_1 to the observer, leaving only $S_{Tr}(E, l_1)$ outside the integral. We consider the processes behind S in Sec. 3.4.2.

Equation (3.18a) shows that ENA diagnostics, like most optical diagnostics of optically thin regions, only provides LOS integrated information of the product $n_a(l) \cdot f_i(l)$, with an additional ambiguity in the total thickness $(l_2 - l_1)$. Even with this handicap, the ENA images still return the average ion distribution function for each LOS if absolute densities and/or source region thickness are not of immediate interest. Additional information is necessary about two of these three unknowns by alternative means to deduce the third quantitatively. For example, several strategies are available for ions from the heliospheric boundary (Sec. 7.5.2). The density of the surrounding interstellar gas n_a is accessible through complementary observations [47–51]. Also, either *Voyager* provides the density of the ions in the IHS *in-situ* for the relevant energy range [52, 53], or global heliospheric modeling offers the thickness of the IHS [54, 55].

The sensor integrates the differential ENA flux from Eq. (3.18a) over the FOV according to its angular response function $T(\xi)$ (Appx. A). For an ENA flux that smoothly varies with angle, we can replace the integral over $T(\xi)$ by the effective solid angle acceptance $\Delta\Omega$. It is set approximately to the width at half transmission or the FWHM. Likewise, we can take the sensor efficiency $\eta_s(E'_j)$ at the center of each energy passband E'_j (finite number of energy steps) and multiply by $\Delta E'$ to obtain the particle count rate in energy step j from direction (θ', ϕ') in spherical coordinates:

$$R(E'_j, \theta', \phi') = \frac{dJ(E', \theta', \phi')}{dE' d\Omega} \cdot a \cdot \eta_s(E'_j)\Delta E'\Delta\Omega$$

$$= \frac{dJ(E', \theta', \phi')}{dE' d\Omega} \cdot G_s(E'_j) \tag{3.19}$$

As discussed in Sec. 3.2, we can determine the bulk flow speed v_B of a remote plasma ion population along the LOS from the observer as the first moment of the ion distribution $f_i(v, \theta, \phi)$ in that direction. It relates directly to the observed differential ENA flux $J_n(v', \theta', \phi')$:

$$
\begin{aligned}
v_B(\theta', \phi') &= \frac{\int_{v_L}^{v_U} f_i(v, \theta', \phi') v \cdot v^2 dv}{\int_{v_L}^{v_U} f_i(v, \theta', \phi') v^2 dv} \\
&= \frac{m^2 \int_{E_L}^{E_U} \dfrac{dJ(E', \theta', \phi') dE'}{\sigma_{aiCX}(E') dE' d\Omega}}{\left(\dfrac{m}{2}\right)^{\frac{1}{2}} \int_{E_L}^{E_U} \dfrac{dJ(E', \theta', \phi') dE'}{\sigma_{aiCX}(E') dE' d\Omega E'^{1/2}}}
\end{aligned}
\tag{3.20}
$$

The integration boundaries are the lowest to the highest energy (E_L to E_U) of the observation range. The result is more accurate if most of the ENA distribution falls inside that range. Otherwise, corrections for the missing part of the distribution are necessary, resulting in a substantial systematic uncertainty.

The result of Eq. (3.20) depends on the energy resolution of the ENA sensor. While it is comparable to that of ion sensors for $E \geq 20$ keV, it is typically relatively poor compared to ion sensors at lower energies, as we will see in Chaps. 4 and 5. The main reason is that converting the ENAs into ions is necessary, with substantial energy loss and straggling before their analysis for E. The energy resolution is often commensurate with the general features of the ion velocity distributions in many plasmas. However, the precision determination of the bulk flow velocity and/or temperature of remote distributions at moderate or low temperatures requires additional information.

The extreme example is the derivation of the interstellar parameters and the characteristics of the plasma flow around the heliosphere from the respective ANA observations, which takes advantage of Keplerian trajectories in the Sun's gravitational field as discussed in Secs. 7.2 and 7.4. It suffices here to note that a widely applicable method to obtain the bulk speed v_B and simultaneously the temperature T from the observed ENA spectra and image (or angular distribution) involves a χ^2–fit to a convected Maxwellian distribution in the source region. This method has been successful in obtaining the ISN flow bulk parameters [56–59] and those of the secondary neutrals, which emerge from the ion distribution in the OHS [60, 61].

Alternatively, an analytical model can describe the interstellar bulk flow because most of the observations occur near the perihelion of the flow trajectories, which substantially simplifies the mathematical description [62, 63]. This method facilitates the extraction of the main physical processes that constrain the interstellar parameters (speed, direction, and temperature) from the observations [64]. The weighting of the observations in the χ^2-fit enables including the uncertainties and estimating the overall measurement uncertainties [57]. These techniques can reveal the plasma flow pattern around the heliosphere and how it is shaped by the ISMF [61, 65, 66]. The physical interpretation of the ENA images requires global models of the heliospheric shape [54, 67–69], of the production of the secondary neutrals, and their trip to the observer [70]. These pre-requisites will be developed further in connection with the analysis of the observations in Chap. 7.

3.4. Propagation of ENAs and Observational Limits

In Chap. 1, we coined *"neutral-atom astronomy"* for ENA diagnostics. Neutral atoms follow trajectories unimpeded by $\vec{E}$ and $\vec{B}$, so they can be used for imaging remote ion populations and sampling neutral gas coming from afar, based on the conservation of the original particle momentum and PSD. However, in transit, the ENA trajectories may be altered by gravitational fields of the Sun [62, 63] and the planets [71], solar radiation pressure on H [72–75], and collisions with atoms or ions. Also, extinction due to ionization by UV photons, electron impact, and CX with ions may reduce the ENA flux. These interactions modify the analysis and set limitations for ENA imaging.

3.4.1. *ENA trajectories in gravitational fields*

For ENA observations made inside the solar system, we must consider the motion of ENAs under solar and planetary gravitation. In most observations, a single celestial body is dominant, and only unbounded trajectories are of interest. First, we consider the solar gravitational effect, including radiation pressure, on ENAs of extra-heliospheric

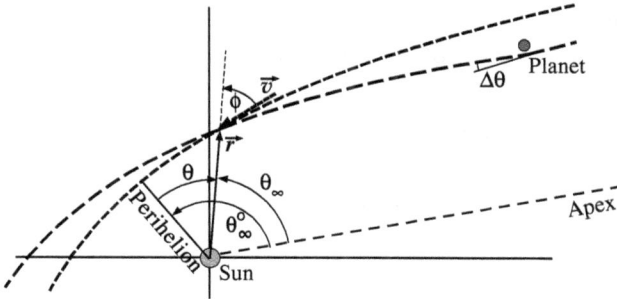

Fig. 3.5.   The Keplerian trajectory of an atom arriving at a location $\vec{r}$ from its apex (thick short dash) is shown with its velocity vector $\vec{v}$ and relevant angles, along with a trajectory deflected by the angle $\Delta\theta$ during an encounter with a planet (thick long dash).

origin, then the deflection in the vicinity of a planet, both depicted in Fig. 3.5.

In the solar gravitational field, ENAs of mass m follow Keplerian trajectories (Fig. 3.5) with velocity $\vec{v}$ and radius vector $\vec{r}$ from the Sun under conservation of total energy W

$$W = \frac{mv^2}{2} - \frac{k}{r} \qquad (3.21a)$$

and angular momentum $|\vec{L}|$ along their path [62, 63].

$$|\vec{L}| = mrv \cdot \sin\phi \qquad (3.21b)$$

The angle ϕ between $\vec{r}$ and $\vec{v}$ is defined by $\sin\phi = |\vec{r} \times \vec{v}|/(rv)$. The combined solar field scales with $k = GmM(1 - \mu)$. G is the gravitational constant, M the solar mass, and μ the relative strength of the radiation pressure, which is only relevant for H. The radiation pressure is approximated as independent of the atom speed relative to the Sun [76]. This notation allows a constant radiation pressure independent of the particle velocity and thus scales as $1/r^2$ from the center of the source of gravity and light. Under these conditions, a neutral particle follows a hyperbolic trajectory [63]:

$$\frac{1}{r} = \frac{mk}{L^2}(1 + \varepsilon \cdot \cos\theta) = \frac{mk}{L^2}(1 + \varepsilon \cdot \cos(\theta_\infty - \theta_\infty^0)) \qquad (3.22)$$

$\varepsilon > 1$ is the eccentricity. θ is the angle from the perihelion, θ_∞ from the hyperbola's asymptote (apex) to $\vec{r}$, and θ_∞^0 from the asymptote to the perihelion, also known as the *true anomaly* of the hyperbolic trajectory. These angles are not to be confused with the polar angle θ in Secs. 3.2 and 3.3.

Section 7.2 contains a more detailed treatment of the ISN trajectories that arrive in the inner solar system and their analysis for the physical parameters of the surrounding ISM.

Inside the planet's gravitational well, *i.e.*, in a magnetosphere, low-E ENAs may be on bound orbits ($\varepsilon < 1$) similar to low-E H in the geocorona. In this case, integration over multiple revolutions must be considered, causing reduced flux due to ionization losses. For ENAs from distant sources, *e.g.*, the heliospheric boundary, these orbits are forbidden and can be readily excluded from the analysis.

For $\varepsilon > 1$ trajectories entering an Earth-orbiting ENA sensor, the deflection by the Earth or a giant planet close to the ENA trajectories can be estimated as follows. The disparity between the planetary scale (Fig. 3.5) and the solar distance makes the solar gravitation a constant and the planetary effect a small perturbation [71]. We describe the perturbation as the total angular deflection $\Delta\theta$ assuming it as very small for all practical cases. In analogy to the treatment of Coulomb collisions during the penetration of matter by charged particles [77], we consider the ENA trajectory within b before and after the nearest approach, where b is the impact parameter of the encounter (Fig. 3.6).

For small deflections, or $\Delta\theta \ll 1$, we can approximate b by the closest distance or $r_{Min} \approx b$. In the following, we assume that the gravitational force at r_{Min} is effective over distance $2b$

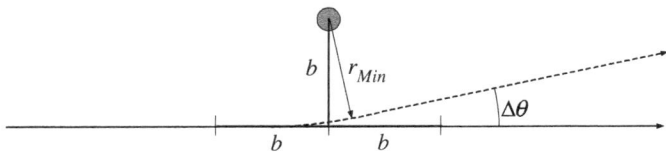

Fig. 3.6. Schematic representation of a neutral atom trajectory with a slight deflection $\Delta\theta$ near a planet. Most of the deflection occurs near the minimum distance r_{Min}, close to the impact parameter b for small deflections.

in Fig. 3.6. This simplification returns the same total change in the momentum $\Delta|\vec{P}|$ as an integration of the variable force over the entire trajectory [77]. Therefore, we integrate the maximum deflecting force $F_{\text{Max}} = F(r_{\text{Min}})$ over the time interval $\Delta t = 2b/v \approx 2r_{\text{Min}}/v$. The factor of 2 includes both the inbound and outbound leg of the ENA trajectory. The change Δv perpendicular to $\vec{v}$, due to the gravitational pull of a planet, yields:

$$\Delta v = 2\frac{\Delta|\vec{P}|}{m} \approx 2\frac{\overline{F \cdot \Delta t}}{m} = 2\frac{GM_P}{r_{Min}^2} \cdot \frac{r_{Min}}{v} = 2\frac{g_P R_P^2}{r_{Min}v} \qquad (3.23)$$

where M_P and R_P are the mass and radius of the planet, g_P the acceleration at the planet's surface, and v the ENA speed in the planet's rest frame. The deflection after the closest approach is $\Delta\theta = \Delta v/v$.

The above estimation solely uses the ENA speed v in the planet rest frame, and the minimum distance r_{Min} from the planet center for ENA detection without interference by the Earth's magnetosphere or a distant planet's atmosphere. As an illustrative example, we estimate the deflection of the ISN gas in early February when the Earth rams into the flow at $v \approx 80\,\text{km/s}$ — a pivotal analysis discussed in more detail in Sec. 7.2. To take observations outside the Earth's magnetosphere $r_{\text{Min}} = 15\,R_E$ is required. With these values, the maximum deflection of the ISN gas flow by Earth's gravity, mostly in ecliptic longitude, is $\Delta\theta \leq 0.08°$ [64, 71], which is substantially smaller than the current measurement uncertainties of $\approx 0.5°$.

3.4.2. *Extinction of ENAs from the source to the observer*

In addition to deflection by gravitation and radiation pressure, ENAs in transit are vulnerable to ionization by UV and X-ray photons, electron impact, and CX that reverts an ENA to an ion in partially ionized plasmas, as well as elastic collisions. These processes will be discussed here and in Sec. 3.4.3.

Ionization of ENAs in transit is a loss process leaving behind positive PUIs implanted into the plasma that the ENAs traverse.

Of the three ionization processes, only CX produces other neutral atoms, but with the velocity distribution of the intervening plasma, which may be starkly different from that of the source plasma of the ENAs. Also, the newly generated neutral atoms spread over 4π, which reduces their flux compared with the original ENA flux. When deducing the source ENA flux from the observation, the total ionization rate ν_{Ion} accounts for the combined losses along the way to the observer, according to:

$$\frac{dj_{ENA}(v,t)}{dt} = -\nu_{Ion}(t) \cdot j_{ENA}(v,t) \tag{3.24a}$$

After integration along the ENA trajectory, the observed ENA flux $j_{ENA}(v)$ relates to the ENA flux at the source $j_{ENA0}(v)$ through the survival probability $S(v)$ according to:

$$j_{ENA}(v) = j_{ENA0}(v) \cdot \exp\left(-\int_0^{t_{Obs}} \nu_{Ion}(t)dt\right) = j_{ENA0}(v) \cdot S(v) \tag{3.24b}$$

ν_{Ion} may vary along the trajectory. Depending on the situation, either its total spatio-temporal variation from observations must be considered, or an approximation can be applied. Angular momentum conservation can be invoked if the ionization scales as $1/r^2$ (for solar photon and CX with SW ions) and the rate is approximately constant along the Keplerian ENA trajectory. Then the ionization rate ν_{IonObs} at the observer distance from the Sun r_{Obs} is relevant according to:

$$j_{ENA}(v) = j_{ENA0}(v) \cdot \exp\left(-\frac{\nu_{IonObs}r_{Obs}}{v}\theta_\infty\right). \tag{3.24c}$$

For heliospheric ENAs, we can safely use θ_∞ for the integration because the ionization is negligible at large distances from the Sun.

Generally, the ionization rate ν_{Ion} includes photoionization ν_{Ph}, electron impact ionization ν_e, and charge exchange ν_{CX}, combined as $\nu_{Ion} = \nu_{Ph} + \nu_e + \nu_{CX}$. For most of the ENA species of interest, photoionization is an essential loss process. In fact, it is the most significant loss for He and Ne. ν_{Ph} depends on the local UV photon flux as a function of energy $I(E)$ according to [78]:

$$\nu_{sPh} = \int_{E_{sI}}^{\infty} \sigma_{sPh}(E)I(E)dE \tag{3.25}$$

where $\sigma_{sPh}(E)$ is the energy-dependent photoionization cross-section, and E_{sI} is the ionization energy for the respective species s. There is a compilation of photoionization rates for H, He, O, and Ne at 1 AU over the recent years [78].

Electron impact ionization is often a small contribution to the overall ionization rate in the SW at and outside 1 AU. However, it is more substantial inside 1 AU, in the IHS (where the SW distribution is heated substantially), and in planetary magnetospheres. The ionization rate ν_{se} at a location $\vec{r}$ reads [79]:

$$\nu_{se}(\vec{r}) = \frac{8\pi}{m_e^2} \int_{E_{sl}}^{\infty} \sigma_{se}(E)f_e(E, \vec{r})EdE \tag{3.26}$$

$\sigma_{se}(E)$ is the E-dependent cross-section for electron impact ionization of species s and $f_e(E)$ the energy distribution of the local electrons. For photoionization and electron-impact ionization, ν_{Ion} is the relevant quantity for the loss of ENAs in transit. The total ENA loss scales with their exposure time to ionization. Hence the integration over time in Eq. (3.24b). Therefore, slower ENAs are affected substantially. Note, that the relevant energy in $\sigma(E)$ and $f(E)$ is not that of the ENA. It is that of the photon in Eq. (3.25) and the electron in Eq. (3.26). The electron speed dominates the relative energy in the center-of-mass frame.

However, for CX loss of ENAs, the *relative speed* $|\vec{v}_n - \vec{v}_i|$ between the ENAs traveling at velocity $\vec{v}_n$ and a local ion of velocity $\vec{v}_i$ is crucial. The rate $\nu_{siCX}(\vec{v}_n)$ yields [80]:

$$\nu_{siCX}(\vec{v}_n) = \int_0^{\infty} \sigma_{siCX}(|\vec{v}_n - \vec{v}_i|) \cdot |\vec{v}_n - \vec{v}_i| \cdot f_i(\vec{v}_i)d\vec{v}_i \tag{3.27a}$$

where $\sigma_{siCX}(v) = \sigma_{siCX}(|\vec{v}_n - \vec{v}_i|)$ is the speed-dependent CX cross-section between neutral species s and ion species i. Depending on the ENA energy and the ion distribution in the intervening medium, approximations are possible. For very low-energy ENAs, the ion speed is dominant, and the ENAs may be considered stationary. For more energetic ENAs traversing a relatively cold plasma, the speed of the ENAs prevails, and Eq. (3.27a) simplifies to:

$$\nu_{siCX}(v) = \sigma_{siCX}(v)vn_i \qquad (3.27\text{b})$$

The ion density n_i replaces the integral over the ion velocity distribution. Here, the CX loss may be evaluated in the spatial rather than the time domain using the mean free path for CX $\lambda_{siCX}(v)$.

$$\lambda_{siCX}(v) = 1/(\sigma_{siCX}(v)n_i) \qquad (3.27\text{c})$$

This notation eliminates the explicit dependence on ENA speed, and only the v-dependence in the cross-section remains.

How to treat these losses depends on the specific ENA observations. We will return to these issues connected with particular types of observations in Chaps. 6 and 7.

3.4.3. *Effects of elastic collisions on the ENA diagnostics*

Another significant interaction for ENAs that affects the ENA imaging comes from elastic collisions in transit. After ENAs leave their source region, we can treat ionizing and CX collisions as part of the overall loss of ENAs along their path. Here, we focus on elastic collisions between ENAs and atoms or ions ("collisions" from here on).

At first glance, we might also describe elastic collisions as a loss process because they can remove ENAs from the original ENA flux. However, ENAs on trajectories that would never reach the observing sensor could also be deflected on a course into the sensor, in this case originating from a different part of the source region. If the observed ENA flux is isotropic, the only effect of elastic collisions would be modifying the original ENA energy distribution in the observation. However, if the source is a localized region in the sky, collisions will blur its image.

To estimate the collisional effect, we must compare the distance of the source with the mean free path λ_{Coll} for collisions of the ENAs in the intervening environment.

$$\lambda_{Coll} = 1/(n_c \cdot \sigma_{El}) \tag{3.28a}$$

n_c is the density of the relevant collision partner species on the way, and σ_{El} is the cross-section for elastic collisions. If several species contribute substantially to the collisions, Eq. (3.28a) needs to be replaced by:

$$\lambda_{Coll} = 1 \bigg/ \sum_s (n_{sc} \cdot \sigma_{sEl}) \tag{3.28b}$$

The contributions of all types of collision partners sc add up, which reduces the overall mean free path. It is important to reemphasize that CX collisions contribute to the extinction of the original ENA flux and add secondary ENAs. At very low energies, they even alter the ENA momentum from that of its parent ion [6, 7], thus contributing to the blurring of the ENA image similar to elastic collisions.

For ENA sources at distances $>\lambda_{Coll}$, the undisturbed flux from their original direction is reduced by $1/e$ for each λ_{Coll}. At the same time, the background of diffuse ENAs that had at least one collision on their trajectory increases accordingly. This effect compares to optical observations where photons from a distance larger than their mean free path are scattered accordingly, deeming the medium "optically thick". (See Sec. 1.2.1 for the UV survey of Earth's exosphere.) In analogy, the environment, which the ENAs traverse with more than one collision, may be called "optically thick for ENAs". In this case, ENA images degrade substantially with a drastically reduced spatial resolution of the source region. Therefore, ENA observations are typically limited to sources closer than λ_{Coll}.

To estimate this limit for the heliosphere and its immediate galactic neighborhood, we consider the most abundant ENAs, *i.e.*, H. Adopting $\sigma_{HEl} \approx 2 \times 10^{-15} \, \text{cm}^2$ for $E = 1 \, \text{keV}$ [80] and an interstellar neutral density $n_H = 0.1 \, \text{cm}^{-3}$ around the heliosphere [49, 81] yields $\lambda_{Coll} \approx 330 \, \text{AU}$. This value is more than three times the distance of the TS ($\approx 90 \, \text{AU}$) [82] and more than

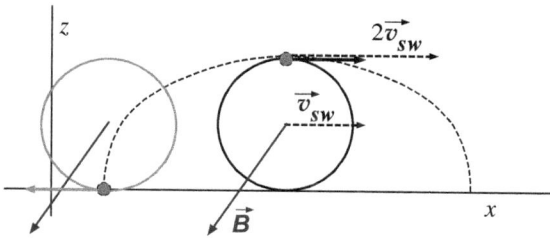

Fig. 3.7. For $\vec{B}$ perpendicular to $\vec{v}_{sw}$, freshly ionized PUIs start in the SW rest frame (solid lines) with $-v_{sw}$, where they gyrate about $\vec{B}$. In the observer frame (dashed lines), PUIs perform a cycloid motion with a minimum speed $v = 0$ and maximum speed $v = 2v_{sw}$.

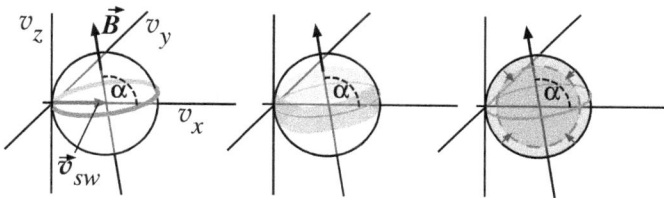

Fig. 3.8. Evolution of a PUI velocity function (from left to right) after initial injection into the SW plasma at pitch angle α relative to the IMF (left). Fluctuations in $\vec{B}$ cause pitch angle scattering that broadens the ring in velocity space (center) and finally leads to an isotropic spherical shell distribution (right). The radial expansion of the SW leads to adiabatic cooling of the distribution indicated by the shrinking of the shell (right). At the same time, additional PUIs enter the outermost shell at v_{sw}.

The PUI motion is that of the valve of a bicycle wheel that rolls along on the road, which a cycloid can describe.

$$x(t) = x_o + r_{ci} \cdot (\omega_{ci} t - \sin(\omega_{ci}t)) \quad z(t) = r_{ci} \cdot (1 - \cos(\omega_{ci}t))$$

$$(3.29c)$$

r_{ci} is the gyro radius of the PUIs.

Equation (3.29a) through c describe the PUI motion in a steady IMF. Under these conditions, the PUI distribution is a narrow ring in velocity space (Fig. 3.8, left panel). Typically, magnetized plasmas like the SW have a variety of Alfvén waves embedded, which scatter PUIs in pitch angle under conservation of their kinetic energy in the

SW frame into a wider ring (Fig. 3.8, center) or even into an isotropic shell with radius v_{sw} (Fig. 3.8, right).

Also, a ring distribution in velocity space is susceptible to instabilities that generate plasma waves and thus cause scattering of the PUIs if intrinsic waves in the background plasma do not scatter the PUIs effectively enough [85]. At arbitrary α, the PUIs form a ring-beam distribution, which drives electromagnetic instabilities [86, 87]. There are two extreme situations. A field-aligned beam for $\alpha = 0°$ produces a two-stream instability [88], and strictly perpendicular ion injection at $\alpha = 90°$ excites electromagnetic cyclotron instabilities under most plasma conditions [89, 90]. At very low plasma pressures, ion Bernstein-mode instabilities at the harmonics of ω_{ci} may become prevalent [91].

In the SW, which expands radially with distance from the Sun, the PUI distributions are subject to adiabatic cooling, as indicated by the shrunk shell with a dashed line in the right panel of Fig. 3.8. Because additional PUIs continuously arrive in the outermost shell, a filled spherical distribution develops [92]. Typically, the PUI distribution has a substantially higher effective temperature than the background plasma and thus contributes significantly to its heat balance and pressure [93]. Because the PUIs stem from gas with a substantially different bulk velocity than the plasma, a momentum exchange occurs with noticeable mass-loading of the plasma [94].

In some cases, PUIs form a distinct, separate distribution in their new environment that produces secondary ENAs for remote detection in CX with the surrounding ANAs. Examples are radiation belt protons, which stem from decaying cosmic ray albedo neutrons (a nuclear CX of sorts) and then became H ENAs as introduced in Sec. 2.3. The *IBEX* Ribbon, which most likely forms through the injection of neutral SW, generates PUIs in the OHS that release ENAs from a final CX with the ISN gas, as discussed in Sec. 7.3. Finally, PUIs can substantially alter the state of the parent plasma where they are injected, which becomes visible in the ENA spectra that emerge from the plasma.

3.5. Summary

Having considered the mechanisms of production and extinction of ENAs, Eq. (3.18a), which connects the observed differential ENA flux of species i in direction $\hat{l}$ of the LOS to the differential parent ion flux $j_i(E, \hat{l})$ between the farthest point at l_{max} and the detector, can now be generalized to include all relevant processes and participants to

$$\frac{dJ_i(E, \hat{l})}{d\Omega dE} \approx \frac{R_{ni}}{\eta(E)G\Delta E} = \int_0^{l_{max}} S_i(\bar{E}, l) j_i(\bar{E}, l) \sum_s [\sigma_{isCX}(\bar{E}) n_s(l)] dl$$

(3.30)

R_{ni} is the count rate ascribed to the ENA flux, E the center of the window ΔE, $\eta(E)$ and G are the sensor detection efficiency and geometric factor discussed in Chaps. 4 and 5. The summation is over ANA species s that produce ENAs. The survival probability $S_i(E, l)$ consists of various extinction rates (Secs. 3.4.2 and 3.4.3):

$$S_i(E, l) = \prod_j S_{ij}(E, l) = \exp\left\{-\sum_j \left[\int_l^0 \nu_{ij}(E, l)\frac{dl}{v_i}\right]\right\}$$

(3.31)

The final goal of ENA diagnostics is to extract the parent ion flux $j_i(E, l)$ on the RHS of Eq. (3.30), from the detected ENA flux on the LHS.

Having discussed the generation of ENAs in remote plasmas and their passage through space, we related the measured ENA spectra to the ion velocity distributions. We now turn to the ENA measurement requirements and the basic sensor design in Chap. 4. Their implementation follows in Chap. 5. We are now prepared to feature the deconvolution of Eq. (3.30), the core challenge in the ENA analysis and interpretations of planetary magnetospheres and at the edge of the heliosphere, in Chaps. 6 and 7, respectively.

References

1. Gruntman, M. (1997). Energetic neutral atom imaging of space plasmas, *Rev. Sci. Instr.*, **68**(10), 3617–3656. DOI: 10.1063/1.1148389.

2. Hsieh, K. C. (2015). Detection of energetic neutral atoms in and out of the heliosphere, *Chin. J. Space Sci.*, **35**(3), 253–292. DOI: 10.11728/cjss2015.03.253 (in Chinese).

3. Möbius, E., Galvin, A. B., Kistler, L. M., Kucharek, H., & Popecki, M. A. (2016). Time-of-flight mass spectrographs — from ions to neutral atoms, *J. Geophys. Res.*, **121**(12), 11,647–11,666. Doi: 10.1002/2016JA022553.

4. Brown, S. C. (1959). *Basic Data of Plasma Physics* (MIT Technology Press and J. Wiley & Sons, New York).

5. Massey, H. S. W., & Gilbody, H. B. (1974). *Electronic and Ionic Impact Phenomena*, **Vol. IV**, Chap. 23&24 (Clarendon, Oxford).

6. Schultz, D. R., Ovchinnikov, S. Y., Stancil, P. C., & Zaman, T. (2016). Elastic, charge transfer, and related transport cross sections for proton impact of atomic hydrogen for astrophysical and laboratory plasma modeling, *J. Phys. B*, **49**(8), 084004. Doi: 10.1088/0953-4075/49/8/084004.

7. Swaczyna, P., McComas, D. J., Zirnstein, E. J., & Heerikhuisen, J. (2019). Angular scattering in charge exchange: issues and implications for secondary interstellar hydrogen, *Astrophys. J.*, **887**(2), 223. Doi: 10.3847/1538-4357/ab5440.

8. Medley, S. S., Donne, A. J. H., Raita, R., Kislyakov, A. I., Petrov, M. P., & Roquemore, A. L. (2008). Invited Review Article: Contemporary instrumentation and application of charge exchange neutral particle diagnostics in magnetic fusion energy experiments, *Rev. Sci. Instr.*, **79**, 011101. Doi: 10.1063/1.2823259.

9. Hintz, E., & Schweer, B. (2013). Plasma edge diagnostics by atomic beam supported emission spectroscopy — status and perspectives, *Plasma Phys. and Contr. Fus.*, **37**(11A), A87–A101.

10. Challis, C. D. (1995). Review of neutral beam heating on JET for physics experiments and the production of high fusion performance plasmas, *Fusion Engineering and Design*, **26**(1–4), 17–28. Doi: 1016/0920-3796(94)00167-6.

11. Speth, E. (1989). Neutral beam heating of fusion plasmas, *Reports on Progr. in Phys.*, **52**(1), 57–121. Doi: 10.1088/0034-4885/52/1/002.

12. Stacey Jr., W. M., & Sigmar, D. J. (1979). Impurity control by neutral-beam injection, *Nucl. Fus.*, **19**(12), 1665–1673. Doi: 0.1088/0029-5515/19/12/010.

13. McElroy, M. B., & Yung, Y. L. (1975). The atmosphere and ionosphere of Io, *Astrophys. J.*, **196**(1), 227–250.

14. Cloutier, P. A., Daniell, R. E., Dessler, A. J., & Hill, T. W. (1978). A cometary ionosphere model for Io, *Astrophys. & Space Sci.*, **55**(1), 93–112. Doi: 10.1007/BF00642582.

15. Ripken, H. W., & Fahr, H.-J. (1983). Modification of the local interstellar gas properties in the heliospheric interface, *Astron. Astrophys.*, **122**(1–2), 181–192.

16. Ripken, H. W., & Fahr, H.-J. (1984). The physics of the heliospheric interface and its implications for LISM diagnostics, *Astron. Astrophys.*, **139**(2), 551–554.

17. Müller, H.-R., & Zank, G. P. (2004). Heliospheric filtration of interstellar heavy atoms: Sensitivity to hydrogen background, *J. Geophys. Res.*, **109**(A7), A07104. Doi: 10.1029/2003JA010269.

18. Izmodenov, V., Malama, Yu., Gloeckler, G., & Geiss, J. (2004). Filtration of interstellar H, O, N atoms through the heliospheric interface: Inferences on local interstellar abundances of the elements, *Astron. & Astrophys.*, **414**, L29–L32. Doi: 10.1051/0004-6361:20031697.

19. Desai, M. I., Allegrini, F., Dayeh, M. A. *et al.* (2015). Latitudinal and energy dependence of energetic neutral atom spectral indices measured by the Interstellar Boundary Explorer, *Astrophys. J.*, **802**(2), 100. Doi: 10.1088/0004-637X/802/2/100.

20. Desai, M. I., Dayeh, M. A., Allegrini, F. *et al.* (2016). Latitude, energy, and time variations in the energetic neutral atom spectral indices measured by the Interstellar Boundary Explorer (IBEX), *Astrophys. J.*, **832**(2), 116. Doi: 10.3847/0004-637X/832/2/116.

21. Livadiotis, G., McComas, D. J., Randol, B. M. *et al.* (2012). Pick-up ion distributions and their influence on ENA spectral curvature, *Astrophys. J.*, **751**(1), 64. Doi: 10.1088/0004-637X/751/1/64.

22. Gurnett, D. A., & Bhattacharjee, A. (2005). *Introduction to Plasma Physics* (Cambridge University Press).

23. Piel, A. (2010). Plasma Physics — An Introduction to Laboratory, Space, and Fusion Plasmas (Springer, Berlin-Heidelberg).

24. Vasyliunas, V. M. (1968). A survey of low-energy electrons in the evening sector of the magnetosphere with OGO 1 and OGO 3, *J. Geophys. Res.*, **73**(9), 2839–2884. Doi: 10.1029/JA073i009p02839.

25. Christon, S. P. (1987). A comparison of the Mercury and Earth magnetospheres: electron measurements and substorm time scales. *Icarus*, **71**(3), 448–471. Doi: 10.1016/0019-1035(87)90040-6.

26. Collier, M R., & Hamilton, D. C. (1995). The relationship between kappa and temperature in energetic ion spectra at Jupiter. *Geophys. Res. Lett.*, **22**(3), 303–306. Doi: 10.1029/94GL02997.

27. Mauk, B. H., Mitchell, D. G., McEntire, R. W. *et al.* (2004). Energetic ion characteristics and neutral gas interactions in Jupiter's magnetosphere, *J. Geophys. Res.*, **109**(A9), A09S12. Doi: 10.1029/2003JA010270.

28. Ogasawara, K., Angelopoulos, V., Dayeh, M. A. *et al.* (2013). Characterizing the dayside magnetosheath using energetic neutral atoms: IBEX and THEMIS observations, *J. Geophys. Res.*, **118**(6), 3126–3137. Doi: 10.1002/jgra.50353.

29. Chotoo, K., Schwadron, N., Mason, G. *et al.* (2000). The suprathermal seed population for corotating interaction region ions at 1 AU deduced from composition and spectra of H^+, He^{++}, and He^+ observed by wind, *J. Geophys. Res.*, **105**(A10), 23107–23122. Doi: 10.1029/1998JA000015.

30. Mann, G., Classen, H. T., Keppler, E., & Roelof E. C. (2002). On electron acceleration at CIR related shock waves, *Astron. Astrophys.*, **391**, 749–756. Doi: 10.1051/0004-6361:20020866.

31. Marsch, E. (2006). Kinetic physics of the solar corona and solar wind, *Living Rev. Sol. Phys.*, **3**(1), 1–100. Doi: 10.12942/lrsp-2006-1.

32. Yoon, P. H., Rhee, T., & Ryu, C. M. (2006). Self-consistent formation of electron κ distribution, 1: theory, *J. Geophys. Res.*, **111**(A9), A09106. Doi: 10.1029/2006JA011681.

33. Decker, R. B., & Krimigis, S. M. (2003). Voyager observations of low energy ions during solar cycle 23, *Adv. Space Res.*, **32**(4), 597–602. Doi: 10.1016/S0273-1177(03)00356-9.

34. Decker, R. B., Krimigis, S. M., Roelof, E. C. *et al.* (2005). Voyager 1 in the foreshock, termination shock, and heliosheath, *Science*, **309**(5743), 2020–2024. Doi: 10.1126/science.1117569.

35. Heerikhuisen, J., Pogorelov, N. V., Florinski, V., Zank, G. P., & Le Roux, J. A. (2008). The effects of a κ-distribution in the heliosheath on the global heliosphere and ENA flux at 1 AU, *Astrophys. J.*, **682**(1), 679–689. Doi: 10.1086/588248.

36. Heerikhuisen, J., Pogorelov, N. V. Zank, G. P. *et al.* (2010). Pick-up ions in the outer heliosheath: a possible mechanism for the interstellar boundary explorer ribbon, *Astrophys. J.*, **708**(2), L126–L130. Doi: 10.1088/2041-8205/708/2/L126.

37. Zank, G. P., Heerikhuisen, J., Pogorelov, N. V., Burrows, R. & McComas, D. J. (2010). Microstructure of the heliospheric termination shock: implications for energetic neutral atom observations, *Astrophys. J.*, **708**(2), 1092–1106. Doi: 10.1088/0004-637X/708/2/1092.

38. Livadiotis, G., McComas, D. J., Dayeh, M. A., Funsten, H. O., & Schwadron, N. A. (2011). First sky map of the inner heliosheath temperature using IBEX spectra, *Astrophys. J.*, **734**(1), 1. Doi: 10.1088/0004-637X/734/1/1.

39. Livadiotis, G., McComas, D. J., Schwadron, N. A., Funsten, H. O., & Fuselier, S. A. (2013). Pressure of the proton plasma in the inner heliosheath, *Astrophys. J.*, **762**(2), 134. Doi: 10.1088/0004-637X/762/2/134.

40. Tsallis, C. (2009). *Introduction to Non-Extensive Statistical Mechanics: Approaching a Complex World* (Springer, New York).
41. Livadiotis, G., & McComas, D. J. (2013). Understanding Kappa Distributions: A Toolbox for Space Science and Astrophysics, *Space Sci. Rev.*, **175**, 183–214. Doi: 10.1007/s11214-013-9982-9.
42. Möbius, E., Hovestadt, D., Klecker, B. *et al.* (1985). Direct observation of He$^+$ pick-up ions of interstellar origin in the solar wind, *Nature*, **318**(6045), 426–429. Doi: 10.1038/318426a0.
43. Isenberg, P. A. (1987). Evolution of interstellar pickup ions in the solar wind, *J. Geophys. Res.*, **92**(A2), 1067–1073. Doi: 10.1029/JA092iA02p01067.
44. Gloeckler, G., Geiss, J., Balsiger, H. *et al.* (1993). Detection of interstellar pick-up hydrogen in the solar system, *Science* **261**(5117), 70–73. Doi: 10.1126/science.261.5117.70.
45. Compton, A. H., & Getting, I. A. (1935). An apparent effect of galactic rotation on the intensity of cosmic rays, *Phys. Rev.*, **47**, 818–821.
46. Ipavich, F. M. (1974). The Compton-Getting effect for low-energy particles, *Geophys. Res. Lett.*, **1**, 149–152.
47. Möbius, E. (1993). Gases of non-solar origin in the solar system, *Landoldt-Börnstein, Numerical Data and Functional Relationships in Science and Technology*, **VI/3A** Chapter 3.3.5.1, 184–188.
48. Möbius, E., Bzowski, M., Chalov, S. *et al.* (2004). Synopsis of the interstellar He parameters from combined neutral gas, pickup ion and UV scattering observations and related consequences, *Astron. Astrophys.*, **426**(3), 897–907. Doi: 10.1051/0004-6361:20035834
49. Bzowski, M., Möbius, E., Gloeckler, G., Tarnopolski, S. & Izmodenov, V. (2008). Density of neutral interstellar hydrogen at the termination shock from Ulysses pickup ion observation, *Astron. Astrophys.*, **491**(1), 7–19. Doi: 10.1051/0004-6361:20078810.
50. Richardson, J. D., Liu, Y., Wang, C., & McComas, D. J. (2008). Determining the LIC H density from the solar wind slowdown, *Astron. Astrophys.*, **491**(1), 1–5. Doi: 10.1051/0004-6361:20078565.
51. Pryor, W., Gangopadhyay, P., Sandel, B. *et al.* (2008). Radiation transport of heliospheric Lyman-alpha from combined Cassini and Voyager data sets, *Astron. Astrophys.*, **491**(1), 21–28. Doi: 10.1051/0004-6361:20078862.
52. Krimigis, S. M., Mitchell, D. G., Roelof, E. C., Hsieh, K. C., & McComas, D. J. (2009). Imaging the interaction of the heliosphere with the interstellar medium from Saturn with Cassini, *Science*, **326**(5955), 971–973. Doi: 10.1126/science.1181079.
53. Fuselier, S. A., Allegrini, F., Bzowski, M. *et al.* (2012). Heliospheric neutral atom spectra between 0.01 and 6 keV from IBEX, *Astrophys. J.*, **754**(1):14. Doi: 10.1088/0004-637X/754/1/14.

54. Zank, G. P., Pogorelov, N. V., Heerikhuisen, J. *et al.* (2009). Physics of the solar wind-local interstellar medium interaction: role of magnetic fields, *Space Sci. Rev.*, **146**(1–4), 295–327.

55. Izmodenov, V. V., & Alexashov, D. B. (2015). Three-dimensional kinetic-MHD model of the global heliosphere with the heliopause-surface fitting, *Astrophys. J. Suppl.*, **220**(2), 32. Doi: 10.1088/0067-0049/220/2/32.

56. Bzowski, M., Swaczyna, P., Kubiak, M. A. *et al.* (2015). Interstellar neutral helium in the heliosphere from Interstellar Boundary Explorer Observations III. Mach number of the flow, velocity vector, and temperature from the first six years of measurements, *Astrophys. J. Suppl.*, **220**(2), 28. Doi: 10.1088/0067-0049/220/2/28.

57. Swaczyna, P., Bzowski, M., Kubiak, M. A. *et al.* (2015). Interstellar neutral helium in the heliosphere from Interstellar Boundary Explorer observations I. uncertainties and backgrounds in the data and parameter determination method, *Astrophys. J. Suppl.*, **220**(2), 27. Doi: 10. 1088/0067-0049/220/2/27.

58. Schwadron, N. A., Moebius, E., Leonard, T. *et al.* (2015). Determination of interstellar He parameters using 5 years of data from the Interstellar Boundary Explorer — beyond closed form approximations, *Astrophys. J. Suppl.*, **220**(2), 25. Doi: 10.1088/0067-0049/220/2/25.

59. Schwadron, N. A., Moebius, E., Lee, M. A. *et al.* (2016). Determination of interstellar O parameters using the first 2 years of data from the Interstellar Boundary Explorer, *Astrophys. J.*, **828**(2), 81. Doi: 10. 3847/0004-637X/828/2/81.

60. Kubiak, M. A., Bzowski, M., Sokol, J. M. *et al.* (2014). Warm breeze from the starboard bow: a new population of neutral helium in the heliosphere, *Astrophys. J. Suppl.*, **213**(2), 29. Doi: 10.1088/0067-0049/213/2/29.

61. Kubiak, M. A., Swaczyna, P., Bzowski, M. *et al.* (2016). Interstellar neutral helium in the heliosphere from IBEX observations. IV. Flow vector, mach number, and abundance of the warm breeze, *Astrophys. J. Suppl.*, **223**(2), 25. Doi: 10.3847/0067-0049/223/2/25.

62. Lee, M. A., Kucharek, H., Möbius, E., Wu, X., Bzowski, M. & McComas, D. J. (2012). An analytical model of interstellar gas in the heliosphere tailored to IBEX observations, *Astrophys. J. Suppl.*, **198**(2), 10. Doi: 10.1088/0067-0049/198/2/10.

63. Lee, M. A., Möbius, E. & Leonard, T. (2015). The analytical structure of the primary interstellar helium distribution function in the heliosphere, *Astrophys. J. Suppl.*, **220**(2), 23. Doi: 10.1088/0067-0049/220/2/23.

64. Möbius, E., Bzowski, M., Frisch, P. C. *et al.* (2015). Interstellar flow and temperature determination with IBEX: robustness and sensitivity to systematic effects, *Astrophys. J. Suppl.*, **220**(2), 24. Doi: 10.1088/ 0067-0049/220/2/24.

65. Park, J., Kucharek, H., Möbius, E. *et al.* (2015). Statistical analysis of the heavy neutral atoms measured by IBEX, *Astrophys. J. Suppl.*, **220**(2), 34. Doi: 10.1088/0067-0049/220/2/34.

66. Park, J., Kucharek, H., Möbius, E. *et al.* (2016). IBEX observations of secondary interstellar helium and oxygen distributions, *Astrophys. J.*, **833**(2), 130. Doi: 10.3847/1538-4357/833/2/130.

67. Izmodenov, V. V., Alexashov, D. B., & Myasnikov, A. V. (2005). Direction of the interstellar H atom inflow in the heliosphere: role of the interstellar magnetic field, *Astron. Astrophys.*, **437**, L35–L38. DOI: 10.1051/0004-6361:200500132.

68. Isenberg, P. A., Forbes, T. G., & Möbius, E. (2015). Draping of the interstellar magnetic field over the heliopause: a passive field model, *Astrophys. J.*, **805**(2), 153. Doi: 10.1088/0004-637X/805/2/153.

69. Zirnstein, E. J., Heerikhuisen, J., Funsten, H. O. *et al.* (2016). Local interstellar magnetic field determined from the Interstellar Boundary Explorer ribon, *Astrophys. J. Lett.*, **818**(1), L18. Doi: 10.3847/ 2041-8205/818/1/L18.

70. Bzowski, M., Kubiak, M. A., Czechowski, A., and Grygorczuk, J. (2017). The helium warm breeze in IBEX observations as a result of charge-exchange collisions in the outer heliosheath, *Astrophys. J.*, **845**(1), 15. Doi: 10.3847/1538-4357/aa7ed5.

71. Kucharek, H., Wurz, P., Möbius, E. *et al.* (2015). Impact of planetary gravitation on high precision neutral atom measurements, *Astrophys. J. Suppl.*, **220**(2), 35. Doi: 10.1088/0067-0049/220/2/35.

72. Lallement, R. (2002). The interaction of the heliosphere with the interstellar medium, in: *The Century of Space Science*, **1**, 1191–, eds. Bleeker, A. M., Geiss, J., and Huber, M. C. E. (Kluwer Acad. Publ.).

73. Schwadron, N. A., Moebius, E., Kucharek, H. *et al.* (2013). Solar radiation pressure and local interstellar medium flow parameters from interstellar boundary explorer low energy hydrogen measurements, *Astrophys. J.*, **775**(2), 86. Doi: 10.1088/0004-637X/775/2/86.

74. Bzowski, M., Sokół, J. M., Tokumaru, M. *et al.* (2013). Solar parameters for modeling interplanetary background, Chap. 3 in: *Cross-Calibration of Past and Present Far UV Spectra of Solar System Objects and the Heliosphere*, ISSI Scientific Report Series **13**, 67–138 (Springer Science Business Media, New York).

75. Rahmanifard, F., Möbius, E., Schwadron, N. A. *et al.* (2019). Radiation pressure from interstellar hydrogen observed by IBEX through solar cycle 24, *Astrophs. J.*, **887**(2), 217. Doi: 10.3847/1538-4357/ab58ce.

76. Rucinski, D., & Bowski, M. (1995). Modulation of interplanetary hydrogen density distribution during the solar cycle, *Astron. Astrophys.*, **296**, 248.

77. Allkofer, O. C. (1971). Teilchen-Detektoren — Particle Detectors (Karl Thiemig KG, Munich).

78. Bochsler, P., Kucharek, H., Möbius, E. *et al.* (2014). Solar photoionization rates for interstellar neutrals in the inner heliosphere: H, He, O, and Ne, *Astrophys. J. Suppl.*, **210**(1), 12. Doi: 10.1088/0067-0049/210/1/12.

79. Owocki, S. P., Holzer, T. E., & Hundhausen, A. J. (1983). The solar wind ionization state as a coronal temperature diagnostic, *Astrophys. J.*, **275**, 354–366. Doi: 0.1086/161538.

80. Lindsay, B. G., & Stebbings, R. F. (2005). Charge transfer cross sections for energetic neutral atom data analysis, *J. Geophys. Res.*, **110**(A12), A12213. Doi: 10.1029/2005JA011298.

81. Frisch, P. C., Bzowski, M., Grün, E. *et al.* (2009). The galactic environment of the Sun: interstellar material inside and outside of the heliosphere, *Space Sci. Rev.*, **146**(1), 235–273. Doi: 10.1007/s11214-009-9502-0.

82. Stone, E. C., Cummings, A. C., McDonald, F. B., Heikkila, B., Lal, N., & Webber, W. R. (2005). Voyager 1 explores the termination shock region and the heliosheath beyond, *Science*, **309**(5743), 2017–2020. Doi: 10.1126/science.1117684.

83. Gurnett, D. A., Kurth, W. S., Burlaga, L. F., & Ness, N. F. (2013). In Situ Observations of Interstellar Plasma with Voyager 1, *Science*, **341**(6153), 1489–1492. Doi: 10.1126/science.1241681.

84. Stone, E. C., Cummings, A. C., McDonald, F. B., Heikkila, B., Lal, N., & Webber, W. R. (2013). Voyager 1 observes low-energy galactic cosmic rays in a region depleted of heliospheric ions, *Science*, **341**(6142), 150–153. Doi: 10.1126/science.1236408.

85. Gary, S. P. (1991). Electromagnetic ion/ion instabilities and their consequences in space plasmas: a review, *Space Sci. Rev.*, **56**, 373–415.

86. Wu, C. S., & Davison, R. C. (1972). Electromagnetic instabilities produced by neutral-particle ionization in interplanetary space, *J. Geophys. Res.*, **77**, 5399–5406.

87. Wu, C. S., & Hartle, R. E. (1974). Further remarks on plasma instabilities produced by ions born in the solar wind, *J. Geophys. Res.*, **79**, 283–285.

88. Briggs, R. J. (1971). Two-stream instabilities, In: *Advances in Plasma Physics*, pp. 43–78, **Vol. 4**, A. Simon & W. B. Thompson eds., Wiley, New York.

89. Florinski, V., Zank, G. P., Heerikhuisen, J., Hu, Q., & Khazanov, I. (2010). Stability of a pickup ion ring-beam population in the outer heliosheath: implications for the IBEX ribbon, *Astrophys. J.*, **719**, 1097–1103.

90. Joyce, C. J., Smith, C. W., Isenberg, P. A., Gary, S. P., Murphy, N., Gray, P. C., & Burlaga, L. F. (2012). Observations of Bernstein waves excited by newborn interstellar pickup ions in the solar wind, *Astrophys. J.*, **745**, ID112.

91. Liu, K., Gary, S. P., & Winske, D. (2011). Excitation of magnetosonic waves in the terrestrial magnetosphere: Particle-in-cell simulations, *J. Geophys. Res.*, **116**, A07212.

92. Vasyliunas, V. M., & Siscoe, G. M. (1976). On the flux and the energy spectrum of interstellar ions in the solar system, *J. Geophys. Res.*, **81**, 1247–1252.

93. Isenberg, P. A., Smith, C. W., Matthaeus, W. H., & Richardson, J. D. (2016). Turbulent heating of the distant solar wind by interstellar pickup protons in a decelerating flow, *Astrophys. J.*, **719**, 716–721.

94. Szegö, K., Glassmeier, K.-H., Brinca, A. *et al.* (2000). Physics of mass loaded plasmas, *Space Sci. Rev.*, **94**, 429–671.

Chapter 4

ENA Instrumentation: General

"To exhaust your farthest vision, climb up the next storey!"
From the poem *On the Stork Tower* by Wang Zhihuan
688–742 CE

This chapter bridges the physics of ENA diagnostics in Chap. 3 and the specific ENA sensor designs in Chap. 5. The goal for *in-situ* or remote observations of space plasmas is to obtain spatial and temporal variations of the plasma velocity (v) distributions, if possible, separate by species or mass, m. These objectives require careful consideration of the mission design, its orbit or trajectory, the S/C, and instrumentation. Section 4.1 considers the requirements on the S/C orbit or trajectory, attitude control, and pointing capability. Section 4.2 describes the critical subsystems of a generic particle instrument for ions and ENAs, emphasizing the *sensor head*, which selects, analyzes, and registers incident particles. Finally, in Sec. 4.3 we address the ENA-specific challenges for the instrument design.

4.1. Choosing Vantage Points for ENA Observations

Any scientific observation requires an accessible vantage point with favorable observing conditions. Planning a space mission (S/C, orbit, and operations) to observe specific space plasmas is like finding the best site for a new observatory. Because of the inherent difference between *in-situ* charged-particle measurements and ENA diagnostics from a distance and their complementarity, let us start with the

71

constraints on *in-situ* measurements and their requirements on the mission design.

Ions move on helical paths along $\vec{B}$. Hence, the measured ion flux is an instantaneous sample of the plasma population in the local magnetic flux tube. Knowing the direction of $\vec{B}$, either measured simultaneously or by modeling, the arrival directions of charged particles provide their pitch-angle (α) distribution. A typical *in-situ* measurement provides single-point sampling along the S/C path, which *cannot distinguish spatial variations from temporal evolution in the plasma under investigation*. Consequently, the S/C must visit regions of interest repeatedly to ascertain spatial structures, such as plasma boundary crossings, or temporal responses to changing external conditions, such as SW or IMF variations. To analyze small-scale or short-time variations, a cluster of S/C orbiting in a formation is essential. Conversely, ENA imaging from afar provides an overview of the plasma structures and their evolution in time, albeit with limited depth perception. To illustrate the choice of vantage points for ENA diagnostics, we pick four missions: *SOHO* at the Lagrangian point L1 and *IBEX* in an Earth orbit viewing the heliospheric boundary and the *interstellar gas*, then *IMAGE* and *TWINS* viewing Earth's magnetosphere from distinct orbits.

4.1.1. *Viewing the heliosphere: SOHO and IBEX*

SOHO is a joint *ESA/NASA* mission launched in 1995 to observe the Sun from its interior to its extended corona and the heliosphere [1]. The 3-axis-stabilized S/C was put into a halo orbit around the Lagrangian point L1 on the Sun-Earth line at 1.5×10^6 km from Earth. This placement allows *SOHO* to continuously observe the Sun and the heliosphere without disturbances from Earth's magnetosphere.

The High Suprathermal-energy Time-Of-Flight (HSTOF) sensor, an add-on to the *SOHO* Charge ELement Isotope Analysis System (CELIAS) [2] (Sec. 5.3.1), seized the opportunity for remote sensing the heliosphere in ENAs. HSTOF's ion-deflecting collimator provides a FOV of $4°$ in the ecliptic and $\pm 17°$ above and below. By pointing $37°$ west of the Sun-S/C line (Fig. 4.1), it avoids the SW and the

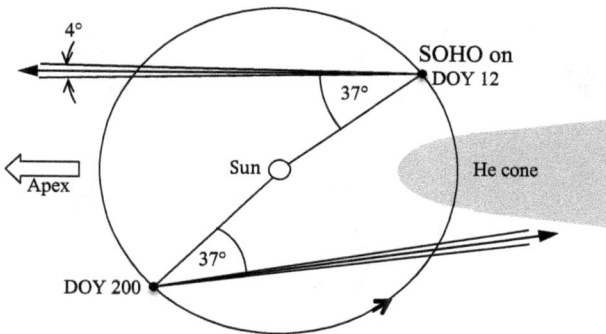

Fig. 4.1. *SOHO* orbits around the Sun in the ecliptic plane with the HSTOF FOV pointing 37° west of the Sun. HSTOF scans for heliospheric ENA emissions once a year along the ecliptic, including the Apex (direction of the Sun's motion relative to the ISM) on Day of Year (DOY) 12 and the interstellar He cone (enhancement of the ISN density by gravitational focusing) in the opposite direction [3]. (© CJSS. Reproduced with permission).

Parker spiral (the "guidewire" for solar energetic ions). As *SOHO* orbits the Sun, HSTOF scans ENAs from a 34°-wide strip centered on the ecliptic once a year (Fig. 4.1) [3].

IBEX is a *NASA* Small Explorer mission, the first ENA-only satellite to view the heliosphere and the ISM [4]. The *IBEX*-Hi and -Lo sensors are described in Secs. 5.3.2 and 5.3.3, respectively. On Oct 19, 2008, a Pegasus rocket launched *IBEX* into a highly elliptical ecliptic orbit with an apogee of $\approx 50\,\mathrm{R_E}$ to provide favorable viewing of the heliospheric boundary regions least affected by the magnetospheric foreground. Its initial orbit being highly sensitive to lunar disturbances, *IBEX* carried enough fuel to maintain the S/C attitude and orbit for at least five years. The initial orbit had a ≈ 7-day period and a $\approx 2.5\,\mathrm{R_E}$ perigee (geocentric). After the first three years of operation, the *IBEX* perigee was raised to $\approx 8\,\mathrm{R_E}$ to achieve a 9.1-day lunar synchronous orbit [5]. This orbit is stable (perigee slowly oscillating between 6 and $16\,\mathrm{R_E}$) for more than 50 years and entirely above the outer radiation belt. The remaining fuel extends the *IBEX* operations beyond two solar cycles.

IBEX is a Sun-pointing spinner (Fig. 4.2, left) with two oppositely pointing ENA sensors scanning a 6.5°-wide (FWHM) circular

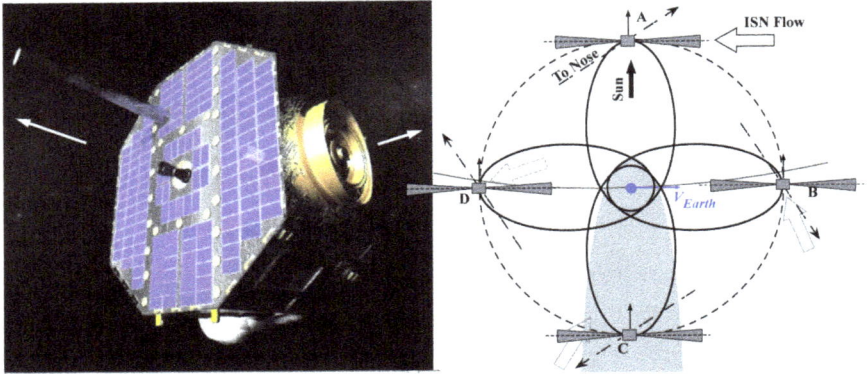

Fig. 4.2. *IBEX* mission. Left: Artist rendition of the *IBEX* S/C. The Sun-pointing spin axis (left arrow) is normal to the solar panel. The *IBEX*-Lo sunshade and collimator are visible with the boresight (right arrow) and one of the two hydrazine tanks as the lower white dome. *IBEX*-Hi is on the opposite side. (© Springer. Reproduced with permission, adapted.) Right: *IBEX* FOVs (grey cones) relative to the ISN flow (open arrow) and the heliospheric nose (dash-dotted arrow) in the GSE frame. The Sun's gravity deflects the ISN flow from its original arrival direction on its way to 1 AU. The dashed circle is the locus of the apogee. At position A around Jan 31, the ISN flow points into the *IBEX*-Lo FOV when the Earth rams into the flow. A, B, C, and D indicate S/C locations three months apart on Jan 31, May 1, Jul 31, and Oct 30, respectively. The ISN flow for the other three S/C orientations is shown in grey and dashed for D when the ISN flow arrives at Earth after its perihelion (not accessible to *IBEX*). The best viewing of the heliosphere occurs when the apogee is away from the magnetosphere (dark grey area) [6].

swath of the sky per S/C spin to avoid sunlight and SW effects optimally. Initially, the spin-axis reorientation to point at the Sun occurred at each perigee, and after the orbit change, at perigee and apogee. This configuration allows each sensor to generate an all-sky map every six months. In the *Geocentric Solar Ecliptic* (GSE) frame, the orientation of the major axis of the *IBEX* orbit drifts around the Earth relative to the magnetosphere, with the apogee pointing toward the Sun in early January when the Earth rams exactly into the ISN flow, as shown in Fig. 4.2 (right, position A).

Analyzing the ISN flow (Sec. 7.2) requires precision control and knowledge of the pointing. Therefore, *IBEX*-Lo includes a star sensor, co-aligned with its boresight, which provides $\approx \pm 0.1°$

precision in spin angle and elevation as verified in-flight relative to the S/C attitude information [7, 8]. In turn, the precision analysis of the ISN flow has even helped uncover a hidden shift in the attitude of *IBEX* by 0.6° after the S/C star tracker recovered from a problem with a software update. This apparent shift was indeed a S/C issue and not related to the ISN flow or its analysis, as validated with the independent attitude determination by the star sensor [9].

4.1.2. *Viewing Earth's magnetosphere: IMAGE and TWINS*

IMAGE was the first mission dedicated to remote sensing Earth's magnetosphere and its interaction with the Sun-driven IP plasma in ENAs, especially during magnetic storms and substorms [10]. Three ENA imagers mapped the ENA producing regions in three energy domains, augmented by radar and UV imaging. The High Energy Neutral Atom imager (HENA, Sec. 5.3.1) for 30–500 keV covered the *ring current* (RC); the Medium Energy Neutral Atom imager (MENA, Sec. 5.3.2) for 1–30 keV the RC, plasma sheet and cusp; the Low Energy Neutral Atom imager (LENA, Sec. 5.3.3) for 10–750 eV the ionospheric outflow. Simultaneously, the GEOcorona photometer (GEO) scanned the geocorona in resonantly scattered HI121.6 nm (H Lyα) and OI135.6 nm from the two dominant ANAs that produce the ENAs. Simultaneously, radar, far UV, and extreme UV monitored SW-driven changes in the MP, aurora, and plasmasphere [11].

Viewing from apogee at high latitudes provides an overview of the magnetosphere structure (Sec. 1.2), especially the inner magnetosphere with the entire auroral zone and the RC. Viewing the same regions at closer distances near perigee complements with shorter LOS from different angles. Both aspects can be combined with a spin-stabilized S/C on an elliptical polar orbit, as described in Fig. 4.3 [12].

Angular momentum conservation keeps the orbital plane fixed in *Geocentric Inertial Coordinates* (GIC). In GSE coordinates, the *IMAGE* apogee sweeps the entire 24 hours of Local Time (LT) over the one-year orbit of Earth, as indicated by the solar illumination

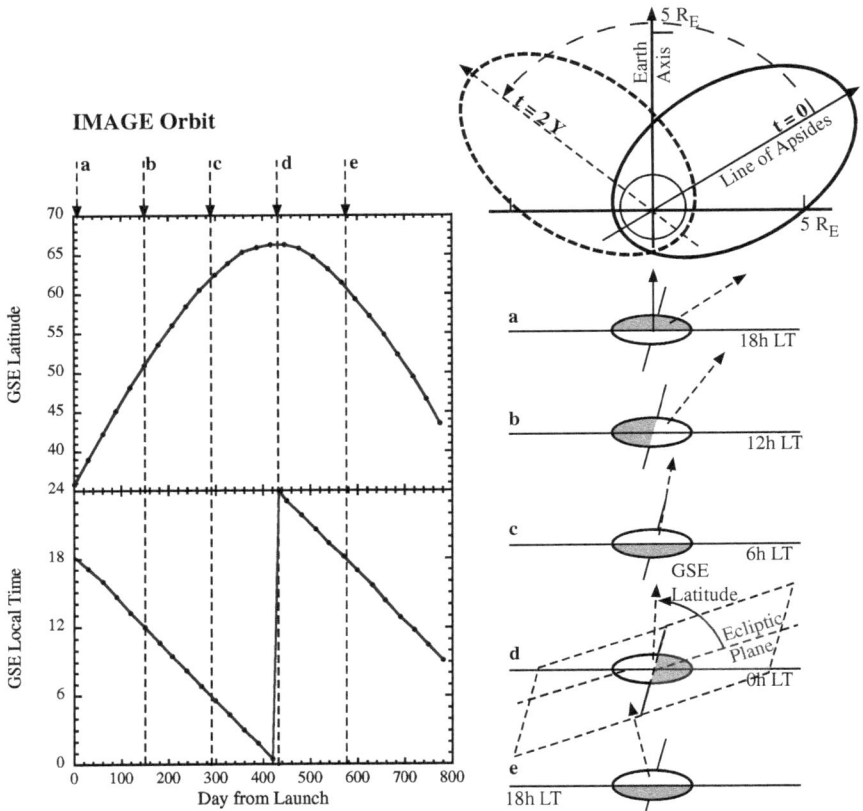

Fig. 4.3. *IMAGE* orbit configuration. Upper right: The orbital plane is stable in GIC, but the line-of-apsides drifts across the north pole during the mission. Lower right (insets **a** through **e**): Earth's motion around the Sun causes the *IMAGE* orbital plane and its line-of-apsides (dashed arrow) to drift across LT. Inset **d** shows the relation between equatorial and GSE latitude. Left: Apogee in GSE Latitude and LT vary as functions of days into the mission. The days marked **a** through **e** correspond to the insets on the right [12]. (© Springer. Reproduced with permission).

of the Earth in Fig. 4.3 (lower right, **a** through **e**). At the same time, the line-of-apsides of the orbit precesses steadily in latitude, starting from initially about 40° latitude over the nominal two-year mission, due to Earth's oblateness (Fig. 4.3 upper-right). As can be gleaned from Fig. 4.3 (left), the *IMAGE* orbit evolution for a two-year mission, starting with a nominal Jan 1, 2000, launch, covers the magnetosphere from all sides. It provides ≈70% more time on the dayside, appropriate for studying SW effects on the magnetosphere. The actual launch on Mar 25, 2000, only shifted the phase of the variations by about three months.

The *IMAGE* sensors are mounted on the sides of its octagonal cylinder to scan the magnetosphere as the S/C spins. The objectives required a spin rate of 0.50 ± 0.01 rpm with the spin axis normal to the orbital plane. Because a S/C inherits its initial spin from the stabilizing spin of the launcher in the direction of its orbit-insertion velocity, the spin axis must be reoriented and the spin rate lowered. A single magnetic torque rod (a solenoid) mounted normal to the spin axis and powered by a controlled current, achieved this goal. Its dipole moment interacts with Earth's intrinsic $\vec{B}$ to produce the required torque. The spin axis was within 1° of the desired direction in the inertial frame within eight perigee passes, as monitored by the S/C's magnetometer, star tracker, and Sun sensor.

After *IMAGE*'s success, *NASA* launched its first mission to view the polar magnetosphere and active RC in 3D from two vantage points in 2008. The *TWINS* mission employs two 3-axis stabilized nadir-pointing S/C, each on a *Molniya* orbit (Fig. 4.4a). Each S/C has two identical ENA spectrographs (Sec. 5.3.2) mounted on the nadir-pointing side, each with a 110° × 4° instantaneous FOV or 140° × 4° combined (Fig. 4.4b). The sensors swivel 180° on a pivot platform to provide a 140° × 180° FOV, centered ≈10° off the nadir minimizing solar exposure and reducing S/C obstruction. The principal axes of the two orbits are 180° apart in longitude, and the S/C are different in phase by a few hours (Fig. 4.4c) [13].

4.2. General Considerations for Space-Borne Particle Instruments

After discussing the mission requirements on the orbits and S/C for ENA observations by four missions focused on two key regions of space, we now turn to the instrumentation. We start with the principles of space-borne particle instruments for ions and electrons,

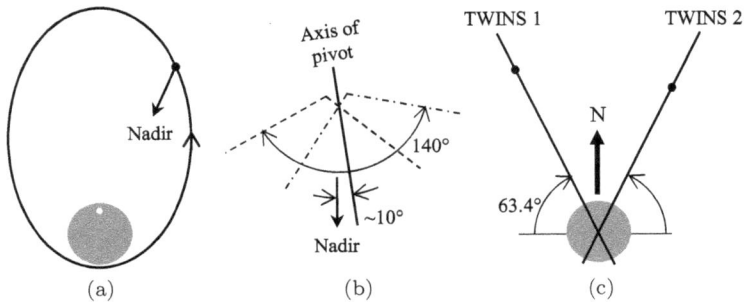

Fig. 4.4. *TWINS* orbits and viewing geometry. (a) Top view of a *Molniya* orbit, scaled to Earth. The north pole (open circle) is tilted forward to show the orbit inclination. The Soviet communication satellites *Molniya* (*Lightning* in Russian) of the 1960s featured: inclination 63.4°, eccentricity 0.74, apogee ≈7.2 R$_E$, period ≈12 h, and argument of perigee 270°. In this orbit, a satellite is unaffected by Earth's oblateness so that its line-of-apsides does not precess like that of *IMAGE* (Fig. 4.3). (b) Each S/C is nadir pointing and contains a pair of ENA imagers with a combined instantaneous FOV of 140° × 4°. Swiveling the sensor by 90° in and out of the page about the pivot axis (≈10° off the nadir) extends the FOV to 140° × 180°. (c) Side view of the two orbits: Their principal axes are separated by 180° in longitude and the S/C on their orbits by a few hours in phase. This configuration provides 3D viewing for several hours per orbit. [http://twins.swri. edu/mission.jsp] [13].

which can be modified for ENAs, before turning to ENA-specific challenges in Sec. 4.3.

Figure 4.5 shows a generic block diagram for space-borne particle instruments. The *sensor head* (or *sensor*), whose purpose and design define and often name the instrument, sorts and detects incident particles of choice. Its *aperture* defines the FOV (or the *collimator* for narrower FOV) and selects the particle *arrival direction* $\hat{l}$ (Appx. A). Next, an *analyzer* selects qualified entrants for the *detectors* to register and the *detector electronics* to produce signals for further processing and transmission. Here, *pulse-height analysis* (PHA) of the detected events that contain values for E and/or v provides the most detailed information about individual particles that arrive in a stochastic manner. At the same time, counters accumulate all incident particles into count rates. All other subsystems support the sensor's operation. The *signal processor & data storage* subsystem sorts and accumulates the PHA events into arrays in $(E, m, q, \hat{l})$,

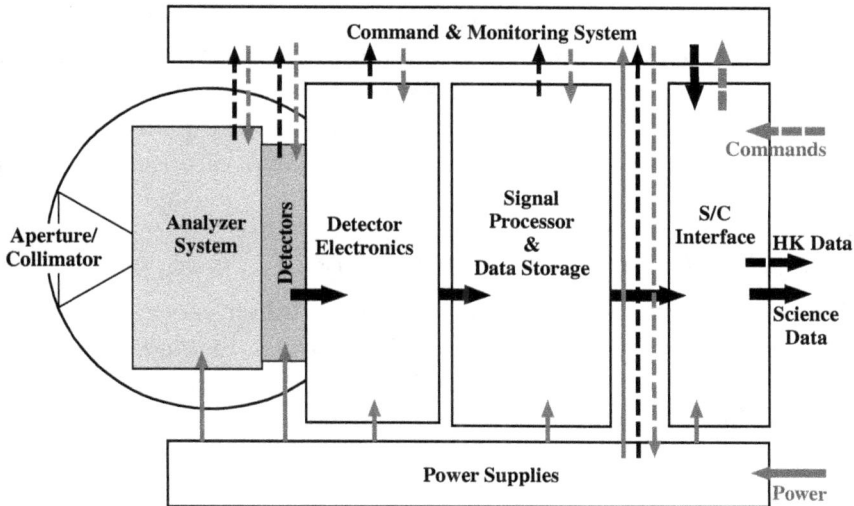

Fig. 4.5. Generic structure of a space-borne particle instrument with its main subsystems. The arrows indicate the flow of science data (solid black), housekeeping (HK) information (dashed black), power (solid grey), and commands (dashed grey). The lines are harnesses with multiple wires or, in some cases, optical transmission lines for data and commands.

stores samples of them at full resolution according to pre-defined priorities, re-adjusts the cadence of the count-rate readout, and compresses the digital information, thus organizing and buffering the data for the telemetry. The *command & monitoring* subsystem relays ground commands to the appropriate instrument subsystems for effective operation and collects vital *housekeeping* (HK) data. The *S/C interface* links the instrument to the S/C, and the *power supplies* provide filtered power to all subsystems at required levels, including high voltage for analyzers and detectors.

A particle instrument, operating as a system, measures the differential fluxes of particles of interest over a selected range of particle variables $(E, m, q, \hat{l})$ as a *spectrograph* and delivers the data to the investigators in the sequence and format according to their design.

The interpretation of the data occurs in their analysis on the ground, where the judgments of knowledgeable investigators are crucial. *Data analysts, therefore, must have intimate knowledge of the instrument characteristics*, such as energy response and resolution, pointing direction and precision, geometric factor, angular response, and S/C operational status to deduce the physical quantities accurately from the observables. Errors can have dire consequences (Appx. A).

4.2.1. *Particle selection by analyzers*

Charged particles can be sorted according to their energy (E), mass (m), and charge (q) by a force acting *normal to their velocity* $\vec{v}$ to bend them towards detection. The available forces are $\vec{F} = q\vec{E}$ (electric) and $\vec{F} = q\vec{v} \times \vec{B}$ (magnetic) or a combination of both. Finding the field configurations and strengths that guide the incident ions to suitably located detectors is the art of *ion optics* [14] using well-tested computer codes [15] that simulate ion trajectories in given $\vec{E}$ and $\vec{B}$ configurations.

The bending force *is* the centripetal force, and thus

$$q|\vec{E}| = mv^2/r \quad \text{for electric and} \tag{4.1a}$$

$$qv|\vec{B}| = mv^2/r \quad \text{for magnetic fields} \tag{4.1b}$$

where r is the curvature radius. (We use $|\vec{E}|$ for the magnitude of $\vec{E}$ to avoid confusion with the energy E.) Rearranging Eqs. (4.1a) and (4.1b) yields

$$|\vec{E}|r = mv^2 q = 2(E/q) \tag{4.2a}$$

$$|\vec{B}|r = mvq = (|\vec{P}|/q) \tag{4.2b}$$

Thus, *electrostatic analyzers* (ESA) sort by *energy-per-charge* (E/q) and *magnetostatic analyzers* (MSA) by *momentum-per-charge* $(|\vec{P}|/q)$. An ion is characterized by its charge $q = Qe$, with charge state Q and unit (protonic) charge e, and its mass $m = A$Da, with mass number A and unit atomic mass Da (1 dalton set to be 1/12 of a C atom). To sort ENAs by $\vec{E}$ or $\vec{B}$ requires ionizing them first (Sec. 4.4.3).

E/q or $|\vec{P}|/q$ alone cannot uniquely identify a particle — they are ratios of the sought-after quantities. To uniquely identify a particle

Table 4.1. Particle identification by one or two analyzers

ESA	MSA	SSD	TOF	Analyzers
E/q	$m/q, v$	E, q	$m/q, v$	ESA
Measured Parameters	$|\vec{P}|/q$	$\sqrt{2m}/q,\ E$	$m/q, v$	MSA
• Diagonal: Single analyzer		E	m, v	SSD
• Off-diagonal: Combination of two analyzers			v	TOF

by m and q, either E and/or v must be measured separately. After selection by an ESA and/or MSA, the particle must be detected. Using a *solid-state detector* (SSD) for this purpose (Sec. 4.2.2) provides E. A *time-of-flight* (TOF) analyzer (Sec. 4.2.3) measures the time it takes a particle across a known distance, which gives v. Both devices also work for ENAs. Combining SSD and TOF measurements identifies the particles uniquely by m, without $\vec{E}$ or $\vec{B}$. Hence no need for ionizing the ENAs first.

Table 4.1 summarizes which parameters can be obtained by a single device or any combination of two. To fully identify m, v, and q of ions, one can add an ESA to a TOF-SSD combination [16], which is unnecessary in an ENA sensor because only $Q = 0$ or 1 can occur. Alternatively, an ESA-TOF combination can identify m and v of ENAs after converting them to ions with $Q = \pm 1$. Chap. 5 discusses specific ENA sensor designs.

4.2.2. Particle detectors

Electron multipliers (EMs) and SSDs are two types of particle detectors readily available for ENA sensors. In the following, we will discuss their operational principles, along with intrinsic advantages and disadvantages.

4.2.2.1. Electron multipliers

The first EMs were part of photomultipliers encased in evacuated glass tubes, consisting of a chain of discrete dynodes placed between the photocathode and anode. The dynodes, coated with an alkali metal, *e.g.*, Cs, to enhance the *secondary electron* (SE) yield, are biased at increasing *electric potentials* (Φ), typically $\Delta\Phi \approx 1\,\mathrm{kV}$ between photocathode and anode. When a photon strikes the photocathode, the emission and acceleration of SEs start a cascade with increasing numbers of SEs at each dynode. Up to $\approx 10^6$ SEs reach the anode, sufficient for a detectable signal. An open dynode chain without the photocathode and glass casing operated under vacuum is a valuable particle detector (Fig. 5.3). For relatively large EMs, even a moderate $\vec{B}$ can alter the SE trajectories, thus reducing the gain. A smaller channel multiplier ($\approx 2\,\mathrm{mm}$ in diameter and $\approx 2\,\mathrm{cm}$ in length), made of heavily Pb-doped glass to sustain $\Delta\Phi \approx 1\,\mathrm{kV}$, was developed to reduce size and mass. Like the dynodes, the inner wall of the tube is coated for high SE yield. Further innovations led to smaller and ruggedized tubes [17]. For example, a funneled spiraltron enabled the detection of 0.6–2.0 keV *neutral atoms* [18].

Eventually, the *MicroChannel Plate* (MCP) developed for night vision became available for ENA imaging [19–22]. A 1–2 mm thick plate of tightly packed microchannels (Fig. 4.6, left), each an EM, facilitates large circular or rectangular position-sensitive detectors

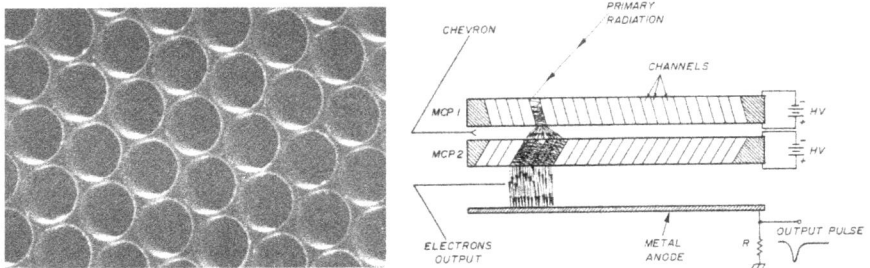

Fig. 4.6. *MicroChannel Plate* (MCP). Left: Magnified view of MCP surface (courtesy J. P. Boutot). Pulling a hexagonal bundle of mm-diameter Pb-glass tubes under heat produces a stretched bundle with starkly reduced diameters. Chemical etching removes the cores, leaving a bundle of microchannels, each 10–25 μm in diameter. Fusing several similar bundles forms a block, from which 1–2 mm thick MCPs are cut [19]. Right: Schematic of an MCP-pair in chevron configuration (not to scale). An incident particle starts a cascade of SEs in one microchannel of the first plate. This avalanche seeds several microchannels of the second plate. The final SE cloud is accelerated to the anode to preserve its size and location on a position-sensing anode [20]. (© Elsevier. Reproduced with permission).

up to $>100 \, \text{cm}^2$ in area. The interstitial space reduces the sensitive area to 83–16%, depending on channel and wall sizes. A typical bias across a plate of $\approx 1 \, \text{kV}$ provides a maximum gain of $\approx 10^4$ per channel without ion feedback due to *secondary ions* (SIs). Released by SE impacts near the exit and accelerated up the channel, they start a new SE cascade. Stacking two plates with their slanted channels, forming a chevron (Fig. 4.6, right) allows higher gains [20]. The lateral spread of the SE cloud from the front plate seeds several channels in the rear plate, yielding gains of $\approx 10^7$.

At gains $\geq 10^6$, an incident particle can deposit $\geq 10^{-13} \, \text{C}$ on the anode within $<200 \, \text{ps}$, yielding a perfect signal for precise timing and position marking. The SE cloud exiting a microchannel cluster at the rear plate (Fig. 4.6, right) allows accurate determination of the position of the cloud centroid on a 2D position-sensitive anode, as illustrated in Fig. 4.7. Shown is a three-electrode so-called "wedge-and-strip" scheme, better suited for simultaneous timing applications than a resistive anode whose ohmic resistance slows down the signals. In this implementation (Fig. 4.7, center, not to scale), the pattern is etched on a printed circuit board with ≈ 30-μm

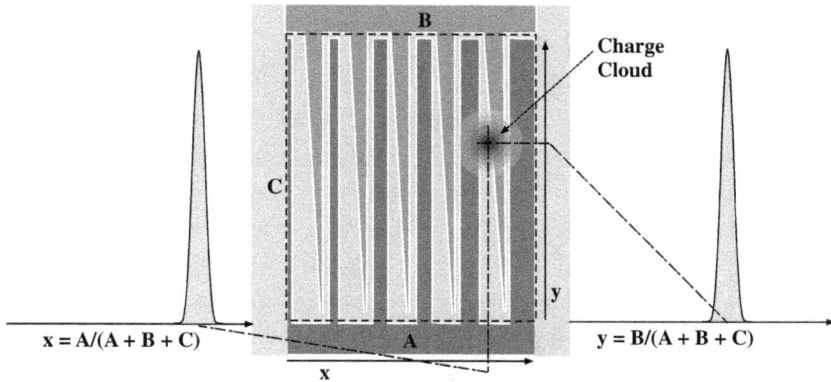

Fig. 4.7. 2D position-sensing wedge-and-strip anode (not to scale). Center: Schematic view of the anode (as for the setup in Fig. 4.6, right). The charge cloud is distributed among anodes A, B, and C according to the fractional anode areas underneath. In the sensitive area (dashed rectangle), normalizing the A and B signal with the total charge (A + B + C) yields the x- (left) and y-position (right) [23].

wide gaps separating three anodes [23]. A is a series of rectangular strips with *increasing width* along the x-axis, B a series of *identical* wedges, and C fills the space between A and B. The (x, y)-position of the centroid of the charge cloud exiting the MCP is computed by normalizing the x-dependent signal A and the y-dependent signal B by the total charge: $x = A/(A + B + C)$ and $y = B/(A + B + C)$. The exact matching of the three amplifier gains is crucial for the x and y linearity.

4.2.2.2. *Solid-state detectors*

Unlike MCPs, SSDs, developed for nuclear and particle physics with the advent of the transistor and other semiconductor devices, provide energy resolution, making them desirable E-analyzers for particles. A semiconductor is a crystal of tetravalent atoms (Si or Ge) with an energy gap E_g (1.16 eV for Si and 0.75 eV for Ge) between the valence and conduction bands. Electrons in the conduction band and holes in the valence band are the carriers of opposite charges. With their larger E_g, Si SSDs can operate at room temperature.

SSDs function as reverse-biased n-p junction diodes and their variants. The n-Si crystals are doped with pentavalent "donor" atoms to introduce excess electrons only $\approx 10^{-2}$ eV "below" the conduction band to facilitate conduction. Conversely, p-Si crystals are doped with trivalent "acceptor" atoms to introduce excess holes only $\approx 10^{-2}$ eV "above" the valence band. With charge equilibrium established, layers of electrons on the p-side and positive ions on the n-side form, making the junction a depletion zone, void of free charge carriers. The two oppositely charged layers provide an $\vec{E}$-field across this zone, with the p-side potential below that of the n-side by U_0. When reversely biased by U_B, the potential difference increases to $(U_0 + U_B)$, and the depletion zone widens at the expense of the p- and n-zones at opposite ends.

The *stopping power* of the particle — energy loss per unit thickness in (keV g^{-1}cm^2) — integrated along its path generates electron-hole pairs with quantum efficiency ε per pair ($\varepsilon = 3.63$ eV for Si and 2.96 eV for Ge) [24]. These charge carriers are accelerated by $(U_0 + U_B)$ to the positive and negative borders, respectively, generating a signal proportional to the energy lost. The widening of the depletion zone to the full SSD thickness d maximizes the ion path in this zone and thus the generated signal.

Due to interactions with the electrons in the SSD, only the electronic stopping power contributes to the signal. For $E <$ 1 MeV/nucleon, the incident ion is not fully stripped. Thus, direct Coulomb interactions with the nuclei in the SSD become important [25], causing more energy loss to the incident particle [26–28]. This *nuclear stopping power* creates crystal-lattice vibrations or phonons without an electric signal, thus leading to a *pulse-height defect* in the SSD signal. Likewise, energy loss in the Au or Al contact layers at the surfaces contribute to this defect also [29]. Modeling provides an initial evaluation of the stopping power using, *e.g.*, the tested and timely updated software *SRIM* [30], and calibration provides the actual value.

While the pulse height of an SSD signal depends on the energy loss across d, its energy resolution and threshold depend mainly on the SSD capacitance ($\propto a/d$ with the SSD area a). Thus, minimizing a increases the resolution and lowers the threshold but also reduces the collecting power. Photolithography of small-scale structures on up-to 15-cm diameter Si wafers, along with highly integrated electronics, has lowered the capacitance for large detection-area SSDs by dividing them into pixels, each a separate detection unit. In addition, improving the purity of the Si-base material and precision doping with ion implant techniques reduced the base noise. Furthermore, dead layers of SSDs that do not contribute to the signal have been minimized. Table 4.2 illustrates this progress.

Pixelated SSDs lend themselves to imaging. Recent progress in integrating transistors, *e.g.*, depleted field-effect transistors [34],

Table 4.2.   Noise level for three SSD examples at room temperature

Host Instrument	Year	Effective area (cm^2)	Noise (keV)
Cosmic-ray telescope [31]	1965	1 detector at 1.2	≤ 50
HSTOF on *SOHO* [32]	1995	192 pixels each at 0.5	8
STE on *STEREO* [33]	2006	4 pixels each at 0.09	0.7

on a pixelated array led to low-noise, energy-resolving, and position-sensing SSDs. Their first space use is for the Mercury Imaging X-ray Spectrometer on *ESA*'s *Bepi-Colombo* S/C [35]. Such an imager can be used for ENAs if filtered against charged particles and photons. A compendium on SSDs may be found in the textbook *Semiconductor Radiation Detectors* [36]. When used as the stop detector in a TOF spectrograph (Sec. 4.2.3), the SSD also measures E to complete the particle identification (Table 4.1).

4.2.3. *Principles of TOF measurements*

TOF refers to the time τ a particle takes to pass a known distance l, whence the particle speed $v = l/\tau$ is derived. The adaptation of the TOF technique, using SE emission from thin C-foils as start signals [37], from nuclear mass spectrography [38] to space plasma physics, enabled the expansion of ion species identification into suprathermal energies ($<1\,$MeV). An early account of TOF sensors for use in space with an SSD demonstrated the capability to identify ^{1}H through ^{40}Ar ions in the range of 2–400 keV/nucleon, including possibly ENAs [39]. For overviews on TOF techniques in space research, we refer the reader to recent reviews [40–42]. Figure 4.8 shows two generic TOF configurations.

An incident particle passes the TOF system from left to right. SEs emitted from a thin C-foil at the entrance and those from a

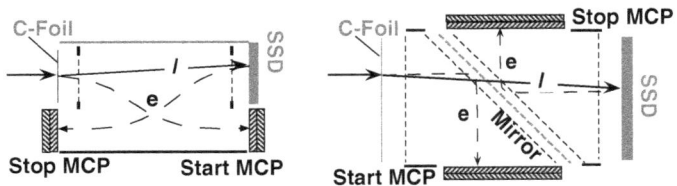

Fig. 4.8.   TOF schematics. Particles enter from the left through a thin C-foil, traverse the TOF path l and stop in an SSD, which can also be an MCP. Note: The path length l may depend on the angle of traversal between the start and stop. The structure in black is typically held at 1–2 kV positive relative to that in grey. Left: Start and stop SEs cross paths before reaching the respective MCP. Right: Dual electrostatic mirrors deflect SEs by 90° [42]. (© AGU. Reproduced with permission).

plate or SSD at the end are guided by appropriate $\vec{E}$ configurations to provide a *start* and *stop* timing signal at the respective MCP with rise times of a few ns or less. The scheme in Fig. 4.8 (left), with the *start* and *stop* SEs crossing paths, has been implemented in *AMPTE* CHEM [16] and SULEICA [43] and in *ACE* and *Ulysses* SWICS [44, 45]. The design in Fig. 4.8 (right) has positively biased grids facing the foil and the SSD to accelerate the SEs immediately into a symmetric dual electrostatic mirror. This configuration assures equal flight paths for the *start* and *stop* SEs emitted from any position with minimal time dispersion; hence the equal transit-time across an *isochronous* mirror [46]. Although used before [47], its merit was only fully appreciated in *SOHO* STOF [2] and *ACE* ULEIS [48] by using the 2D positioning capability for the start and stop signals, giving a more accurate determination of l and v. The multiple layers of grids, however, reduce the sensor efficiency η.

With v alone, the energy per atomic unit mass E_{TOF}/A in the TOF section is known and is related to the original particle energy E as:

$$\frac{E \cdot (1 - \alpha_1)}{A} = \frac{E_{TOF}}{A} = \frac{1\,\mathrm{Da}}{2} \left(\frac{d}{\tau}\right)^2 \qquad (4.3)$$

where α_1 reflects the energy loss in the foil, which can be determined through simulation [30] augmented by final calibration [49]. To obtain both E and A, however, TOF must be combined with another measurement.

4.2.3.1. *Combination of TOF and SSD measurements*

If the stop detector at the end of the TOF section is an SSD (Fig. 4.8), the residual energy E_{Res} of the particle is measured. Using Eq. (4.3), A and E can be determined independently as:

$$A = \frac{E_{Res}}{(1 - \alpha_2) \cdot E_{TOF}/A} = \frac{E_{Res}}{(1 - \alpha_2) \cdot (\frac{1\,\mathrm{Da}}{2})(\frac{d}{\tau})^2}$$

$$E = \frac{A}{(1 - \alpha_1)} \frac{1\,\mathrm{Da}}{2} \left(\frac{d}{\tau}\right)^2 \qquad (4.4)$$

where α_2 describes the pulse-height defect of the SSD. Like α_1, α_2 depends on E and A and is determined by simulation and calibration.

4.2.3.2. *Combination of TOF and ESA measurements*

For ions and, as we will see, ions from converted neutral atoms, the TOF technique can also be combined with an ESA, which obtains E/Q through electrostatic deflection between oppositely charged curved plates [50–52]. Combining E/Q and E/A yields A/Q of the ion:

$$\frac{A}{Q} = \frac{E \cdot (1 - \alpha_1)/Q}{E_{TOF}/A} \tag{4.5}$$

or with post-acceleration by the voltage U_{PAC} for low-energy particles:

$$\frac{A}{Q} = \frac{E \cdot \frac{1-\alpha_1}{Q} + eU_{PAC}}{E_{TOF}/A} \tag{4.5a}$$

This combination is sufficient if most of the ions are singly charged ($Q = 1$), almost exclusively the case for converted neutral atoms. Examples of ESA-TOF combinations for ions include *FAST* TEAMS [53] and *Cluster* CODIF [54]. Combining ESA, post-acceleration, TOF, and SSD allows the independent determination of A, Q, and E, as in *ACE* and *Ulysses* SWICS [44, 45] and *STEREO* PLASTIC [55].

4.2.3.3. *Intrinsic noise-suppression capability of TOF sensors*

To extract low-flux particle events from an overwhelming background among the detected signals requires effective noise suppression. With the coincidence of two or more independent signals from the same particle within a time window $\Delta \tau_C$, as for TOF, the suppression comes naturally.

Let R_S be the *signal* or *true event rate*, and R_B the *background* or *noise event rate*. When collected by two detectors of respective efficiencies η_1 and η_2, the *detected signal rate* R_{SD} is:

$$R_{SD} = R_S \cdot \eta_1 \cdot \eta_2 \tag{4.6}$$

Let us now compare R_{SD} with the chance coincidence rate R_{BC} of a much higher background event rate R_B. If the mean separation of two consecutive noise events, $\Delta\tau_B = 1/R_B$, is much longer than $\Delta\tau_C$, or $R_B \cdot \Delta\tau_C \ll 1$, then the chance coincidence rate of two uncorrelated noise events is:

$$R_{BC} = R_B^2 \cdot \Delta\tau_C \cdot \eta_1 \cdot \eta_2 \tag{4.7a}$$

$$\frac{R_{BC}}{R_{SD}} = \frac{R_B}{R_S} R_B \cdot \Delta\tau_C \tag{4.7b}$$

The double-coincidence increased the *signal-to-noise ratio* $(S/N) = (R_B/R_S)^{-1}$ by $(R_B \cdot \Delta\tau_C)^{-1}$. A typical coincidence window $\Delta\tau_C = 200\,\text{ns}$ suppresses a relatively high background rate of $R_B = 5 \times 10^4\,\text{s}^{-1}$ by a factor of 100. Adding another coincidence with a similar $\Delta\tau_C$ changes Eq. (4.7b) to

$$\frac{R_{BC}}{R_{SD}} = \frac{R_B}{R_S}(R_B \cdot \Delta\tau_C)^2 \tag{4.7c}$$

squaring the already huge suppression factor. Triple coincidence, first used on a cosmic-ray instrument on *NASA*'s *Explorer VI* [56], allows detecting infrequent events amid high noise levels with confidence. Such an assuring feature is innate to all TOF spectrographs!

4.2.3.4. *TOF spectrographs for neutral atoms*

TOF ion mass spectrographs work equally well for ENAs, because the SE yield of a thin foil does not depend on Q of the incident particle, and the average Q upon exit only depends on its E and A [57]. In fact, most ENA sensors use double- or triple-coincidence TOF schemes. However, ENA-specific challenges must be met *before* benefitting from the intrinsic noise-suppressing power of TOF sensors.

4.3. ENA-Specific Observational Challenges

Besides adopting the skills and tools developed for *in-situ* space-plasma investigations, remote-sensing ENA observations need solutions for ENA-specific challenges. The ENA flux, given by Eqs. (3.18a) and (3.18b), is typically low because of small CX cross-sections ($\sigma_{iaCX} < 2 \cdot 10^{-15}\,\text{cm}^2$), low ANA densities ($n_a <$

10^5 cm^{-3}), and possible extinction *en route*. In general, the ENA flux is orders of magnitude lower than the local ion flux, except at altitudes <500 km, where n_a is high (>10^5 cm^{-3}) and H ENAs were first discovered (Sec. 2.1). Also, the identification of protons near the equator that stem from H ENAs generated in the inner radiation belt occurred at low altitudes (Sec. 2.3). The imposing local ion flux must not enter the sensor, especially ions with m and E of the anticipated ENAs. Thus, *ion suppression* is the *first challenge*.

The *second challenge* is *EUV attenuation*. EUV photons ($\lambda <$ 124 nm or $E >$ 10 eV) are ubiquitous and intense, especially H Lyα (121.6 nm) and HeI (2^1P $\rightarrow$ 1^1S: 58.4 nm). They are sunlight resonantly scattered at the very ANAs that give birth to ENAs. The Lyα intensity, for example, is 200–600 R (1 Rayleigh = $10^6/4\pi$ photons/(cm^2 sr s)) in the anti-solar direction at 1 AU. These photons are energetic enough to trigger EMs and intense enough to trigger some SSDs by "pile-up". Thus, the entrance system must minimize the entry of EUV into the ENA analyzers.

The first two challenges make ENA observation a search for a needle in a haystack. Addressing them is essential *before* taking advantage of the noise suppression capability of TOF sensors.

The *third challenge* specific to low-E neutral atoms is the need to *convert them to ions*, so they can be analyzed by $\vec{E}$- and/or $\vec{B}$-fields and/or accelerated for effective detection. This challenge escalates with decreasing E, particularly for the very low-E ANAs.

We will address each of these three challenges separately in Secs. 4.3.1, 4.3.2, and 4.3.3. A generic scheme in Fig. 4.9 (as in the *IBEX* [58, 59] and *IMAP* [60] ENA sensors) illustrates how a *single* ENA sensor can meet *all three challenges*.

Finally, calibration of the ENA sensors with a well-behaved neutral-atom beam is essential. Providing such a beam is a challenge addressed in Appx. B.

4.3.1. *Suppression of charged particles*

Rejection of ions with the same m and E of the ENAs of interest can be achieved electrostatically, without affecting ENAs. A $\vec{B}$-field is less effective for ions because of their large gyro radii. However,

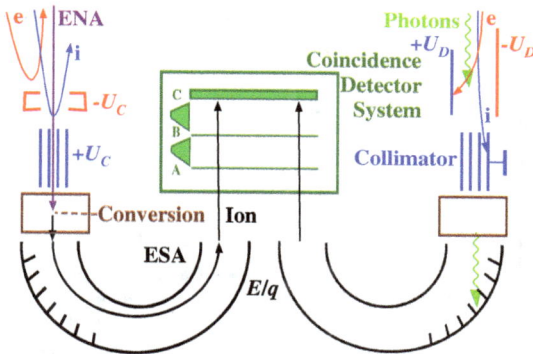

Fig. 4.9. Functional blocks of a generic ENA sensor that meets all three challenges. The entrance electrodes and the collimator provide the $\vec{E}$ field separating ENAs from charged particles by repelling (left) or deflecting (right) them. The collimator defines the arrival direction and angular resolution. The conversion section turns the ENAs into ions for their E/q analysis by an ESA, also stopping photons. The selected ions are then accelerated for identification and detection in a TOF coincidence subsystem.

both help to reject electrons, of which massive amounts could enter a sensor, particularly photoelectrons generated on the S/C surfaces. Depending on the electrostatic design, ions can be *deflected* or *repelled* (Fig. 4.9).

- *Deflection* occurs between the aperture and the sensor (Fig. 4.9, right). The simplest ion deflector is a pair of parallel plates of length l, separated by distance d with a voltage U_D across the gap. Ignoring fringe fields, ions with

$$E/q \leq \frac{U_D \cdot l^2}{4d^2} \qquad (4.8)$$

are deflected towards the negative plate and do not enter the next section of the sensor. Figure 4.10 shows the ion transmission of the parallel-plate ion deflector of CELIAS HSTOF on *SOHO* (Fig. 5.10). The transmission of ions of $E < 88\,\text{keV/e}$ is suppressed to $<10^{-2}$ [61]. The deflector also collimates and defines the sensor FOV and geometric factor.

Instead of a stack of parallel plates with a fixed gap d as in HSTOF (Fig. 5.10), divergent plates, with d decreasing linearly

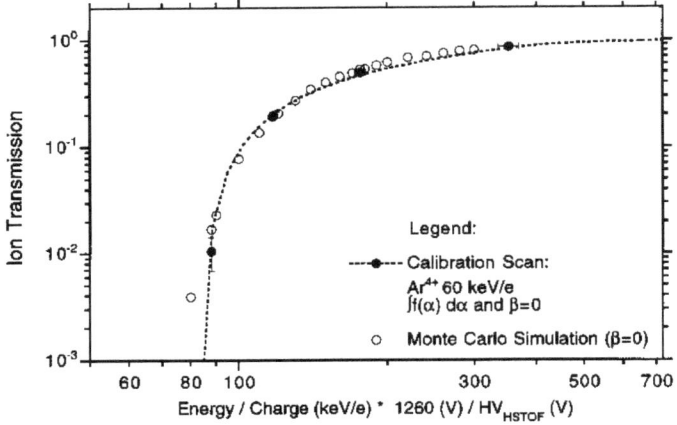

Fig. 4.10. Fraction of ions that pass the ion-deflecting collimator of HSTOF as a function of E/q. Each pair of plates has $l = 10.0\,\text{cm}$, $d = 0.45\,\text{cm}$, and $U_D = 2.52\,\text{kV}$. The arrival is along the boresight ($\beta = 0$) with the shortest and least effective path across the $\vec{E}$ field [61]. (© AAS. Reproduced with permission).

along l, are installed in MIMI INCA on *Cassini* and HENA on *IMAGE* (Sec. 5.3.1). Figure 4.11 shows this configuration, along with trajectories through the detector system (Fig. 4.11, right). The immediate benefit of the divergent-plate design is a substantial increase of the FOV ($90° \times 120°$) and G. Consequently, the TOF assembly and image plane must accommodate the wider FOV. The tradeoff for these advantages is that the higher voltage needed to deflect the ions in the wider entrance gap increases $\vec{E}$ at the narrow exit, where d between the oppositely biased plates is the smallest. Arcing may arise when U_D between the plates exceeds a certain level. A preventive solution is lowering U_D and lengthening l. The INCA collimator deflects ions $\leq 500\,\text{keV/e}$ when operating at $U_D = 12\,\text{kV}$.

The simplicity of the flat plates belies two problems: 1) ions hitting either plate may bounce off as neutral atoms, masquerading as ENAs; 2) particles or photons hitting the negative plate release SEs, which are then accelerated toward the positive plate. Here, they could produce X-rays and trigger an electron avalanche, thus overloading the start detector or even the power supply. Possible

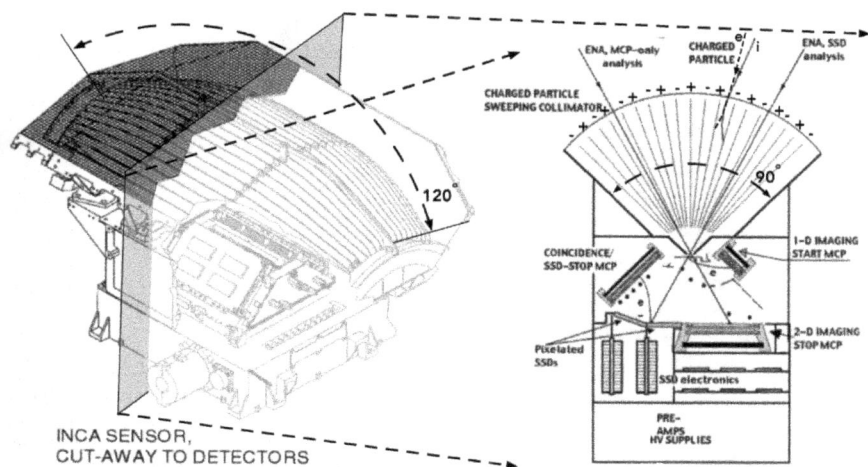

Fig. 4.11. Divergent flat plate collimators for ion deflection. Left: Perspective view of MIMI INCA on *Cassini* and *IMAGE* HENA [62]. Right: Schematic cross-section of HENA [63]. Alternatingly biased fanned-out flat plates provide a 90° × 120° FOV. (© Springer. Reproduced with permission, adapted).

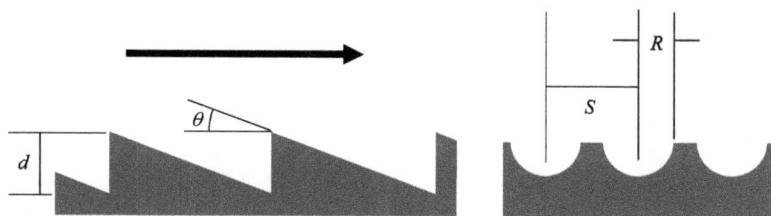

Fig. 4.12. Cross-sectional view of two types of serrations (arrow points away from the entrance). Left: Sawtooth, with θ the maximum half-angle of the FOV (Appx. A), defined by a pair of plates, *e.g.*, in HSTOF and HENA. The serration depth d is typically ≤ 1 mm. Right: Cylindrical troughs as in the *TWINS* stainless steel collimator plates. The serration depth (radius) is $R \approx 50\,\mu$m at a separation of S close to $100\,\mu$m [65].

mitigation strategies are: 1) reducing U_D and d to lower the SE yield while maintaining $\vec{E}$ for ion deflection; 2) serrating the surfaces (Fig. 4.12) or replacing the plates with Venetian-blind blades to trap particles and photons [64]; 3) coating the surface, *e.g.*, with $\approx 0.25\,\mu$m of gold black or dendritic Ebonol-C to further reduce light scattering [65].

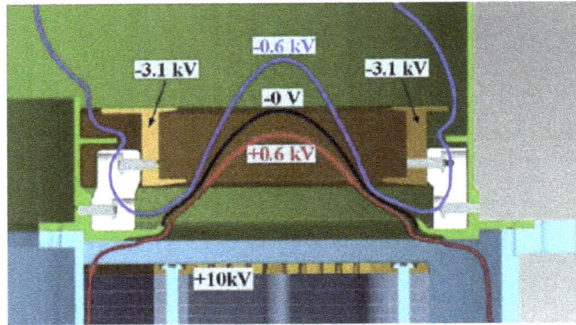

Fig. 4.13. Cross-sectional view of the *IBEX* charged-particle repellers. The electrodes at $-3.1\,\text{kV}$ reject electrons up to $600\,\text{eV}$, and the collimator at $+10\,\text{kV}$ ions up to $10\,\text{keV}$. The equipotential lines are cross-sections of annular surfaces [58]. (© Springer. Reproduced with permission).

- *Repulsion* occurs before the ions reach the collimator (Fig. 4.9, left; Fig. 4.13), thus eliminating the noise-causing byproducts of deflection. Indeed, electrons must be rejected upfront by negative electrodes to prevent an electron-gun effect. Electrons could ionize the rest gas, producing ions internally that could masquerade as ENAs in the sensor (Sec. 5.3.2).

4.3.2. *Attenuation of EUV*

EUV attenuation requires material intervention. In TOF analyzers, thin, low-Z C-foils that commonly provide SEs for the start signals also reduce the EUV flux [66, 67]. These foils are formed by depositing arc-evaporated C on a glass slide coated with a detergent film. After floating the foil off the slide in distilled water, it is lifted with an electroformed Ni or Au mesh of known transmittance, to which it adheres. The foil thickness, given by its surface density σ in $\mu\text{g/cm}^2$, is finally calibrated by the actual ion E-loss in the foil [68]. The best fit to three independent data sets provided the Lyα transmittance of $T(\sigma) = (0.6 \pm 0.2)e^{-(0.60 \pm 0.05)\sigma}$ as a function of the C-foil surface density σ [69].

Composite foils better attenuate EUV but reduce the transmittance of low-E ENAs. Figure 4.14 shows the tradeoff in threshold E (defined by 50% of incident protons, a proxy for H, exiting the foil

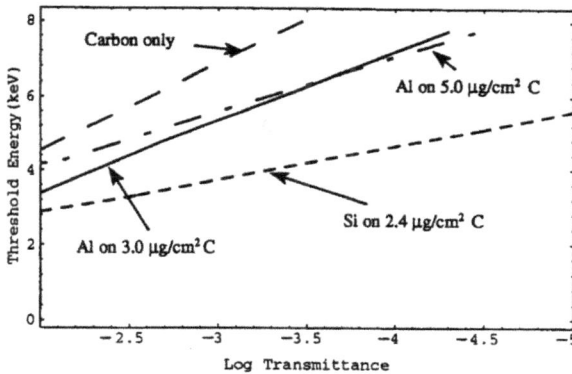

Fig. 4.14. Threshold E *vs.* Lyα transmittance for C, Al/C, and Si/C composite foils [70]. (© Elsevier. Reproduced with permission).

with $E > 1\,\mathrm{keV}$ sufficient to trigger the MCP) *vs.* Lyα transmittance for four foil sets. The Si/C foils seem the best, *e.g.*, a Si/C foil with $14 \pm 1\,\mu\mathrm{g/cm}^2$ Si on $2.4 \pm 0.2\,\mu\mathrm{g/cm}^2$ C reduces Lyα by 10^{-5}, while passing 50% of 5-keV H atoms with $E > 1\,\mathrm{keV}$ [70]. Three early ENA sensors used Si/C foils (Sec. 5.3.1).

An alternative EUV attenuator is the *freestanding transmission grating*, made of rectangular Au bars (Fig. 4.15, left). It was first used in the *SOHO* CELIAS EUV monitor [71] and then proposed and tested for ENA instruments (Sec. 5.3.2) [72, 73]. Closely spaced metal bars attenuate EUV like waveguides without degrading the particle energy. However, they reduce the FOV and G. The *TWINS* grating has 6% ENA transmission [13], far lower than the 50% for a Si/C foil with comparable EUV suppression.

The best way to reduce EUV signals is to convert the incident ENAs to ions, then analyze them by an ESA or MSA. These analyzers bend the ion trajectories into the detectors inaccessible to photons (Fig. 4.9), thus bringing us to the third challenge for low-E ENA observations, the need for conversion.

4.3.3. *Conversion of ENAs into ions before their analysis*

ENAs of E below the detection thresholds of either SSDs or MCPs ($\leq 1\,\mathrm{keV}$) must be converted to ions and accelerated above the

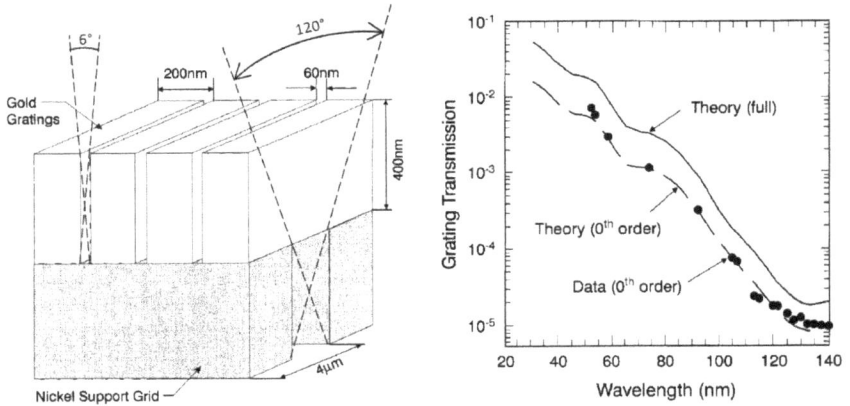

Fig. 4.15.   Freestanding transmission-grating EUV filter. Left: The basic module used in MENA and on *TWINS* (Sec. 5.3.2) sets a $\approx 120°$ FOV in one and $\approx 6°$ in the orthogonal direction. Right: Theoretical and measured transmittance as a function of wavelength [65]. (© Springer. Reproduced with permission).

threshold. Ionization with an electron beam, even *via* a low-power cold cathode [74] as used in gas mass-spectrographs [75], is ineffective for the low fluxes of ENAs or ANAs. Instead, we consider ionization through their interaction with solids, *i.e.*, penetrating a thin foil or reflecting from a solid surface.

4.3.3.1. *Conversion of ENAs to positive ions*

The first space-borne ENA spectrograph (Sec. 5.1.2) and *IBEX-Hi* (Sec. 5.3.2), four decades apart, used a thin C-foil to ionize a fraction of the incident ENAs before analyzing them by an ESA. The Q distribution of particles exiting a C-foil has been studied extensively [76–78]. Figure 4.16 shows the C-foil ionization efficiency as a function of ENA energy [79].

Ions emerging from the foil have lower E than the incident ENAs, but they can trigger detectors with higher E thresholds after acceleration. However, transferring signals from a detector at a potential $\geq 10\,\mathrm{kV}$ to the signal processor at ground potential requires electrical insulation and optical or inductive coupling. As indicated in Fig. 4.9, ion analysis using an ESA also provides an effective EUV trap.

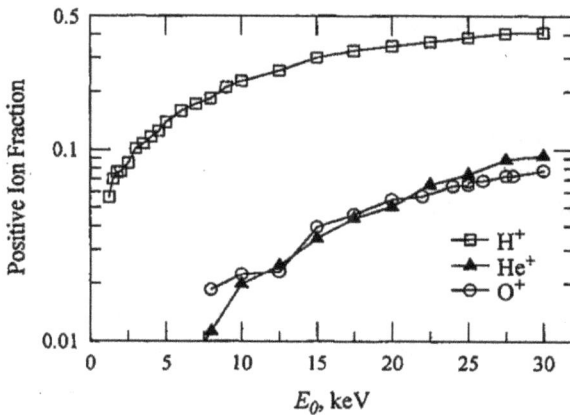

Fig. 4.16. The measured fraction of incident H, He, and O ENAs, emerging from a $1.1\,\mu g/cm^2$ C-foil as $Q = +1$ ions, increases with energy [79]. (© AIAA. Reproduced with permission).

4.3.3.2. Conversion of low-energy ENAs or ANAs to negative ions

Figure 4.16 suggests that the ionization efficiency for ENAs that penetrate a C-foil drops precipitously with decreasing energy. Thus, detecting ENAs of $E < 1\,keV$ with concentrations $< 10^3$ atoms/cm^3, e.g., ANAs participating in the production of ENAs, requires an alternative conversion method.

At $E \leq 0.5\,keV$, species that form stable negative ions, such as H and O, produce an increasing fraction thereof after penetrating a thin foil [79]. Similarly, upon impacting a surface, a projectile penetrates a thin surface layer. Thus, striking a surface at a grazing angle, even at the lowest energies, effectively converts ENAs to negative ions with minimum scattering, a process known as surface conversion. At grazing incidence on a very smooth surface (roughness $< 5\,nm$) [80], a large fraction of the atoms leaves the surface under specular reflection as negative ions [81]. Hence, surface conversion is the method of choice at the lowest energies, as proposed earlier [82, 83] and implemented in *IBEX*-Lo and *IMAGE* LENA (Sec. 5.3.3). Negative ions are often used in the first stage of accelerators and are crucial to atom-wall interactions in controlled fusion, thus well known.

Surface conversion of atoms to negative ions works as follows. Atoms or positive ions that touch the surface enter the uppermost layers of the *conversion surface* (CS) and, on average, attain a Q distribution that depends on the particle species and E, as when penetrating a thin C-foil. A fraction of atoms with negative binding energy for electrons (or electron affinity E_{ea} in their vacuum state), leaves as negative ions after reflection. While still at distances below $10a_B$ (Bohr radius), they induce a positive image-charge below the surface, leading to a deepening and broadening of E_{ea} to overlap with electrons up to the Fermi level ϕ_F below the surface. Here, resonant electron transfer between the surface and the escaping atom occurs. Thus, materials with low ϕ_F are conducive to negative ion production [84]. It is also evident that the probability of keeping an extra electron decreases rapidly with distance from the surface as E_{ea} approaches the vacuum value. Racing through this zone beyond $10a_B$ may increase the likelihood to retain the extra electron and thus explain why the negative ion fraction increases with the escape velocity perpendicular to the surface, as illustrated in Fig. 4.17 [85].

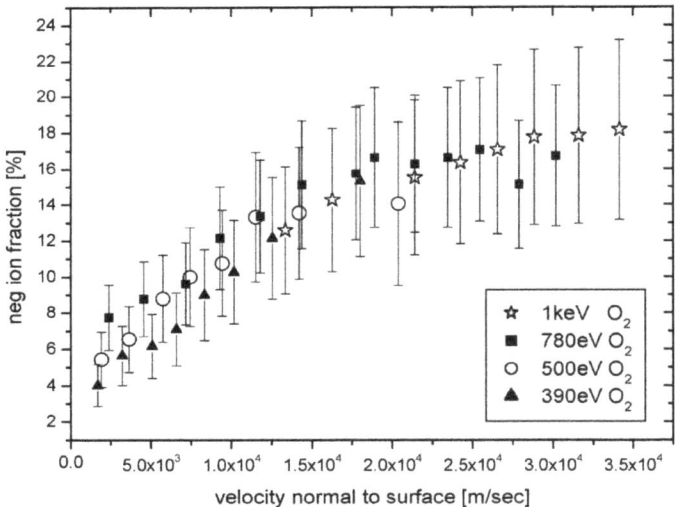

Fig. 4.17.   Negative ion fraction as a function of the outgoing velocity normal to the CS, measured for O_2 on a diamond-like C surface [85]. (© AIP. Reproduced with permission).

While converting ENAs and ANAs to negative ions on the CS maintains the species identity, the CS also emits inadvertently sputtered particles, usually of the CS constituents and fragments of adsorbed H_2O. A similar fraction emerges as negative ions and adds a challenge. Sputtered ions from ENAs at higher E contribute to the signal at lower E, thus requiring additional calibration steps [59]. However, these sputter products enable the diagnostics of He and Ne atoms, which do not produce negative ions due to their low E_{ea}. Identification occurs *via* the unique m and E dependent abundances in the resulting sputter products [59, 86]. Effective mass analysis and detection of the ions must follow the ESA selection and post-acceleration across several kV [87].

Having reviewed the general requirements for ENA instrumentation, we move on to specific ENA sensor designs in Chap. 5. We will discuss especially how they address the ENA-specific challenges. We hope the material presented here will awaken the creativity in our readers to develop even more ingenious designs for future missions.

References

1. Special issue on SOHO (1995). *Solar Phys.*, **162**(1–2).
2. Hovestadt, D., Hilchenbach, M., Bürgi, A. *et al.* (1995). CELIAS — charge, element and isotope analysis system for SOHO, *Sol. Phys.*, **162**(1–2), 441–481. DOI:10.1007/BF00733436.
3. Hsieh, K. C. (2015). Detection of energetic neutral atoms in and out of the heliosphere, *Chin. J. Space Sci.*, **35**(3), 253–292. DOI:10.11728/cjss2015.03.253. (In Chinese).
4. Special issue on IBEX (2009). *Space Sci. Rev.*, **146**(1–2).
5. McComas, D. J., Carrico, J. P., Hautamaki, B. *et al.* (2011). A new class of long-term stable lunar resonance orbits: Space weather applications and the interstellar boundary explorer, *Space Weath.*, **9**(11), S11002.DOI:10.1029/2011SW000704.
6. McComas, D. J., Allegrini, F., Bochsler, P. *et al.* (2009a). IBEX — interstellar boundary explorer, *Space Sci. Rev.*, **146**(1–2), 11–33. DOI: 10.1007/s11214-009-9499-4.
7. Bzowski, M., Möbius, E., Kucharek, H. *et al.* (2012). Precision pointing of IBEX-Lo observations, *Astrophys. J. Supp.*, **198**, art ID 12, DOI:10.1088/0067-0049/198/2/9

8. Hlond, M., Bzowski, M., Möbius, E. *et al.* (2012). Precision pointing of IBEX-Lo observations, *Astrophys. J. Suppl.*, **198**(9).

9. Swaczyna, P., Kubiak, M. A., Bzowski, M. *et al.* (2022). Very Local Interstellar Medium Revealed by Complete Solar Cycle of Interstellar Neutral Helium Observations with IBEX, *Astrophys. J. Suppl.*, **259**(42).

10. Special issue on IMAGE (2000). *Space Sci. Rev.*, **91**(1–2).

11. Fuselier, S. A., Burch, J. L., Lewis, W. S., & Reiff, P. H. (2000). Overview of the IMAGE science objectives and mission phases, *Space Sci. Rev.*, **91**(1–2), 51–66.

12. Gibson, W. C., Burch, J. L., Scherrer, J. R. *et al.* (2000). The IMAGE observatory, *Space Sci. Rev.*, **91**(1–2), 15–50.

13. McComas, D. J., Allegrini, F., Baldonado, J. *et al.* (2009b). The two wide-angle imaging neutral-atom spectrometers (TWINS) NASA mission-of-opportunity, *Space Sci. Rev.*, **142**(1–4), 157–231. DOI: 10.1007/s11214-008-9467-4.

14. Dahl, P. (1973). *Introduction to Electron and Ion Optics* (Academic Press, New York).

15. Dahl, D. A. (2000). SIMION for the personal computer in reflection, *Int. J. Mass Spectr.*, **200**(1–3), 3–25. DOI:10.1016/S1387-3806(00)00305-5.

16. Gloeckler, G., Ipavich, F. M., Stüdemann, W., Wilken, B., *et al.* (1985). The charge-energy-mass spectrometer for 0.3–300 keV/e ions on the AMPTE CCE, *IEEE Trans. Geosci. & Remote Sens.*, **GE-23**(3), 234–240.

17. Evans, D. S. (1965). Low energy charged particle detection using the continuous channel electron multiplier. *Rev. Sci. Instr.*, **36**(3), 375–382.

18. Gruntman, M. A., & Kalinin, A. P. (1977). Registration characteristics of neutral particles with power 0.6–2.0 keV channel electron multiplier with funnel, *Acad. Sci. USSR Inst. Space Res. Moscow Report*, **Pr-311**(1–23) (*NASA Report* No. TM-75540, Sep. 1978)

19. Boutot, J. P., & Audier, M. (1977). Les detecteures et ensembles de detection utilizes en physique nucleaire, *Journéee de Saclay* (in French).

20. Wiza, J. L. (1979). Microchannel plate detector, *Nucl. Instr. & Meth.*, **162**(1–3), 587–601.

21. Colson, W. B., McPherson, J., & King, F. T. (1973). High-gain imaging electron multiplier, *Rev. Sci. Instr.*, **44**(12), 1679–1696.

22. Lampton, M. (1981). The microchannel image intensifier, *Sci. Amer.*, **245**(5), 62–71.

23. Martin, C., Jelinsky, P., Lampton, M., Malina, R. F., & Anger, H. O. (1981). Wedgeandstrip anodes for centroid fining position-sensitive photon and particle detectors, *Rev. Sci. Instr.*, **52**(7), 1067–1074.

24. Alig, R. C., Bloom, S., & Struck, C. W. (1980). Scattering by ionization and phonon emission in semiconductors, *Phys. Rev. B*, **22**(12), 5565–5582.

25. Sigmund, P. (1997). Charge-dependent electronic stopping of swift nonrelativistic heavy ion, *Phys. Rev. A*, **56**(5), 3781–3793.

26. Lindhard, J., & Scharff, M. (1961). Energy dissipation by ions in the keV region, *Mod. Phys. Rev.*, **124**(1), 128–130.

27. Lindhard, J., Scharff, M., & Schiøtt, H. E. (1963). Range concepts and heavy ion ranges, *Mat. Fys. Medd. Dan. Vid. Selsk.*, **33**, 1–42.

28. Northcliffe, L. C., & Schilling, R. F. (1970). Range and Stopping-power Tables for Heavy Ions, *Nucl. & Atom. Data*, **A7**, 233–463.

29. Ipavich, F. M., Lundgren, R. A., Lambird, B. A., & Gloeckler, G. (1978). Measurements of pulse-height defect in Au-Si detectors from H, He, C, N, O, Ne, Ar, Kr from ≈2 to ≈400 keV/nucleon, *Nucl. Instr. & Meth.*, **154**(2), 291–294.

30. Ziegler, J. F., Biersack, J. P., & Ziegler, M. D. (2008). *SRIM the Stopping Range of Ions in Matter*, Lulu Press Co., Morrisville, NC, USA.

31. Tuzzolino, A. J., Perkins, M. A., & Kristoff, J. (1965). Stabilization of lithium-drifted silicon surface-barrier detectors, *Nucl. Instr. & Meth.*, **37**, 204–216.

32. Hilchenbach, M. (2002). Space-borne mass spectrometer instrumentation, *Intl. J. Mass Spectrom.*, **215**(1–3), 113–129.

33. Tindall, C. S., Palaio, N. P., Ludewig, B. A. *et al.* (2008). Silicon detectors for low energy particle detection, *IEEE Trans. Nucl. Sci.*, **55**(2), 797–801.

34. Lutz, G. (2005). DEPFET development at the MPI semiconductor laboratory, *Nucl. Instr. & Meth. in Phys. Res. A*, **549**(1), 103–111.

35. Majewski, P., Andricek, L., Christensen, U., Hilchenbach, M. *et al.* (2010). DEPFET maccropixel detectors for MIXS: first electrical qualification measurements, *IEEE Trans. Nucl. Sci.*, **57**(4), 2289–2396.

36. Lutz, G. (2007). *Semiconductor Radiation Detectors: Device Physics*, Springer-Verlag, Berlin, Germany. (eBook available).

37. Dietz, E., Bass, R., Reiter, A., Friedland, U., & Hubert, B. (1971). Time-of-flight spectrometer for mass identification of heavy ions, *Nucl. Instr. Meth.*, **97**(3), 581–586. DOI: 10.1016/0029-554x(71)90261-8.

38. Schneider, W. F. W., Kohlmeyer, B., & Bock, R. (1970). Mass-identification of alpha-partices and heavy ions by time-of-flight methods, *Nucl. Instr. and Meth.*, **87**(2), 253–259.

39. Gloeckler, G., & Hsieh, K. C. (1979). Time-of-flight technique for particle identification at energies from 2–400 keV/nucleon, *Nucl. Instr. Meth.*, **165**(3), 537–544.

40. Gruntman, M. (1997). Energetic neutral atom *imaging of space plasma*, *Rev. Sci. Instrum.*, **68**(10), 3617–3656.

41. Wüest, M. (1998). Time-of-Flight Ion Composition Measurement Technique for Space Plasmas, in: *Measurement Techniques in Space Plasmas — Particles*, Eds. Pfaff, R. F., Borovsky, J. E. and Young, D. T., *Geophys. Monograph*, **102**, 141–156.

42. Möbius, E., Galvin, A. B., Kistler, L. M., Kucharek, H., & Popecki, M. A. (2016). Time-of-flight mass spectrographs — from ions to neutral atoms, *J. Geophys. Res.*, **121**(12), 11,647–11,666.

43. Möbius, E., Hovestadt, D., Klecker, B. *et al.* (1985). The time-of-flight spectrometer SULEICA for ions of the energy range 5–270 keV/charge on the AMPTE/IRM, *IEEE Trans. on Geosci. El.*, **GE-23**(3), 274–279.

44. Gloeckler, G., Geiss, J., Balsiger, H. *et al.* (1992). The solar wind ion composition spectrometer, *Astron. Astrophys. Suppl.*, **92**(2), 267–289.

45. Gloeckler, G., Cain, J., Ipavich, F. M., & Tums, E. O. (1998). Investigation of the composition of solar and interstellar matter using solar wind and pickup ion measurements with SWICS and SWIMS on the ACE spacecraft, *Space Sci. Rev.*, **86**(1–4), 497–539.

46. Wilken, B., & Stüdemann, W. (1984). A compact time-of-flight mass-spectrometer with electrostatic mirrors, *Nucl. Instr. and Meth. in Phys. Res.*, **222**, 587–600.

47. McEntire, R. W., Keath, E. P., Fort, D. E., Lui, A. T. Y., & Krimigis, S. M. (1985). The medium-energy particle analyzer (MEPA) on the AMPTE CCE spacecraft, *IEEE Trans. on Geosci. El.*, **GE-23**(3), 230–233.

48. Mason, G. M., Gold, R. E., Krimigis, S. M. *et al.* (1998). The ultra-low-energy isotope spectrometer (ULEIS) for the ACE spacecraft, *Space Sci. Rev.*, **86**(1–4), 409–448.

49. Allegrini, F., Ebert, F. W., & Funsten, H. O. (2016). Carbon foils for space plasma instrumentation, *J. Geophys. Res. Space Physics*, **121**(5), 3931–3950.

50. Gosling, J. T., Asbridge, J. R., Bame, S. J., & Feldman, W. C., (1978). Effects of a long entrance aperture upon the azimuthal response of spherical section electrostatic analyzers, *Rev. Sci. Instrum.*, **49**(9), 1260.

51. Carlson, C. W., Curtis, D. W., Paschmann, G., & Michael, W. (1983). An instrument for rapidly measuring plasma functions with high resolution, *Adv. Space Res.*, **2**(7), 67–70.

52. Carlson, C. W., & McFadden, J. P. (1998). Design and application of imaging plasma instruments, in: *Measurement techniques in Space Plasms: Particles*, Eds. Pfaff, R. F., Borovsky, J. E., Young, D. T., *AGU Monograph*, **102**, 125–140.

53. Klumpar, D. M., Möbius, E., Kistler, L. M. *et al.* (2001). The time-of-flight energy, angle, mass spectrograph (TEAMS) experiment for FAST, *Space Sci. Rev.*, **98**(1–2), 197–219.

54. Rème, H., Bosqued, J. M., Sauvaud, J. A. *et al.* (1997). The CLUSTER ion spectrometry experiment, *Space Sci. Rev.*, **79**(1–2), 303–350.

55. Galvin, A. B., Kistler, L. M., Popecki, M. A. *et al.* (2008). The plasma and suprathermal ion composition (Plastic) investigation on the STEREO observatories, *Space Sci. Rev.*, **136**(1–4), 437–486.

56. Fan, C. Y., Meyer, P., & Simpson, J. A. (1960a). Cosmic radiation intensity decreases observed at the earth and in nearby planetary medium, *Phys. Rev. Lett.*, 4(8), 421–423.

57. Gonin, M., Buergi, A., Oetliker, M., & Bochsler, P. (1992). Interaction of solar wind ions with thin carbon foils: Calibration of time-of-flight spectrometers, *Proc. of the First SOHO Workshop: Coronal Streamers, Coronal Loops, and Coronal and Solar Wind Composition*, ESA-SP **348**, 381–384.

58. Funsten, H. O., Allegrini, F., Bochsler, P. *et al.* (2009). The Interstellar Boundary Explorer High Energy (IBEX-Hi) Neutral Atom Imager, *Space Sci. Rev.*, **146**(1–4), 75–103.

59. Fuselier, S. A., Bochsler, P., Chornay, D. *et al.* (2009). The IBEX-Lo sensor, *Space Sci. Rev.*, **146**(1–4), 117–147.

60. McComas, D. J., Christian, E. R., Schwadron, N. A. *et al.* (2018). Interstellar Mapping and Acceleration Probe (IMAP): A New NASA Mission, *Space Sci. Rev.*, **214**(116).

61. Hilchenbach, M., Hsieh, K. C., Hovestadt, D. *et al.* (1998). Detection of 55–80 keV hydrogen atoms of heliospheric origin by CELIAS/HSTOF on SOHO, *Astrophys. J.*, **503**(2), 916–922.

62. Krimigis, S. M., Mitchell, D. G., Hamilton, D. C. *et al.* (2004). Magnetosphere Imaging Instrument (MIMI) on the Cassini Mission to Saturn/Titan, *Space Sci. Rev.*, **114**(1–4), 233–329.

63. Mitchell, D. G., Jaskulek, S. E., Schlemm, C. E. *et al.* (2000). The high-energy neutral atom (HENA) imager for the IMAGE mission, *Space Sci. Rev.*, **91**(1–2), 67–112.

64. Hsieh, K. C., Zurbuchen, T. H., Orr, J., Gloeckler, G., & Hilchenbach, M. (2004). A collimator design for monitoring heliospheric energetic neutral atoms at 1 AU, *Adv. Space Res.*, **34**(1), 213–218.

65. Pollock, C. J., Asamura, K., Baldonado, J. *et al.* (2000). Medium energy neutral atom (MENA) imager for the IMAGE mission, *Space Sci. Rev.*, **91**(1–2), 113–154.

66. Hsieh, K. C., Keppler, E., & Schmidtke, G. (1980). Extreme ultraviolet induced forward photoemission from thin carbon foils, *J. App. Phys.*, **51**(4), 2242–2246.

67. McComas, D. J., Allegrini, F., Pollock, C. J. *et al.* (2004). Ultrathin (≈10 nm) carbon foils in space instrumentation. *Rev. Sci. Instrum.*, **75**(11), 4863–4870.

68. Stoner, J. O. (1969). Accurate determination of carbon-foil surface densities, *J. App. Phys.*, **40**(2), 707–709.

69. Hsieh, K. C., Sandel, B. R., Drake, V. A., & King, R. S. (1991). H Lyman α transmittance of thin C and Si/C foils for keV particle detectors, *Nucl. Instr. Meth.*, **B61**(2), 187–193.

70. Drake, V. A., Sandel, B. R., Jenkins, D. G., & Hsieh, K. C. (1992). Transmittance of Al/C foils at 1216 Å, *Nucl. Instr. & Meth. in Phys. Res.*, **B72**(2), 153–158.

71. Ogawa, H. S., McMullin, D. R., Judge, D. L., & Korde, R. (1993). Normal incidence spectrophotometer with high-density transmission grating technology and high-efficiency silicon photodiodes for absolute solar extreme-ultraviolet irradiance measurements, *Opt. Engr.*, **32**(12) 3121–3125.

72. Gruntman, M. A. (1995). Extreme-ultraviolet radiation filtering by freestanding transmission gratings, *Appl. Opt.*, **34**(25), 5732–5747.

73. Gruntman, M. A. (1997a). Transmission grating filtering of 54–140 nm radiation, *Appl. Opt.*, **36**(10), 2203–2205.

74. Curtis, C. C., & Hsieh, K. C. (1986). Spacecraft mass spectrometer ion source employing field emission cathodes, *Rev. Sci. Instrum.*, **57**(5), 989–990.

75. Keppler, E., Afonin, V. V., Curtis, C. C. *et al.* (1986). Neutral gas measurements of comet Halley from Vega 1, *Nature*, **321**, 273–274.

76. Allison, S. K. (1958). Experimental results on charge-changing collisions of hydrogen atoms and ions at kinetic energies above 0.2 keV, *Rev. Mod. Phys.*, **30**(4), 1137–1168.

77. Kallenbach, R., Gonin, M., Bochsler, P., & Bürgi, A. (1995). Charge exchange of B, C, O, Al, Si, S, F and Cl passing through thin carbon foils at low energies: Formation of negative ions, *Nucl. Instr. Meth. in Phys. Res. B*, **103**(2), 111–116.

78. Gonin, M., Kallenbach, R., Bochsler, P., & Bürgi, A. (1995). Charge exchange of low energy particles passing through thin carbon foils: Dependence on foil thickness and charge state yields of Mg, Ca, Ti, Cr and Ni, *Nucl. Instr. Meth. in Phys. Res. B*, **101**(4), 313–320.

79. Funsten, H. O., McComas, D. J., & Scime, E. E. (1995). Low-energy neutral-atom imaging techniques for remote observations of the magnetosphere, *J. Spacecraft & Rockets*, **32**(5), 899–904.

80. Moore, T. E., Chornay, D. J., Collier, M. R. *et al.* (2000). The low-energy neutral atom imager for IMAGE, *Space Sci. Rev.*, **91**(1–2), 155–195.

81. Wurz, P. (2000). Detection of energetic neutral atoms, In: The outer heliosphere: beyond the planets. Based on the spring school "Die äußere Heliosphäre — Jenseits der Planeten", in Bad Honnef, Germany, 12–16 April 1999. K. Scherer, H. Fichtner, E. Marsch, (editors). Katlenburg-Lindau, Germany: Copernicus-Gesellschaft. ISBN 3-9804862-3-0, 251–288.

82. Gruntman, M. A. (1993). A new technique for *in situ* measurement of the composition of neutral gas in interplanetary space, *Planet. Space Sci.*, **41**(4), 307–319.

83. Wurz, P., Schletti, R., & Aellig, M. R. (1997). Hydrogen and oxygen negative ion production by surface ionization using diamond surfaces, *Surf. Sci.*, **373**(1), 56–66.

84. Van Amersfoot, P.W., Geerlings, J. J. C., Kwakman, L. F. Tz. *et al.* (1985). Formation of negative hydrogen ions on a cesiated W(110) surface; the influence of hydrogen implantation, *J. App. Phys.*, **58**(9), 3566–3572.

85. Wahlström, P., Scheer, J. A., Wurz, P., Hertzberg, E., & Fuselier, S. A. (2008). Calibration of charge state conversion surfaces for neutral particle detectors, *J. Appl. Phys.*, **104**(3), 034503.

86. Bochsler, P., Petersen, L., Möbius, E. *et al.* (2012). Estimation of the neon/oxygen abundance ratio at the heliospheric termination shock and in the local interstellar medium from IBEX observations, *Astrophys. J. Supp.*, **198**(2), 13.

87. Wieser, M., Wurz, P., Bochsler, P., Moebius, E., Quinn, J. *et al.* (2005). NICE: an instrument for direct mass spectrometric measurement of interstellar neutral gas, *Meas. Sci. & Technol.*, **16**(8), 1667–1676.

Chapter 5

ENA Sensor Implementations

"Measure what can be measured and make measurable what cannot
be measured."

Galileo Galilei, 1564–1642

This chapter will show how specific ENA instruments meet their
requirements for different scientific objectives, with emphasis on
ENA-specific challenges (Sec. 4.3). We hope the story will inspire
and lead to even more versatile ENA sensors that solve the remaining
and still emerging problems in space-plasma research.

In Sec. 5.1, we will introduce two ion sensors of particular
relevance to ENAs and the first space-borne ENA spectrograph.
Because the TOF technique is a watershed for space-borne particle
sensors, we will present non-TOF and TOF ENA-sensors next in
Sec. 5.2 and Sec. 5.3. The latter will cover various instruments divided
into three E ranges from high to low, with increasing challenges
toward the lower energies.

5.1. The Road to ENA Instrumentation

To fully appreciate the shared attributes of all space-borne ENA
and non-ENA instruments, we start with two mature non-ENA
instruments in Sec. 5.1.1. First, we present a gas mass-spectrograph
that analyzed the neutral atoms and molecules in the Venusian
atmosphere. Second, we describe an ion spectrometer on a low-
altitude Earth-orbiter that detected protons generated by ENAs from

107

the outer radiation belt (Sec. 2.3). Section 5.1.2 discusses the first ENA spectrometer in space to illustrate the commonalities among these instruments.

5.1.1. *Common root: two non-ENA instruments of relevance*

• *Gas mass-spectrometer on NASA's 1978 Pioneer Venus Multi-probe Mission*: The *in-situ* analysis of neutral gas in the upper atmosphere with mass-spectrometers began with sounding rockets, followed by satellites, and eventually expanded to probes dropped into atmospheres of other planets. Here, we describe one that measured the composition and temperature profile of the Venusian thermosphere on a 10-km/s descent.

This gas spectrograph (Fig. 5.1) operated under ambient pressure up to 10^{-2} mbar as it descended into the Venusian atmosphere [2]. A magnetically focused 75-eV electron beam El ionizes some ambient gas that passes the ion-repeller R. The resulting molecular and atomic ions $(Q = +1)$ are extracted and accelerated through slit

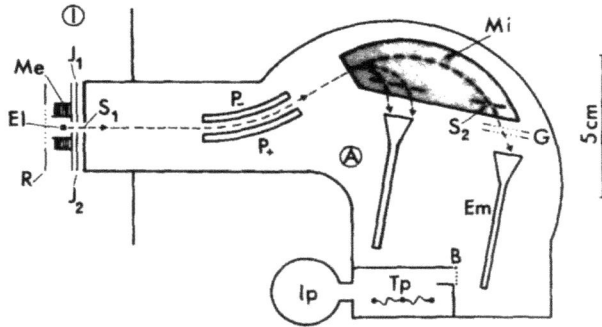

Fig. 5.1. Schematics of a gas spectrograph on *Pioneer Venus*. The main subsystems along the particle trajectories are labeled: ion repeller R, electron beam for ionization El, electron-beam focusing magnets Me, ion-focusing electrodes J_1 & J_2, entrance slit to the analyzer S_1, ESA electrodes P_- & P_+, magnetic analyzer Mi, exit slit S_2, grid assembly G (electrometer for ion collection), electron multiplier Em, baffle system B, titanium-sublimation pump Tp, and ion-getter pump Ip [1]. (© AIP. Reproduced with permission).

S_1 (0.07 × 5.0 mm) into a double-focusing Mattauch-Herzog mass-spectrograph. The ESA with P_+ and P_- selects an almost parallel ion beam from the divergent ion flux within a narrow E/q range. The magnetic analyzer Mi sorts the ions by $|\vec{P}|/q$ (Sec. 4.2.1). For a given E/q, the ion gyroradii obey $r_{ci} \propto (m_i/q_i)^{1/2}$. Thus, the exit location of the ions at Mi depends on m. The two EMs along the exit edge of Mi capture ions of 1–4 Da (H^+ and He^+) and 12–44 Da (C^+ through CO_2^+), respectively. The Em counts only 10% of the C and O group particles exiting slit S_2. The low-transmission grid G, serving as an electrometer, collects another 80%, thus providing a second independent flux measurement. In addition, the ion current is measured in the ionization region El, which doubles as an ion pressure gauge calibrated for the species detected at S_2. The ionization efficiency in El depends on species and pressure, as indicated in Fig. 5.2 by the count rates and Em currents as functions of the pressure. Limited by power, pumps Tp and Ip operated at 15-s intervals to keep the spectrograph operational behind slit S_1 during the descent at the altitudes of interest.

The ion-repelling grid and analyzer-detector system of this gas mass-spectrograph are essential for ENA sensors. However, its

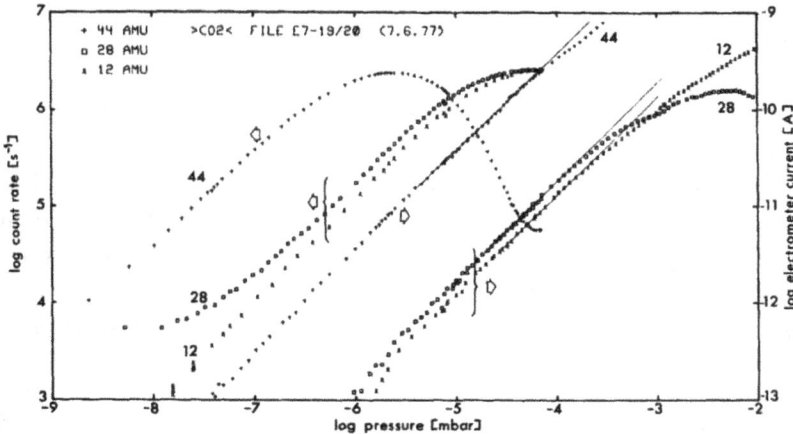

Fig. 5.2. Pressure calibration curves. The Em count rates and currents at G for C, CO, and CO_2 are shown as functions of CO_2 pressure obtained from the ion current at El [1]. (© AIP. Reproduced with permission).

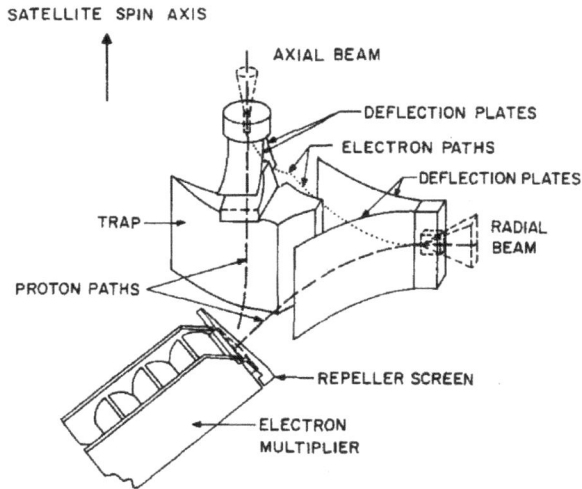

Fig. 5.3. Schematic view of SPS. Particles entering the circular aperture form the "axial beam". Those entering the rectangular aperture that scans the azimuth once per spin form the "radial beam". Each "beam" enters an ESA with a pair of mirrored curved plates separating electrons and protons. The two "beams" time-share a single EM detector per species. The EM for electrons is not shown. The "trap" captures undeflected particles [3]. (© AIP. Reproduced with permission).

aperture and ionization scheme, designed for a high-density gas, is not appropriate for the low-flux ANAs or ENAs, which is the focus of this book.

- *Soft Particle Spectrometer (SPS) on the International Satellite for Ionospheric Studies (ISIS-1)*: Launched on Jan 30, 1969, into a polar orbit of 88.5° inclination, 3500-km altitude apogee, and 570-km altitude perigee, SPS (Fig. 5.3) was designed for 10 eV–10 keV electrons and protons [3]. Not only did SPS lead to the discovery that the equatorial region of Earth's outer radiation belt is a rich H-ENA source (Sec. 2.3), its effective use of sensor resources and observation time is worth describing.

  As shown in Fig. 5.3, each ESA has two curved outer electrodes (+ for electrons and – for ions) that face curved walls of a trap for undeflected particles at ground potential. Biased at 20 logarithmically spaced steps between ±2.3 and ±2,300 V, each ESA guides electrons (dotted line) or protons (dashed line) for 20 $|E/q|$ intervals covering $10\,\mathrm{eV/e}$–$10\,\mathrm{keV/e}$ into one of two EM detectors. The time-sharing between the simultaneous collection of two proton and electron spectra by the radial and the axial ESAs provides four spectra every $20\,\mathrm{s}$ with only two EMs! In front of each EM, a negatively biased repeller rejects low-E electrons. Incident photons, high-E charged particles, and ENAs end up in the traps.

Would adding an ENA detector inside the trap make SPS an ENA sensor? Regrettably, high photon and ion fluxes would drown out the ENAs. Detecting ENAs in space needs more provisions.

5.1.2. *First ENA instrument in space*

Section 2.2 showed the first direct ENA detection in space (Fig. 2.4) on Apr 25, 1968, by an ENA spectrometer on a Nike-Tomahawk sounding rocket to study particle precipitation in aurora [4]. Figure 5.4 shows a schematic view.

Figure 5.4 shows a meridional-cut of the *hemispheric electrostatic analyzer* (HESA) and cylindrical entrance system. A 4-kV/cm $\vec{E}$-field in the 1-cm gap between a pair of concentric plates effectively prevents protons of $E < 50\,\mathrm{keV}$ from entering the HESA, and a rubber magnet with $|\vec{B}| = 10^{-2}\,\mathrm{T}$ achieves the same for all electrons, including locally-produced SEs. Approximately 10% of the incident H^0 emerges as H^+ from the 2-$\mu\mathrm{g/cm}^2$ C-foil at the entrance and enter the HESA. The HESA is biased at the inner shell in five logarithmically spaced steps between -0.1 and $-4.0\,\mathrm{kV}$, thus selecting ENA-generated H^+ in five E/q intervals over 0.6–22 keV. The spiraltron counts them at a 10^{-2}-s cadence. The geometric factor is $G = 4.3 \times 10^{-2}\mathrm{cm}^2\mathrm{sr}$. Including ENA to ion conversion, scattering by the foil, and support grid transmission, the overall detection efficiency η increases with E from 3×10^{-4} to 4×10^{-2} over the sensor range [4].

This first ENA instrument used the best technology at the time to address all ENA-specific challenges (Sec. 4.3). It took full advantage of a hemispherical analyzer: 1) With its analyzer constant $r/2\Delta r \approx 19$, it covers a wide E/q range (0.6–22 keV/e) by stepping through

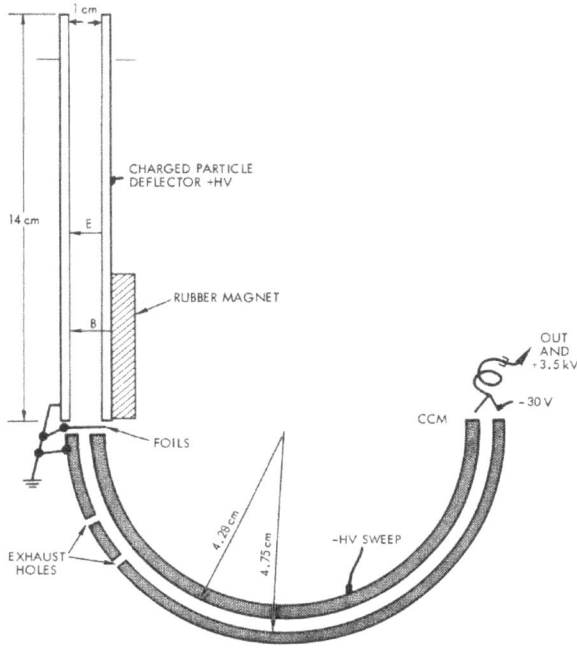

Fig. 5.4.   Schematic meridional cut of the first ENA spectrometer. The two concentric semi-circular arcs mark the inner and outer shells of the HESA with a gap $\Delta r = 4.7$ mm and a mean curvature radius for the ion path $r = 4.52$ cm [4]. (© AGU. Reproduced with permission).

moderate bias voltages $\Delta\Phi$ (0.1–4.0 kV). 2) It focuses the incoming particles over the out-of-plane angle in Fig. 5.4 at the HESA exit and 3) blocks all stray UV. The idea of a HESA is intriguing. Rotating its cross-section about a vertical axis to the right of the exit forms a toroidal ESA increasing G substantially, as used in the two *IBEX* ENA imagers three decades later (Secs. 5.3.2 and 5.3.3).

5.2. Non-TOF ENA Sensors

We now turn to ENA sensors on satellites and space-probes that did not use the TOF technique. We start from the low-E end with the first ANA analyzer that measured the ISN He flux *in situ* in the inner heliosphere and close with SSDs for ENAs at higher E.

5.2.1. *The first ANA instrument: GAS on Ulysses*

The first and successful ANA sensor flown is the ISN gas instrument, GAS on *Ulysses* [5], 13 years before its more versatile successor *IBEX*-Lo (Sec. 5.3.3) [6]. The sole objective of GAS was to obtain the velocity distribution of ISN He in the heliosphere and thus to gain insight into the kinetic properties of the ISN gas. Based on the previously derived ISN bulk flow speed (Sec. 7.2) [7, 8], the He velocity distribution should stand out in the sky maps taken with an instrument sensitive to the impact of ISN He in the *Ulysses* S/C frame. Section 7.2 presents the related analysis and challenges.

To map the entire sky, a detector suitable for ISN He with a collimated FOV mounted on a scan platform needed a ride on a spinning S/C with the appropriate IP trajectory. The *Ulysses* mission [9], with its journey over the poles of the Sun, presented the opportunity. Notably, the GAS experiment almost did not make it because it had been selected for the US probe of the original two-S/C Out-of-Ecliptic mission, which was later canceled due to a significant cut of the *NASA* budget in 1981 [10]. Only *ESA*'s offer to include unique European instruments on the *ESA* probe after the US cancelation and the willingness of the cosmic-ray team on the *ESA* probe to take GAS under their wings brought this instrument to the launch pad.

In principle, both SE and SI emission from a surface upon the impact of neutral atoms with sufficient E would work as suitable detection mechanisms for GAS. However, Lyα photoemission of electrons presents a severe background to the atom detection *via* SEs, and the SI sputter yield is typically rather low. For He, LiF did not show these disadvantages [11]. The sputtered ions of LiF are primarily positive and close in m to that of the impactor. Thus, He sputters Li$^+$ favorably. Also, LiF is mostly transparent to UV light and therefore has a low photoemission yield. Thus, a LiF surface became the key detection element of GAS.

Figure 5.5 shows two orthogonal cuts of GAS. Particles and light enter the sensor from the left through two circular apertures at the end of a light baffle and the entrance to the detector section, forming

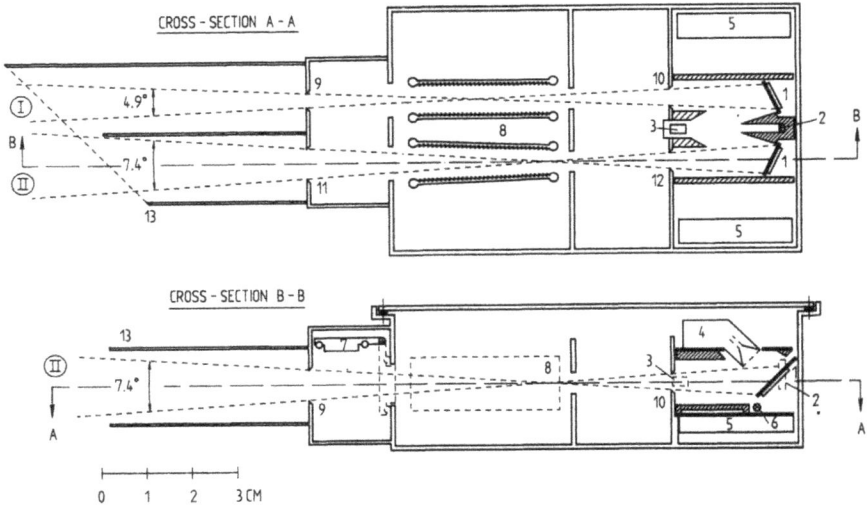

Fig. 5.5. Cross-sectional views of the GAS sensor (top) in the plane that contains the spin axis (pointing up, roughly toward the Sun) and (bottom) in the plane normal to the spin axis. 1: conversion plates with heaters, 2: quartz crystal to monitor LiF deposition, 3: furnace with LiF for refreshing the conversion plates at intervals, 4: CEMs for SE or SI detection, 5: CEM amplifier and electronics, 6: W-filaments for in-flight testing of the CEMs, 7: vacuum-tight cover, opened in flight, 8: electrostatic deflection plates to sweep out charged particles, 9 and 10, 11 and 12: circular apertures that define the 4.9° and 7.4° FOV, 13: Sun-light baffle [5]. (© A&A. Reproduced with permission).

two channels (4.9° FOV in I and 7.4° in II). Charged particles of $E <$ 80 keV/e in channel I and $<$50 keV/e in channel II are diverted out of the FOV by the $\vec{E}$-field between the deflection plates. The remaining neutral atoms and photons in each channel hit a black conductive Pb-glass conversion plate covered by a thin (150 nm) LiF layer. A furnace with a small supply of LiF that faces both conversion plates can refresh the layer repeatedly in flight. Before each refreshment, Ohmic heating to 200°C cleans the Pb-glass substrate. Heated to 800°C, the furnace deposits 8 nm of fresh LiF. A quartz crystal gauges the deposition by a change in its resonance frequency.

The SIs and SEs ejected from each conversion plate are detected by a channel electron multiplier (CEM) collecting SIs (when biased −) or SEs (when biased +). Both GAS channels operate simultaneously in either mode. During the SI mode, the CEMs are immune to SEs, thus minimizing UV-generated background.

The SE and SI efficiencies of LiF as functions of the He energy were measured in a test setup [12]. A prototype of the GAS detector section, similar to the one flown, was used for their calibration [13]. Figure 5.6 shows the results for different times after freshly applied LiF layers, indicating systematic uncertainties of $\approx 40\%$. Even after extended exposure to air, the efficiency decreases at most by 50%. The SI and SE efficiencies are about equal at a value of $\approx 10^{-2}$ for 80 eV particles and drop sharply toward lower E. Therefore, the GAS sensor detects He only above 30 eV in the S/C frame and is completely insensitive to ISN H, distinguishing He clearly from H. In turn, ISN He measurements with *Ulysses* were restricted to the

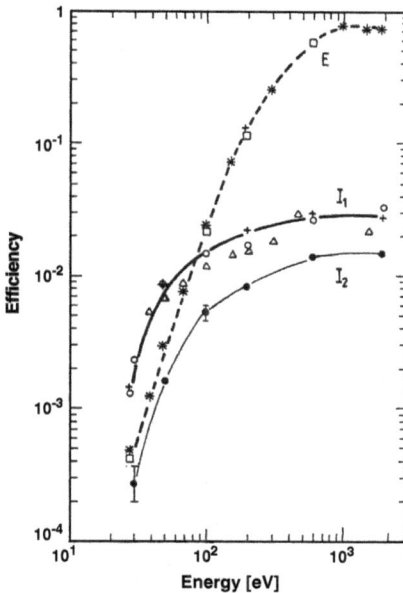

Fig. 5.6. Efficiencies for the production of SEs (E) and SIs (I_1 and I_2) on a LiF surface as functions of the He energy. The values on curve I_1 stem from one day (cross) and six days (triangle) after deposition of 150 nm LiF and two hours (circle) after adding 50 nm of LiF. Curve I_2 was measured after exposing the surface to air for one month and curve E one day (square) and six days (star) after deposition of the 150 nm layer [5]. (© A&A. Reproduced with permission).

first six months after launch and about four months during each fast latitude scan, when the S/C velocity boosted the He E enough for detection [5].

Because of their higher masses, ISN species, such as O and Ne, are also detectable by GAS. However, their relative scarcity renders them at most a minimal effect on the overall ISN He signal [14]. *Ulysses* GAS identified the interstellar He flow with a high degree of certainty. GAS suppressed incoming charged particles and SEs from UV photons. It is insensitive to ISN H, and ISN O and Ne arrive with very low fluxes. GAS accumulated the ISN He maps over 32 bins in azimuth (spin angle) at selectable increments (0.7°, 1.4°, 2.8° and 11.25°) and at $n \cdot \Delta\varepsilon$, where $n = 1, 3, 5, \ldots 31$, and $\Delta\varepsilon$, the solution, can be 1°, 2°, 4°, or 8° in elevation. The higher resolution maps centered on the ISN flow direction [5].

In the SE mode, GAS obtained maps of the backscattered Lyα and thus the ISN H distribution, when the ISN He energy was below the detection threshold. Switching between SI- and SE-modes conveniently adds a much-appreciated heliospheric data set [5].

5.2.2. *ENA sensors with MCP single-pulse detection*

For source regions with high ENA-fluxes, *e.g.*, planetary magnetospheres or ionospheres, and if local charged particle access is effectively blocked, sensors relying solely on single-pulse detection may be adequate for imaging [15]. The first MCP-only ENA sensor was part of the PIPPI experiment on the Swedish microsatellite *Astrid* (Sec. 2.4) for auroral studies [16]. The MCP section, designed for the ENA flux at $E = 0.1$–70 keV, evolved into the Neutral Particle Imager (NPI) of the ASPERA-3 and -4 suite to explore Martian and Venusian ENAs, respectively (Sec. 6.3.2). The other part of PIPPI was an SSD section (Sec. 5.2.3), replaced within ASPERA by the TOF camera NPD (Sec. 5.3.3). Figure 5.7 is a perspective view of the NPI sensor with a sample ENA trajectory [17].

Two parallel flat serrated annular plates, separated by 3 mm, define the aperture and ion-deflector. Setting the upper plate at $+5$ kV by command blocks charged particles of $E/q < 60$ keV/e. Posts divide the 3-mm $\times 2\pi$ aperture into 32 sectors, each facing a section

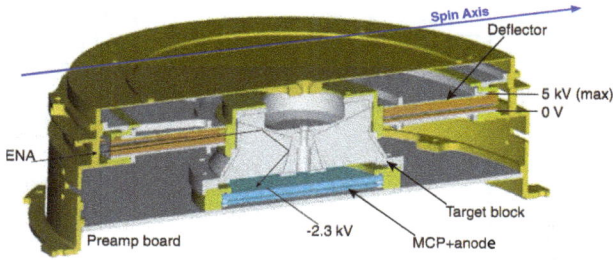

Fig. 5.7. 3D cut-away view of the ASPERA NPI sensor, along with a sample ENA trajectory through the collimator-deflector entrance *via* the reflector plate to the annular MCP and anode at the bottom (in blue) [17]. (© Springer. Reproduced with permission, adapted.)

of a centrally placed cone with two angled reflecting surfaces. An ENA entering from the left in Fig. 5.7 bounces off the two reflecting surfaces at shallow angles (20°) and triggers the MCP whose anode has 32 sectors. Each sector has a 9° × 18° FOV with 4.6° × 11.25° FWHM resolution. The anode sector signal and the time of impact relative to the spin phase mark the ENA arrival direction. The sector facing the Sun is blocked, and its anode monitors the detector noise. With the S/C spin axis in the aperture plane (Fig. 5.7), NPI completes an all-sky map every 1/2 S/C spin.

The reflected particles can have $Q = 0, +1$, or -1. The surfaces can also emit SEs and SIs. NPI selects particles of $Q = +1$ or -1 for detection with the bias polarity at the MCP front surface. In Fig. 5.7, it is set at -2.3 kV to attract $Q = +1$ ions. A layer of resin-based graphite dispersion (DAG 213) over the surfaces yields a 3×10^{-8} and 6×10^{-8} UV suppression in the $Q = +1$ and -1 modes, respectively, including the MCP photon efficiency.

Figure 5.8 shows η of NPI as a function of E for H_2O^+ at different MCP biases. In *Astrid* PIPPI, the Neutral Particle Detector (NPD) complements NPI toward higher energies with SSDs (Sec. 5.2.3), while the NPD in ASPERA uses MCPs (Sec. 5.3.3).

5.2.3. *ENA sensors with SSD single-pulse detection*

For ENAs of $E > 10$ keV, SSDs supercede MCPs because at least H atoms exceed the SSD thresholds. SSDs provide a direct

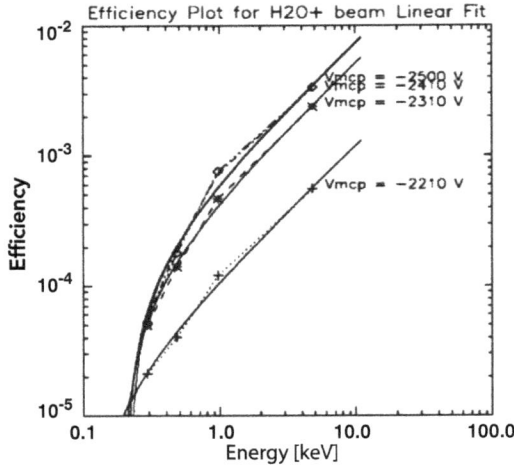

Fig. 5.8.   Detection efficiency η as a function of particle energy for H_2O^+ ions into the ASPERA NPI sensor for different MCP bias voltages [17]. (© Springer. Reproduced with permission).

E-measurement as an additional advantage. In the following paragraphs, we discuss a few examples of ENA sensors that solely rely on single-pulse detection with SSDs.

- *ENA sensor with SSDs on Astrid*: Stacked on top of the MCP-only NPI (Fig. 5.7) was an SSD section covering ENA energies in the range of 13–140 keV (not shown). Its entrance system design is similar to that of NPI but features two sets of serrated collimating plates and 15 SSDs. The 2π-FOV is divided into 15 sectors by plastic spokes so that each SSD has a $2.5° \times 25°$ FOV and $G = 2.5 \times 10^{-3}\,cm^2\,sr$. The 15th SSD has the Sun in its FOV. It is blocked and just monitors background and noise. The oppositely biased collimating plates divert charged particles of $E/q < 140\,keV/e$ from the SSDs. The ENA signals are sorted into 8 E-channels with a resolution $\Delta E/E = 0.3$, with an imaging scheme similar to NPI [16]. The ASPERA suite does not include the PIPPI-SSD section.
- *Advanced ENA sensor for the Double Star*: A version of the PIPPI-SSD section with increased geometric factor is the NeUtral

Atom Detector Unit (NUADU) for the *Chinese-European Double-Star* mission [18]. NUADU features a cylindrical sensor head with apertures staggered in polar angles that provide a $5° \times 15°$ FOV $(G = 1.4 \times 10^{-2}\,\text{cm}^2\,\text{sr})$ per pixel. The S/C spin is divided into 128 sectors providing full 4π coverage every $1/2$ spin. Like in NPI, the FOV-defining parallel collimator plates also serve as ion deflectors, rejecting ions of $E/q < 300\,\text{keV/e}$ when biased at $\pm 5\,\text{kV}$. When turned off, NUADU can evaluate the ion flux. Each pixel contains an ion-implanted Si SSD with a 200-nm Al window, blocking UV but setting a 45-keV threshold, thus providing an ENA E-range of 45–300 keV.

- *IMS HI on CRRES, the first ENA detector in orbit*: The ion mass-spectrometer, IMS-HI, aboard *NASA/USAF CRRES*, launched on Jul 25, 1990, included the first ENA detector in orbit, shown in Fig. 5.9 [19]. A broom magnet at the entrance sweeps away the electrons. A semi-circular magnet with $|\vec{B}| = 7\,\text{kG}$ shapes the incident ion trajectories according to their gyro-radii and guides them to six designated E-resolving SSDs. Because the ions of interest are singly charged, combining the measured E and SSD location yields the ion mass. The incident ENAs and photons are unaffected by $\vec{B}$ and reach SSD-7, a p-type Si diode with

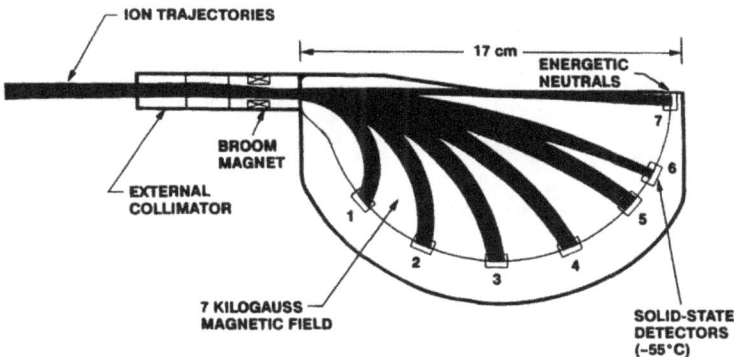

Fig. 5.9. Schematics of *CRRES* IMS-HI. A collimator sets the FOV, and a broom magnet sweeps away electrons of $E < 1\,\text{MeV}$. With seven SSDs strategically located around the edge of a magnetic spectrograph, SSD-7 measures ENAs and photons, unaffected by $\vec{B}$ [19]. (© AIAA. Reproduced with permission).

a 20-μg/cm^2 Al surface barrier to reduce its EUV sensitivity. However, the barrier raises the E-threshold to 40 keV. Passive radiative cooling of the SSDs, pre-amps, and electronic box to $-50°$C, $-12°$C, and $0°$C, respectively, affords a 2-keV (FWHM) resolution of the SSDs [20].

IMS-HI failed to detect ENAs. 1) $G = 10^{-3}$ cm^2 sr, set by the narrow magnet pole gap and the aperture-SSD distance (Fig. 5.9), was too small to detect ENA fluxes, even in the absence of ion foreground at altitudes <900 km. 2) The lack of a coincidence measurement and anti-coincidence protection rendered the rare ENA events invisible in the presence of an overwhelming background from omni-directional energetic ion fluxes at altitudes >10^3 km. However, IMS-HI taught us three lessons: 1) focus on ENA detection, 2) avoid magnetic analyzers because of their constraints on G and large mass, and 3) include a coincidence measurement because ENA fluxes are extremely low.

5.3. ENA Spectrographs with TOF Measurement

Section 4.2.3 discussed in detail the principles of a TOF measurement, its use to identify particles unambiguously in conjunction with other techniques, and its inherent ability to suppress random background. In Sec. 5.3, we describe the implementation of the TOF technique in sample ENA instruments grouped into three E-intervals:10–200 keV, 0.3–20 keV, and 5–2000 eV, in Secs. 5.3.1, 5.3.2, and 5.3.3, respectively.

5.3.1. *ENA cameras for the 10–200 keV range*

Because the SE yield from the TOF entrance foil is comparable for ions and ENAs, TOF spectrographs for ions [21] can work for ENAs (Sec.4.2.3). However, suppression of the ion influx (Sec. 4.3.1) is necessary for unambiguous identification of the ENAs.

- *CELIAS HSTOF on SOHO*: The first sensor of this type is the HSTOF section of the CELIAS STOF sensor on *SOHO* [22]. This

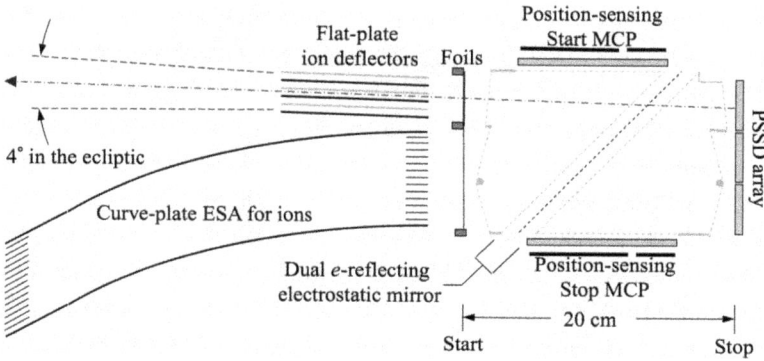

Fig. 5.10. Schematics of the *SOHO* STOF and HSTOF sensor [24]. HSTOF deflects incoming ions in the *E*-range of the ENA observation with its flat collimator plates at alternating polarities and thus is an ENA mass-spectrograph. The grids (grey dots) on both sides of the mirror are shaped and biased to guide the start SEs from the upper foil and the stop SEs from the upper third of the pixelated SSD array to their respective detectors (shorter anodes behind the start and stop MCP) (*cf.* Fig. 4.8, right). (© CJSS. Reproduced with permission).

modification emerged as a target of opportunity after the instrument selection for *SOHO* [23]. Removing a previously selected ion sensor for the 0.5–5 MeV/nuc from the payload due to lack of funds required an "economical" fix to maintain the contiguous energy coverage. Replacing 1/3 of STOF's curved-channel ESA with a set of straight channels reduced the STOF geometric factor by 33% but bridged the gap in energy coverage with only minor changes (Fig. 5.10). The curved deflection plates guide ions of 20–600 keV/e into STOF as proposed. The flat-plate collimator deflects ions of $E/q < 150$ keV/e (Fig. 4.10), making HSTOF the ion sensor for $E/q > 150$ keV/e, thus restoring the 0.5–5 MeV/nuc range left vacant by the deselected sensor. Simultaneously, HSTOF serves as an ENA sensor for $E_{ENA} < 100$ keV/nuc. The long-awaited chance to probe the heliosphere *via* ENAs had arrived!

HSTOF, whose heliospheric viewing is shown in Fig. 4.1, maintains a substantial *E*-overlap with STOF to allow additional validation and identification of ion fluxes that might interfere with the ENA spectrum. A Si/Lexan/C UV-reducing foil (thicknesses 28,

31, and 5 nm, respectively) forms the entrance window. Each ENA event generates two TOF signals yielding v and one SSD pulse-height resulting in E, thus identifying the particle uniquely while suppressing the background with a triple-coincidence. This ingenious last-minute "fix" enabled HSTOF to obtain the first heliospheric energetic H^0 and He^0 spectra (Sec. 7.5.2) [25].

SOHO HSTOF was an add-on with limited ENA-capability, launched on Dec 2, 1995. The first dedicated ENA imager with simultaneous ion capability was the Ion Neutral CAmera (INCA) [26, 27], part of the Magnetosphere Imaging Instrument (MIMI) package [28] on *Cassini* to Saturn and its moons, launched on Oct 15, 1997. On its heel, *NASA*'s *IMAGE* mission, the first to exploit ENA imaging as one of its primary goals, launched on Mar 25, 2000 (Sec. 4.1.2) [29]. We will feature the three *IMAGE* ENA imagers, along with their progenies, starting with HENA.

- *IMAGE HENA and Cassini INCA*: The HENA sensor [30] is a replica of MIMI INCA on *Cassini*, chosen to replace the original HENA after mission selection (Fig. 4.11). INCA was modified with a pixelated SSD, placed adjacent to the reduced position-sensitive stop MCP, to provide E-resolution in part of the sensor for a complete particle identification like in HSTOF.

INCA and HENA operate as ion- and ENA-sensors, subject to the bias on the collimator plates. As noted in Sec. 4.3.1, their collimators broaden the FOV to $120° \times 90°$ (Fig. 4.11). The S/C spin extends HENA's coverage to $120° \times 360°$. In ENA mode, $\vec{E}$ between the plates (biased alternately at $\pm 6\,\text{kV}$) deflects ions of $E \leq 500\,\text{keV/e}$, passing ENAs for TOF analysis.

Like in HSTOF, the composite foil ($6.5\,\mu\text{g/cm}^2\,\text{Si}, 7\,\mu\text{g/cm}^2$ polyimide, and $1\,\mu\text{g/cm}^2\,\text{C}$) at the entrance of HENA and INCA reduces UV intensity and provides the start SEs for the TOF analyzer. However, the INCA and HENA TOF sections have a more intricate design for ENA detection due to their $90° \times 120°$ FOV for imaging (Fig. 4.11). The start detector is a 1D position-sensing MCP along the entrance slit adjacent to the entrance foil (close to the *IMAGE* spin axis). The stop detector is a 2D position-sensing MCP, side-by-side with SSDs for HENA, facing the entrance slit. A second foil immediately above the stop MCP emits SEs toward the coincidence and SSD-stop MCP (Fig. 4.11, right) to provide the TOF stop timing. This MCP is located opposite the start MCP across the entrance slit. For the HENA SSD section, SEs emitted from the SSD front serve that purpose. In HENA, the second C-polymer-C foil ($7\,\mu g/cm^2$ polyimide and $5\,\mu g/cm^2$ C) further protects the stop MCP from the UV flux (higher at Earth than at Saturn). With the position-sensing MCP or the SSD, the start and coincidence MCP signals provide a triple-coincidence measurement. HENA's pixelated SSDs provide E of the incoming particles and thus complete the particle identification. INCA takes advantage of a variation with m and E of the incident ENAs in the number of SEs emitted from both sides of the stop foil. It thus resolves H and O (the dominant species around Saturn) based on the combined pulse height of the coincidence and stop MCP signals [31].

Strategically placed wires and shaped electrodes (dots and line segments in Fig. 4.11) biased at fixed potentials accelerate and guide the SEs from the start and stop foil or from the SSD surface well-separated to the start and stop MCP, respectively. The differences in the flight times of the ENAs and SEs arriving at the coincidence and SSD-stop MCP for varying ENA arrival angles are due to different impact locations on the stop foil (also SSD for HENA). The 2D position information from the stop MCP and the pixelated SSDs allows their correction.

The ion-deflecting collimator with a $90° \times 120°$ FOV, the distributed wire and shaped electrodes in the TOF chamber, and the composite foil in front of the stop MCP made INCA and HENA unique. INCA and HENA achieved angular resolutions of $5°$ for H at $100\,keV$ to $20°$ at $20\,keV$ for the $120°$ FOV ($3°$ to $15°$ for the $90°$ FOV). Scattering in the entrance foil causes larger angular spreads for low-E H [30]. Among others, Sec. 5.3.3 addresses this challenge to ENA imaging at lower E. The malfunction of the HENA SSD segment was a heavy loss for *IMAGE*, only partially compensated by the

MCP pulse-height analysis. Chapters 6 and 7 cover the scientific accomplishments of HENA and INCA, respectively.

5.3.2. *ENA cameras for the 0.3–20 keV range*

The foil at the entrance, intended to reduce the UV-related background, degrades the angular resolution of low-E ENAs due to enhanced scattering. Alternate UV suppression schemes allow thinner foils that provide the start SEs with reduced ENA scattering. This section will feature four designs, all using ultrathin foils for the TOF start signals. Each covers a portion of the 0.3–20 keV range, using different means to improve the angular resolution. We start with the sensors on *IMAGE* [32] and *TWINS* [33] (Sec. 4.1.2), which use *free-standing transmission gratings* (Sec. 4.3) as a UV filter.

- *IMAGE MENA and TWINS*: Since the *TWINS* sensors go back to MENA, we combine their discussion. Figure 5.11 is a schematic view of the center unit of the three identical MENA units. The collimator contains fan-shaped and narrowly spaced parallel plates, with a 4° FOV across the plates and 110° in polar angle. A 20° offset in polar angle between the units gives MENA an instantaneous 4° × 150° FOV with considerable overlap. Each *TWINS* S/C has two units, with a 30° offset in polar angle, resulting in a 4° × 140° instantaneous FOV (Fig. 4.4). The S/C spin enables a 2π scan across the 4° FOV in each case.

The 4° FOV defined by the narrow spacing between the collimator plates is commensurate with the free-standing transmission grating that attenuates UV (Sec. 4.3.2 and Fig. 4.15). The grating is an array of 460-nm tall and 140-nm wide parallel gold strips, separated by 60-nm gaps passing particles while effectively attenuating UV by $< 2 \times 10^{-5}$. Thus, an ultrathin ($0.5 \, \mu g/cm^2$) C-foil serves solely as the start-SE generator with much-reduced angular scattering of the ENAs. The fully biased collimator suppresses ions of $E < 20 \, keV/e$ to $< 10^{-4}$ of their original flux [32].

Figure 5.11 shows an ENA entering the sensor through a 60-nm slit of the grating. It passes the ultrathin C-foil, generating start

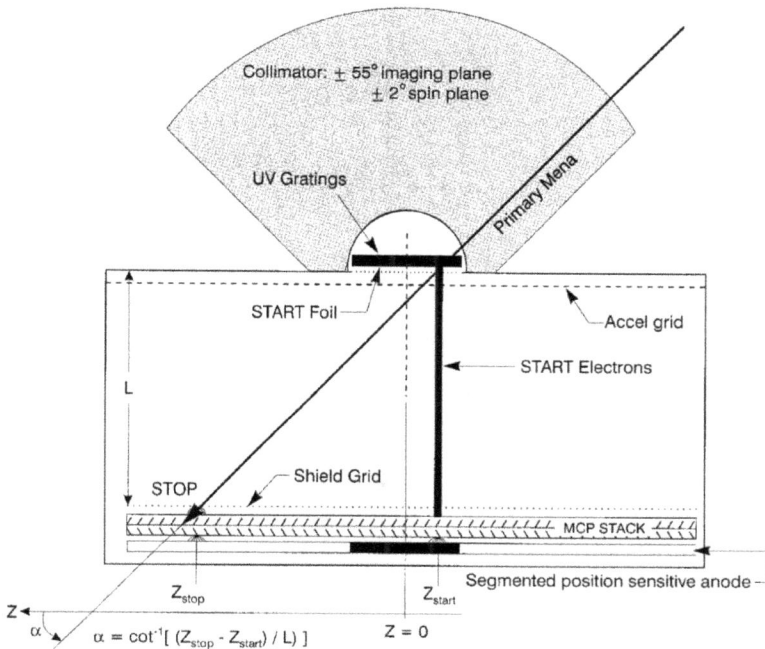

Fig. 5.11. Schematic cut through the center unit of the MENA sensor head. The S/C spins about the z-axis [32]. In the plane of this figure, the adjacent units have a 20° offset to the left and right of the vertical axis normal to the z-axis. (© Springer. Reproduced with permission).

SEs. A uniform $\vec{E}$-field between the foil and the accelerating grid sends the start SEs toward the central portion of the MCP stack, marking the ENA start position. Then, the slower ENA strikes the MCP, completing the TOF analysis and marking the stop location. A segmented anode records the start and stop positions on the MCP. Each position pair yields the arrival polar-angle and the total flight path of the ENA. The measured TOF τ and the event timing in spin phase jointly determine the ENA velocity vector $\vec{v}$ (v and $\hat{l}$).

The angular resolution of MENA in polar angle is $\approx 4°$ for 10 keV H ENAs. The resolution degrades with decreasing energy, and for heavier ENAs, such as O. In the absence of an SSD, heavier ENAs are identified only statistically by their increased pulse heights in the stop-MCP signal, as discussed for HENA and INCA.

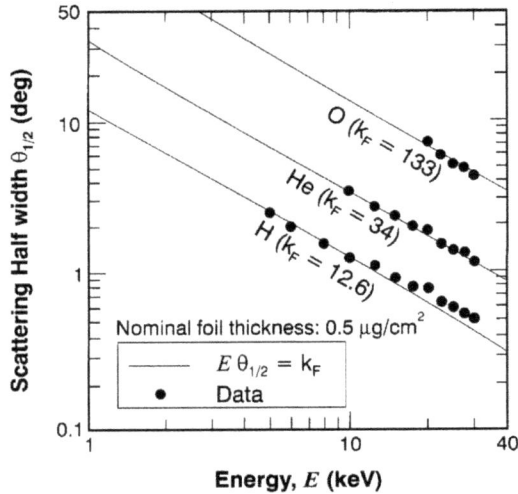

Fig. 5.12. Angular spread of ENAs after emerging from a $0.5\,\mu\mathrm{g\,cm^{-2}}$ C-foil [32]. (© Springer. Reproduced with permission).

Figure 5.12 shows the variation of the angular scattering width with ENA energy and species.

- *JENI for JUNO*: When the ambient UV flux is sufficiently lower than the charged-particle flux, an ultrathin entrance foil can be used for the start SEs [32, 34, 35] to maintain adequate angular resolution for H atoms down to 1 keV. This situation applies to the Jovian Energetic Neutral and Ion (JENI) sensor on the Jupiter orbiter *JUNO*. The UV flux is far below the ion and electron flux based on the *Voyager* and *Galileo* findings [36].

Figure 5.13 shows a cut through JENI in the x-z plane (top) and its start- and coincidence-anode in the x-y plane (bottom). The trajectories of an ENA and the SEs generated from the foils indicate the triple coincidence for each certified ENA event. All MCP anodes are 1D position-sensitive using delay lines [36]. The unique features of INCA and HENA (Sec. 5.3.1) are evident in JENI's design with two symmetric collimator and entrance slit assemblies. The collimator has alternatingly biased fanned-out serrated plates, followed by a narrow slit. An entrance foil and a rear coincidence foil allow position-sensing

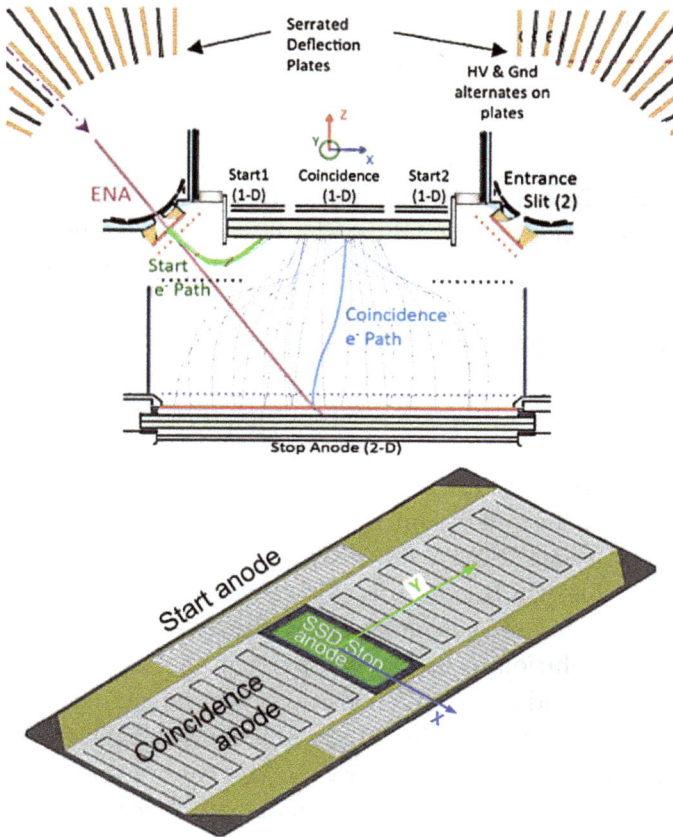

Fig. 5.13. Top: Cut through the JENI sensor in the x-z plane, showing a symmetric collimator/entrance pair sharing one stop plane. Bottom: The anode board in the x-y plane accepts SEs from the two entrance foils, the rear foil, and the SSD, separately [36]. (© AGU. Reproduced with permission, adapted.)

of the SEs with grids and plates at fixed potentials guiding the start and coincidence SEs to their respective MCPs. JENI differs from MENA and TWINS by the absence of the transmission grating.

JENI's dual entrance symmetry simplifies its TOF section over HENA and INCA to a single plane for the start MCP and the coincidence and SSD-stop MCP facing the stop plane (Fig. 5.13, top). Here, a narrow SSD stretches between two broader stop MCP segments of equal length along x (Fig. 5.13, bottom). Like the adjoining MCPs, the pixelated SSD records the ENA stop position. It also measures E to identify the ENA (Sec. 4.2.3.1) for its mass. JENI's partial use of a pixelated SSD is a HENA heritage.

The combined instantaneous FOV is $120° \times 140°$. The spin axis orientation along y provides a $120° \times 360°$ panorama FOV scanned by both entrance fans. Like HENA and INCA, JENI can select between ENA and ion modes by switching the bias on the collimator plates. The two collimator entrance assemblies can operate independently in either mode.

The rate of chance double-coincidences due to charged particles at the expected single rates of start SEs (green track in Fig. 5.13, top, $R_{BStart} = 2 \times 10^5/s$) and stop SEs (blue track, $R_{BCoinc} = 4 \times 10^5/s$) is $\approx 2 \times 10^4/s$ (Eq. (4.7)). Such a high background rate is unacceptable for ENA detection, although these rates are manageable by fast electronics. The solution is the use of triple coincidence for each ENA event (Sec. 4.2.3.3) as in HSTOF, INCA, and HENA. The third signal is that of the SSD or the stop MCP behind the rear foil. The latter restricts the coincidence time window to the TOF resolution for individual particle species, $\Delta\tau_{TOF} = 4$ ns. Also, the start and stop MCP positions are correlated, enabled by the exact electrostatic steering of the SEs and position sensing at the start and stop MCP. Thus, separating the counts into 20 position-designated rates reduces the triple coincidence rates in Eq. (4.7b) by another factor of 20, to $R_{BC} \approx 2.5/s$.

To cope with the dynamic range of the expected ENA and background rates in the Jovian environment, JENI has entrance slits of different widths arranged on a drum. An actuator can rotate the drum to adapt the slit widths, changing the sensor collecting power.

- *The LENI Concept*: The Low Energy Neutral Imager (LENI) concept offers another solution to improve the angular resolution and lower the E-threshold [37]. Figure 5.14 shows a radial cut through the electro-optics. ENAs enter between deflector plates from the left and hit a stop MCP on the right.

The sensor head forms a quadrant, with the entrance at the outer rim on the left. A stop MCP collects the ENAs near the

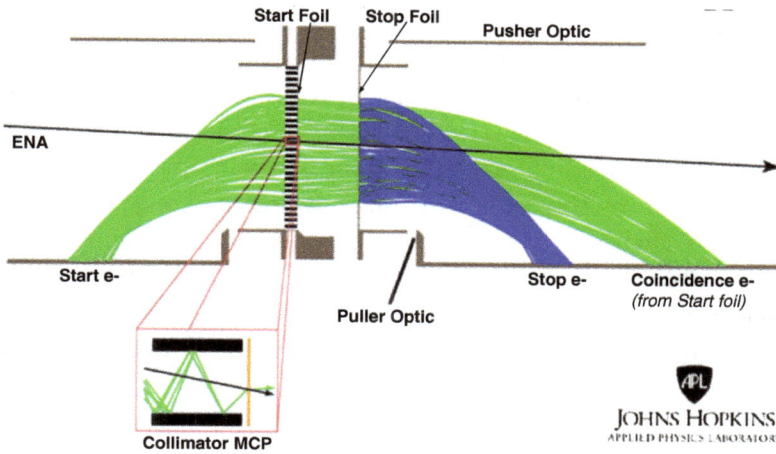

Fig. 5.14. The radial cut of the LENI electro-optics includes a sample ENA trajectory and SE trajectories. Not shown are the deflectors on the left and the stop MCP on the right [37]. (© AGU. Reproduced with permission, adapted).

vertex on the right (not shown). Alternatingly biased parallel plates at the entrance prevent incoming ions and electrons from entering the sensor. A curved MCP strip, with channels of 2° FOV each, serves as a collimator that gives LENI a 2° × 90° instantaneous FOV and reduces the UV flux into the sensor. This collimation scheme turns the ENA arrival direction into a position on the curved MCP with a 2° × 2° resolution. Upon penetrating the ultrathin start C-foil at the MCP exit, the ENA generates SEs (green) on both sides of the foil. The backward-emitted SEs enter the channel, just passed by the ENA. They emerge from the MCP multiplied and reach the imaging start MCP (Fig. 5.14, left). After acceleration, the forward-emitted SEs penetrate the stop foil and reach the far end of the coincidence MCP (Fig. 5.14, right). The ENA penetrates the stop foil and generates stop SEs (blue), which move on to the near end of the coincidence MCP and provide a TOF measurement between the two foils. Striking the stop MCP (not shown), the ENA affords a second TOF measurement. Using consistency between the TOF measurements and the additional coincidence SE signal further suppresses any background. The ENA identity may only be inferred statistically from the MCP pulse-heights as in INCA and the non-SSD sections of HENA and JENI for the lack of an SSD.

The single-channel collimation provides the ultimate angular resolution, regardless of the scattering in the foil that follows the collimating MCP. However, collimation with pixel-size angular resolution reduces the collection power, which can only be partially compensated with an increased sensor aperture, requiring valuable S/C resources, *i.e.*, volume and mass. The alternative is an increase in collection times. The need for scanning to generate the ENA image further reduces the time resolution, thus rendering this concept ineffective for planetary magnetospheres.

- *IBEX-Hi on the first ENA-only S/C*: If the time scales of the expected variations in the ENA images are of the order of months to years, the use of large-area single-pixel sensors may be the superior approach. This scenario arises when taking global ENA images of the heliospheric boundary and sampling the ISN gas flow through the heliosphere. Their expected variations take several solar rotations governed by solar activity. The result of dedicating an entire S/C to heliospheric imaging is the *IBEX* mission [38] with its large-area single-pixel ENA cameras, *IBEX*-Hi (0.5 to 6 keV) and *IBEX*-Lo (0.01 to 2 keV), accommodated on a small satellite. Section 4.1.1 describes the *IBEX* viewing geometry and operations strategy. *IBEX* provides full-sky ENA maps at a 6-month cadence with seasonal viewing of different portions of the heliosphere and Earth's magnetosphere [39]. Figure 4.2 (right) shows how the *IBEX* orbit drifts relative to Earth's magnetosphere over one year and how the ISN flow enters the *IBEX* FOV for a few months when the Earth rams into the oncoming flow. At that time, the apogee of the orbit points toward the Sun.

The left half of Fig. 4.9 shows a radial cut of both sensors with all functional principles. Both cameras are cylindrically symmetric and equipped with identical or similar subsystems as much as possible [6, 40, 41]. Unprecedented, the *IBEX* sensors repel $\leq$3.1-keV electrons and $\leq$10-keV/e ions *before* they reach the collimator with

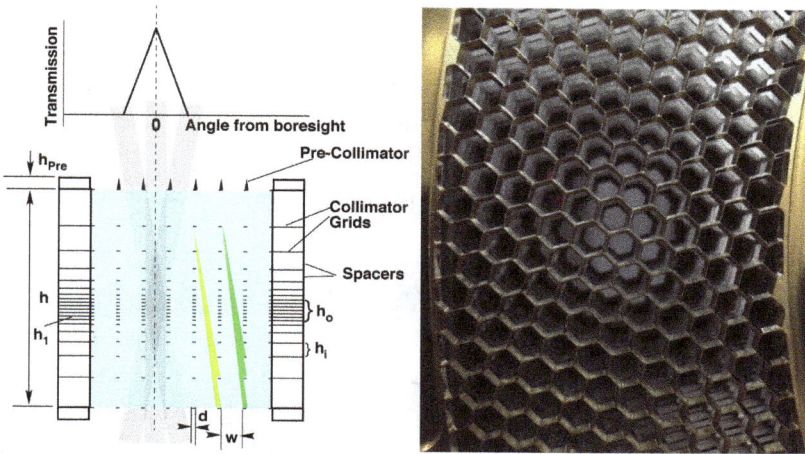

Fig. 5.15. *IBEX* collimator. Left: Angular response and schematic of an annular collimator cut. ENAs enter from above [40]. The pre-collimator reduces the acceptance angle range. The yellow and green rays show how the consecutive grids prevent ENAs from leaking through the adjacent channels. For the vertical grid spacing, see the text (© Springer. Reproduced with permission.). Right: Photograph of a collimator section showing how the exit area is accessible only to ENAs from a narrow angular cone, thus defining the angular transmission (Appx. A.2.1).

the configuration shown in Fig. 4.13. The sensors' $\approx 14° \times 14°$ FOV and $\approx 7° \times 7°$ angular resolution come from a novel collimator design (Fig. 5.15), which constrains the ENA arrival directions along with a precise pointing of the sensor boresight.

The *IBEX* collimators consist of 21 layers of precision photo-etched Al grids with identical hexagonal openings, precisely aligned in a stack (Fig. 5.15, left). The width w of the openings and the total height h between the entrance and exit determine the FOV (Appx. A). The sequence of the spacers between the grids is close to a geometric progression. It starts with the smallest gap h_1 at the center and increasing the gaps h_i toward both ends. To prevent cross-channel ENA leaks past the grid lines of width d into adjacent channels, the spacer heights h_i follow the sequence

$$h_1 < h_{Pre} \cdot d/w \quad h_0 > h_{Pre} \quad h_2 < h_0 \quad h_{i+1} < h_i + h_{i-1} \quad (5.1)$$

h_{Pre} is the height of the pre-collimator, which prevents particles at large angles from entering. The stack uses available Al material thicknesses and is robust against tolerances. Extremely sharp edges at the grid openings ($\approx 5\,\mu$m from the photo-etching process) minimize the scattering of ENAs toward the collimator exit. Keeping all structures external to the sensor outside the FOV suppresses particles that interact with the S/C surface [42].

After passing the collimator, the ENAs enter the conversion subsystem. In *IBEX*-Hi, an ultra-thin C-foil ($0.5\,\mu$g/cm^2) turns an E-dependent fraction of ENAs into positive ions (Fig. 4.16) [40]. *IBEX*-Lo, which covers $E = 0.01$ to $2\,$keV (Sec. 5.3.3), uses grazing incidence on a conversion surface to produce negative ions [6]. These ions are then pre-accelerated into the ESA subsystem of the sensor.

Both *IBEX* sensors use a toroidal ESA to analyze the ENA-derived ions for E/q. Concentric deflector plates, shaped like a Bundt cake pan, turn the ion trajectories through $180°$. Toroidal ESAs have a large entrance aperture at the outer perimeter, maximizing the collection area. The exit area inside concentrates the ion flux onto a small cylindrical coincidence detector section at the center (Fig. 5.16). To enlarge their collecting power, the ESAs accept ions over a wide E-range with $\Delta E/E \approx 0.7\,$FWHM. According

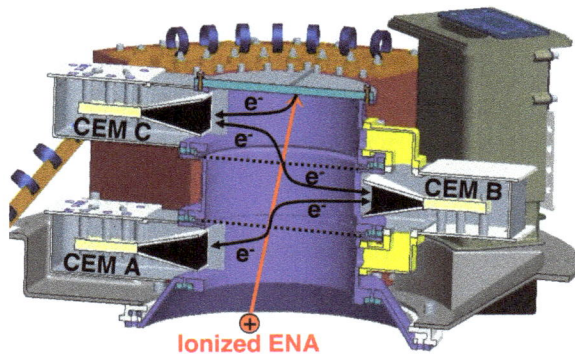

Fig. 5.16.   3D CAD model of the IBEX-Hi detector subsystem with a typical ion trajectory and the collection of secondary electrons, released from two consecutive thin foils and detector endplate, by three CEMs [40]. (© Springer. Reproduced with permission).

to Liouville's Theorem, compressing the ion flux onto a small exit area results in an angular spread increased by the entrance and exit area ratio. However, a post-acceleration voltage, $U_{PAC} = -6\,kV$, accelerates the ions from the ESA exit into the detector section, which reduces their angular spread and increases their detection efficiency.

In the detector section, the post-accelerated ions pass two consecutive thin foils and end on a metal plate. Each foil and the endplate surface generate SEs, collected, and registered by three Channel EMs, CEM A, B, and C, respectively, in the correct timing sequence. The coincidence windows are optimized during the sensor calibrations. Electronic counters accumulate rates of all double-coincidence (A-B, B-C, A-C) and triple-coincidence events, triple events being the cleanest.

IBEX-Hi uses CEMs instead of MCPs because they present a much smaller cross-sectional area to penetrating energetic radiation, minimizing the related background rates. Also, the electrostatic configuration of the detector section collects the SEs very efficiently into each CEM. While *IBEX*-Hi offers no direct species separation, further analysis can deduce the presence of heavy ENA species from different ratios of coincidences between CEM A, B, and C probabilistically. However, because of the more significant energy loss and scattering in the foils, the overwhelming majority of observed ENAs are H.

IBEX has taken ENA imaging to its current limits in obtaining images of the heliospheric boundary [38]. Some pixels of the ENA maps contain only a few counts per day. Thus, utmost care at all steps to suppress background and noise, including triple-coincidence, is essential.

5.3.3. *Neutral atom cameras for the 5–2000 eV range*

At ENA energies below 0.5 keV, their energy loss in even the thinnest feasible foils is so severe that only a tiny fraction of them passes the foil. In addition, the efficiency to produce positive ions drops substantially (Fig. 4.16). However, conversion of ENAs in their interaction with smooth surfaces, especially into negative ions, becomes more advantageous, as proposed for low-energy ENA instruments (Sec. 4.3.3.2) [43, 44].

ENA-surface interactions may lead to products with $Q = 0, +1$, or -1 at varying yields, depending on the ENA species and surface material. The choice of the surface and the means to identify the species of the converted ENA depends on the needed mass-resolution and energy range of the expected ENA composition and flux. Mostly H and O atoms are present in Earth's magnetosphere and planetary ionospheres [45, 46]. The ISN wind inside and outside the heliosphere contains H, N, He, O, and Ne atoms [47, 48]. The reflected and sputtered neutral atoms from airless bodies, such as the Moon and Mercury, contain even wider mass ranges [49–51].

We will feature seven ENA sensors to cover these topics: *IBEX*-Lo, *IMAGE* LENA, NPD of ASPERA on *ESA*'s *Mars* and *Venus Express* missions, CENA on the Indian lunar mission *Chandrayaan-1*, ASANA on the Chinese lunar rover *Yutu-2*, ELENA and STROFIO on *ESA*'s Mercury mission *Bepi-Colombo*. *IBEX*-Lo is first in line for its kinship with *IBEX*-Hi, even though it succeeded *LENA*.

- *IBEX-Lo on the first ENA-only S/C: IBEX*-Lo [6] overlaps with *IBEX*-Hi at 0.5–2 keV and extends the E-range down to 0.01 keV. Besides heliospheric ENAs, *IBEX*-Lo specifically targets the ISN flow through the heliosphere (Sec 7.2) and secondary neutrals from the OHS (Sec. 7.4), which requires high ENA collecting power, superior background suppression, and separation of different species. Its charged-particle rejection scheme, the large-area single-pixel collimator, and the toroidal ESA are identical to *IBEX*-Hi. However, *IBEX*-Lo uses surface conversion to turn ENAs into $Q = -1$ ions and an annular triple TOF section for mass analysis.

Figure 5.17 shows a schematic radial cut of *IBEX*-Lo above the axis of rotational symmetry. The conversion subsystem consists of trapezoidal CS facets, inclined at 15° relative to the sensor boresight and covering a conical ring behind the annular collimator exit. The CS is a vapor-deposited diamond-like carbon layer on a

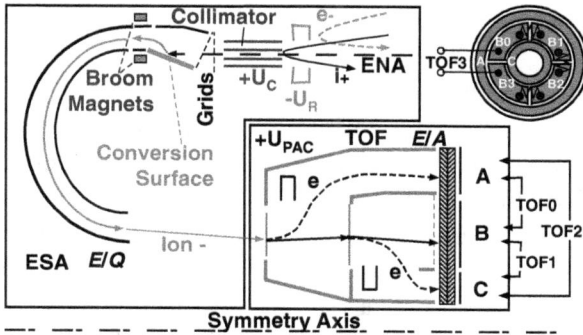

Fig. 5.17. A schematic radial cut of *IBEX*-Lo above the axis of cylindrical symmetry (not to scale) shows the rejection of electrons by $-U_R$ and ions by $+U_C$. An ENA passes the collimator to the CS for conversion to a negative ion. Analysis for E/q in the ESA follows, then post-acceleration, and TOF analysis in the triple-TOF subsystem. Upper right: Front view of the annular anodes A and C, and anode quadrants B0-B3. Including the delay TOF3 for B, *IBEX*-Lo takes four TOF measurements (TOF0, 1, 2, 3) [6]. (© Springer. Reproduced with permission).

highly polished Si wafer, demonstrated as long-lasting, with superior efficiency, and service-free [52].

Relieved of the need to map the positions from the CS into the detector section, the *IBEX*-Lo ion optics only must effectively collect the negative ions through the ESA and focus them into the TOF subsystem. Broom magnets at the ESA entrance sweep aside abundant SEs that emerge from the CS, generated by ENAs and UV, as already implemented in the first ENA instrument (Sec. 5.1.2). These SEs would cause an excessive background for a sensor that collects negative ions. The curved ESA also blocks background-causing particles and UV photons. Still, positive ions, generated from the residual gas in the gap between the positively biased collimator and the slightly negatively biased CS by photo and electron-impact ionization, can be accelerated onto the CS and generate negative ions, thus masquerading as genuine ENAs. A conical grid was placed between the ground grid facing the collimator and conversion surface (Fig. 5.17) to mitigate this "ion-gun" effect. The radial component of the resulting $\vec{E}$-field deflects the unwanted positive ions away from the CS. This measure may have saved *IBEX*-Lo from an excessive background when the positive collimator bias quit working after launch [6, 53].

The annular TOF subsystem (Fig. 5.17, lower right) is adapted from *Cluster* CODIF [54]. Nesting two similar TOF sections achieved a genuine triple coincidence, not merely with three detector signals, but three TOF measurements for superior background suppression and clean species identification! SEs generated by an incoming ion at the first C-foil reach the outer A ring of a single two-stage MCP, while SEs from the second foil strike the inner C ring. The ion ends on the center B ring, which consists of four quadrants (B0–B3). All sections on the signal anode are electrically separated (Fig. 5.17, upper right). TOF measurements are taken from each foil to the MCP (TOF0, TOF1) and between the two foils (TOF2). TOF3, derived from the delay line between B0 and B3, provides the quadrant position for further background characterization.

During the *IBEX*-Lo background characterization, the TOF0, 1, 2 coincidence rates were $<10^{-3}\,\text{s}^{-1}$, indicating a very clean and low-noise TOF section. According to Eq. (4.7c), triple coincidence reduces this background rate by another factor $R_{\text{Start}} \cdot \Delta\tau$, $\Delta\tau$ being the TOF range for H ENAs. Based on the interconnection of the three TOF measurements and the additional delay line measurement, the signals from an ENA-generated negative ion must fulfill the checksum condition: |TOF0 + TOF3 − TOF1 − TOF2| < 1ns. Further requiring consistency between all TOF measurements for a single species, H ENAs are selected within $\Delta\tau$ for TOF0, 1, and 2. Using these criteria in a background characterization run during the *IBEX*-Lo calibration, no triple H count occurred over 2.5 days, thus exceeding even the ambitious goal of <1 count per day [55].

Besides its excellent background suppression, the *IBEX*-Lo TOF section also provides a high detection efficiency that is homogenous over the entire unit. While the triple-TOF coincidence offers the best background suppression, any single TOF measurement (TOF0, 1, or 2) achieves the highest counting statistics. Even with the post-acceleration voltage reduced from 16 kV to 7 kV after a discharge during a long eclipse in 2012, *IBEX*-Lo still fulfills its science objectives, except for the O and Ne ISN flow [56]. The TOF efficiencies for converted and sputtered H are reduced by about a factor of two, and the background is slightly elevated, with no significant impact on the ENA observations. In other words, *IBEX-Lo* still takes very clean observations as of this writing.

- *IMAGE LENA*: Because analyzing the magnetospheric processes requires time resolutions of minutes to hours, LENA uses 1D imaging in polar angle through a 1-cm^2 pinhole with an instantaneous FOV of 8° in azimuth ϕ and 90° in polar angle θ. The S/C spin completes the 360° × 90° coverage.

Figure 5.18 shows a schematic cross-sectional view of LENA in the plane of the S/C spin. A set of collimating plates and the pinhole aperture S1 limit the incoming ENAs to an 8° FOV in azimuth. The plates are biased to electrostatically prevent incoming ions up to 100 keV/e from entering S1, thus suppressing the ion flux by at least two orders of magnitude to detect ENAs in the presence of much higher ion fluxes in the magnetospheric environment [57].

ENAs that arrive within the 90° FOV in θ pass S1 onto one of the four CS facets covering 90° in θ and biased at -20 kV.

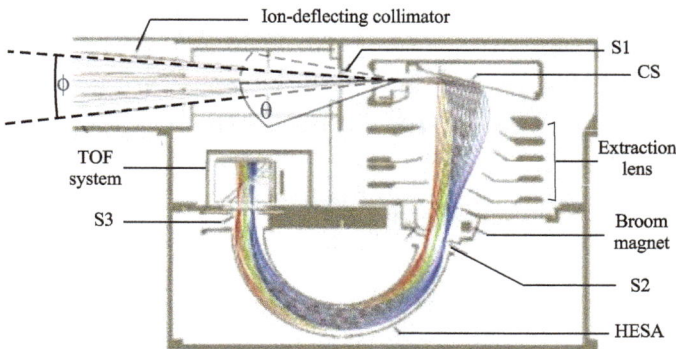

Fig. 5.18. The schematic cut of *IMAGE* LENA perpendicular to the spin axis shows the collimation of incoming ENAs in azimuth (or spin angle) ϕ. The out-of-plane imaging in polar angle θ is indicated by a tilted wedge. Ray-traced trajectories of three energy groups of converted ions from the CS are shown all the way into the imaging TOF section [57]. (© Springer. Reproduced with permission, adapted).

A multi-electrode lens extracts the $Q = -1$ ions and focuses them into the entrance S2 to a truncated hemispherical ESA (near ground potential). The ESA maps the image of the post-accelerated ions at S2 onto its exit plane S3. The focusing characteristics of the ion optics make the resolution of LENA in E and θ largely independent of the scattering at the CS. The radial position (foci of the colored rays) marks $E(\Delta E/E \approx 1)$, and the lateral position (normal to Fig. 5.18) marks $\theta(\Delta\theta \approx 8°)$. A $\approx 2\mu g/cm^2$ C-foil at S3 is the entrance to an electrostatic mirror TOF system (Fig. 4.8, right), which images the start positions on four MCP pairs with 2D position-sensitive wedge-and-strip anodes [58], sorting the ENAs according to E and θ. The ions move on to the stop MCP, completing the TOF measurement.

The ESA has two voltage settings to cover two E-ranges: 10–300 eV and 25–750 eV. Each setting separates ions with a resolution of $\Delta E/E \approx 1$ between different E-ranges, as illustrated by the colored bands in Fig. 5.18. The ESA-TOF combination, which nominally only yields A/Q (Eq. (4.5a)), uniquely provides A with a mass resolution $A/\Delta A \approx 4$, sufficient to resolve O from H, because the CS only forms $Q = -1$ ions [57]. The ion optics is optimized for this resolution by biasing the CS at -20 kV.

The CS is a highly polished bare W surface. The original plan of adding a top layer with low work-function material (Caesium, Cs) to increase the negative ion yield was scrubbed. Its implementation would require flight operation at high temperature and extreme cleanliness to maintain or replenish a clean Cs surface layer for a mere doubling of the negative-ion yield [57]. The CS also emits SEs when struck by particles or UV photons. Broom magnets sweep out the electrons to suppress the SE interference. In addition, Ni black, Cr black (on Al surfaces), or Dow-9 (on Mg surfaces) cover all inner surfaces. The ESA design requires at least three bounces for a UV photon to reach S3, with the outer shell covered with Gold-black and the inner one with Cr black.

We now turn to ENA sensors on planetary and lunar missions.

- *NPD of ASPERA on Mars and Venus Express*: The Analyzer of Space Plasma and Energetic Atoms (ASPERA) on *Mars* and

Venus Express consists of an Electron Spectrometer (ELS), Ion Mass Analyzer (IMA), Neutral Particle Imager (NPI, covered in Sec. 5.2.2), and a Neutral Particle Detector (NPD). Here, we turn to NPD, which covers 0.1–10 keV in E and resolves H and O. Like NPI, NPD relies on reflecting incident ENAs off an optimized surface for analysis. Still, it also uses this surface to start the TOF measurement [17].

Figure 5.19 shows one of two identical NPD sensor heads. The fan-shaped deflection plates and a rectangular pinhole define a $9° \times 90°$ FOV. The $\vec{E}$-field between the plates at ±5 kV blocks the entry of ions with $E < 70$ keV/e. An ENA hits the start surface at an angle of approximately $15°$ and releases SEs. $\vec{E}$-fields guide them via a collection grid into one of the start MCPs depending on their polar incidence angle. Near-specularly reflected ENAs continue to strike the stop surface, emitting SEs into one of the three stop MCPs, completing a TOF measurement with a $\approx30°$ polar angle resolution. The ENAs may attain $Q = 0$ or ±1, but simulations show that ions with $E > 80$ eV reach the stop surface without substantial

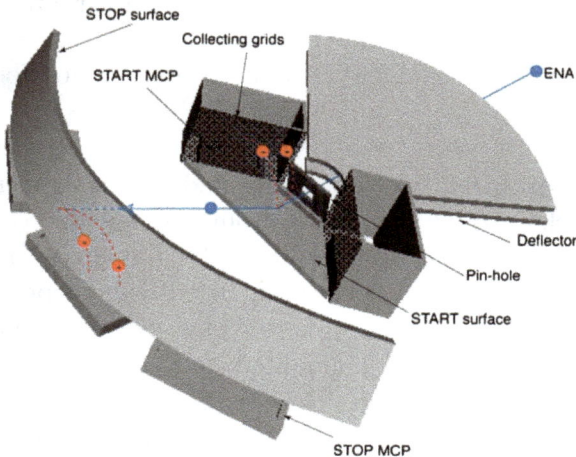

Fig. 5.19. The schematic 3D view of one NPD sensor shows a sample ENA trajectory (solid lines) and the collection of SEs (dashed lines) from the start and stop surfaces by two sets of MCPs [17]. (© Springer. Reproduced with permission, adapted).

disturbance by the electron collection fields. Using the A- and E-dependence of the SE yield at the stop surface provides additional information to resolve O of $E < 4\,keV$ and H of $E > 0.3\,keV$ by their TOF.

Two identical sensors extend the coverage in polar angle to $180°$. Mounted on a scan platform, they sweep out 2π in solid angle.

Unlike for the ENA sensors featured thus far, photons can enter NPD through the pinhole entrance and cause noise. Therefore, the start and stop surfaces must be highly UV-absorptive and low in photoelectron yield, in addition to being highly reflective for atoms with low angular scattering. The start surface has a multi-layer coating of Cr_2O_3, MgF, and WO_2, optimized to absorb Lyα at $15°$ incidence. The stop surface consists of graphite with $\approx$100-nm roughness, covered by $\approx$500 nm of MgO — multiple challenges met!

- *Sensors for ENA emissions from airless bodies*: The sensors presented thus far are designed for remote-sensing plasmas far away from surfaces of orbiting bodies. Analyzing ENAs backscattered from the surfaces of airless bodies, *e.g.*, the Moon and Mercury, requires resolving at least several elemental groups between O and Fe. This challenge calls for new ENA sensor designs. We will illustrate those at the hand of two sensors for the Moon and two for Mercury.

Ion bombardment due to the SW and magnetospheric plasmas causes a backscattering of ENAs from the regolith on the lunar surface. Regolith itself is a product of cumulative energetic ion bombardment. The ENA flux may be altitude-dependent, modified by the thin lunar exosphere. The *Chandrayaan Energetic Neutrals Analyzer* (CENA), mounted on the Indian lunar orbiter *Chandrayaan-1* [49], and the *Advanced Small Analyzer for Neutrals Atoms* (ASANA), riding on the Chinese lunar rover *Yutu-2* [59], may provide complementary data on the surface interaction and the effects of the lunar exosphere on the backscattered ENAs.

CENA observes ENAs with $E = 0.01$–$3.2\,keV$ by converting them into $Q = +1$ ions, followed by E/q selection and a TOF

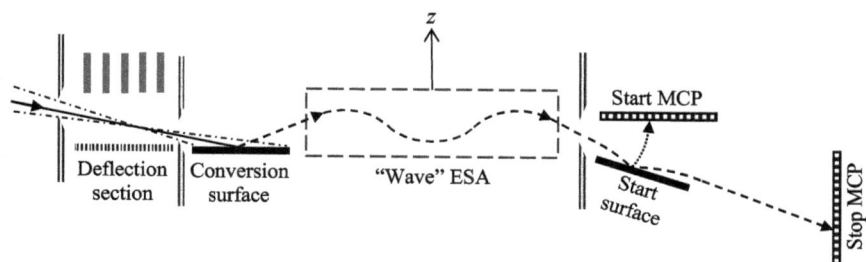

Fig. 5.20. A schematic radial cut of CENA in the x-z plane shows its functional subsystems. Unaffected by the $\vec{E}$-field in the ion deflection section, incident low-E ENAs (solid line) reach the CS, converting them into positive ions (dashed line). After selection by E in the "wave" ESA and post-acceleration, they enter the TOF section. SEs (dotted line) hit the Start MCP, while the reflected ion continues to the Stop MCP. The sensor forms a circular arc about its z-axis that goes through the center of the "wave" ESA [49].

measurement to obtain v. Their combination yields m and E of the incident ENA. Figure 5.20 shows the functional principles and essential subsystems of CENA.

An ENA enters CENA from the left into a $5° \times 160°$ FOV and passes the ion deflector, which blocks ions $<15\,\mathrm{keV/e}$ and traps all SEs with a plate at $+5\,\mathrm{kV}$ facing grounded fins. The CS converts the ENA into a $Q = +1$ ion (trajectory as a dashed line). A wave-like ESA similar to the one used in *SOHO* CELIAS MTOF [22] selects the ions by E/q with a resolution $\Delta E/E \approx 0.5$ using eight logarithmically spaced voltage settings. Post-acceleration of the ions by $-1.5\,\mathrm{kV}$ reduces the angular spread on the way to the start surface. SEs generate a start and position signal at the start MCP allowing a $5°$ resolution in polar angle. After a reflection under $15°$, the ions continue to the stop MCP, completing the TOF analysis regardless of Q. With its surfaces serrated and blackened, the mechanical ESA structures sufficiently suppress photons before reaching the TOF section.

Combining E/q from the ESA and v from the TOF (after $1.5\,\mathrm{kV}$ post-acceleration) in Eq. (4.5a) yields A/Q and thus A with $Q = +1$. CENA resolves H and the element groups CNO, NaMgSiAl, KCa, and Fe. CENA achieves $G = 10^{-2}\,\mathrm{cm^2 sr\,eV/eV}$ at $3.2\,\mathrm{keV}$ and $0.2\,\mathrm{cm^2 sr\,eV/eV}$ at $25\,\mathrm{eV}$, with $\eta = 0.01\%$–1%, varying with A and E.

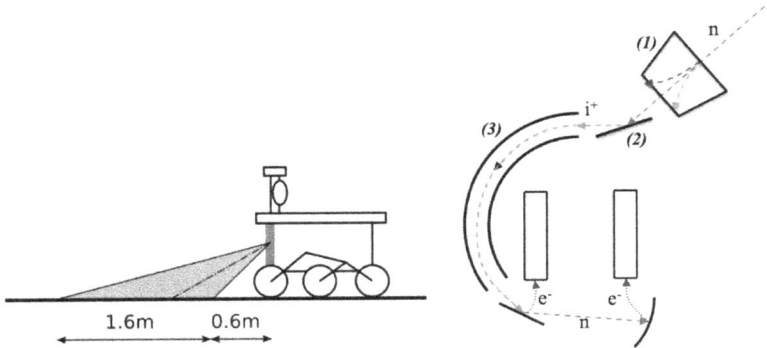

Fig. 5.21. Left: ASANA on *Yutu-2* views the lunar surface from the front of the rover with a 30° FOV at 0.6 m elevation. Right: The sensor schematic shows the path of an ENA (n) from entering the charged-particle deflector (1) to the conversion into an ion (i) at the CS (2). The ion passes the ESA (3) and completes the TOF measurement again as an ENA (n). The deflector (1) diverts incoming electrons and ions [59]. (© Springer. Reproduced with permission).

ASANA is unique among the ENA sensors featured in this book. It is the first to survey the lunar surface in ENAs from a rover on the Chinese *Chang-e 2* mission (Fig. 5.21, left). It shares the standard architecture of the *SWIM family* sensors [60] and differs from CENA (Fig. 5.20) only by its ESA type. As shown in Fig. 5.21 (right), the ENAs pass the charged-particle deflector (1) and strike the W(100) CS (2) at a shallow angle to turn into a $Q = +1$ ion. The positive potential of the CS pushes the newly formed ion into a 127° cylindrical ESA (3). The exiting ion bounces off another CS and hits the stop plate. The SEs emitted from the second CS and the stop plate trigger the start and stop detector of the TOF analyzer, respectively.

ASANA resolves H ENAs from the heavy ENA groups like CENA with 16%–100% resolution for 0.01–10 keV. SW sets the upper limit in E, and sputtered particles extend to the lower limit. With the deflector bias turned off, ASANA analyzes ions in the same E-range with 8% resolution and resolves species with $A/Q = 1, 2, 4, 8, 16,$ and 32.

The Search for Exospheric Refilling and Emitted Natural Abundances (SERENA) instrument on *ESA*'s Mercury mission *Bepi-Colombo* [61] carries two ENA sensors: ELENA (Emitted Low-Energy

Neutral Atoms) for the sputtered and backscattered ENAs (0.02–5 keV) from Mercury's surface and STROFIO ($\sigma\tau\rho o\phi\eta$ for "rotate" in Greek) for < 1 eV neutral atoms up to $A = 64$ to complement ELENA. Both sensors uniquely provide TOF start signals of the ENAs without any material contact.

ELENA is meant to resolve the sputtered ($E < 10^2$ eV) heavy ENAs from the backscattered ($E > 10^2$ eV) SW H ENAs by TOF analysis. Its start section in Fig. 5.22 (left) consists of two 1-μm thick, 1.0-cm^2 Si$_3$N$_4$ films, separated by $\approx 500\,\mu$m. Each contains a 2D array of 260-nm $\times$ 1.4-μm slits that are 1.4 μm apart (10^5 slits/mm^2). The rear array is a shutter that oscillates by $\approx 1\,\mu$m along y with up to 100 kHz, driven by a piezo-electric device. Only ENAs that enter within a time window centered on the phase when all slit pairs are perfectly aligned can reach the 3×10 cm^2 stop MCP, which takes the stop time and position on a 1D discrete anode (Fig. 5.22, right). Start timing at the shutter and the stop time yield the TOF and the z-location on the stop anode the arrival direction of the ENAs within the $5° \times 76°$ FOV. The S/C track across Mercury along y turns the 1D signal into a 2D image [51, 61, 62].

Sadly, due to technical delays, the two sets of slits stay permanently open in an aligned position and solely serve to reduce the UV flux by diffraction. The sensor may still yield total flux maps in y-z but without m and E resolution [61].

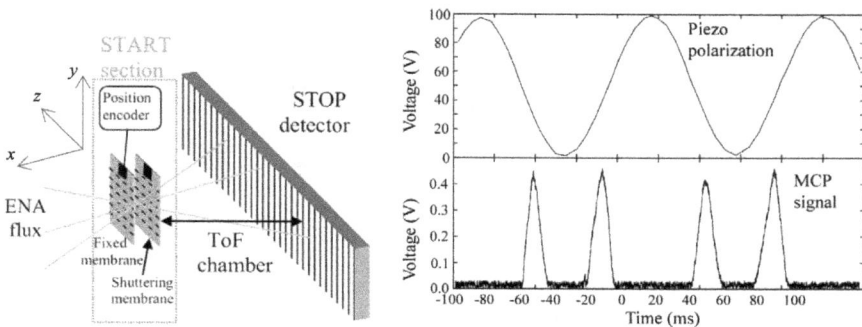

Fig. 5.22.   3D schematic view of the proposed ELENA sensor and its timing response [62]. (© Elsevier. Reproduced with permission).

STROFIO provides high mass resolution ($M/\Delta M \approx 60$) for $A \leq 64$ atoms at $E < 1\,\mathrm{eV}$ in a *rotating $\vec{E}$-field* [50]. This section describes STROFIO's functional principles without its structural complexity [61].

Figure 5.23 illustrates its working principle (an artist's rendition based on private communication with S. Livi). Figure 5.23 (left) shows a radial cut through the rotationally symmetric sensor. Electron impact ionizes ambient gas that rams into the sensor. An electrostatic lens extracts, accelerates, and focuses the newly created ions onto the detector plane before entering the rotating-field (RF) dispersion region with an axial speed v_0. Here, a radial $\vec{E}$-field rotates about the central axis at 195 kHz, generated by synchronized oscillating potentials at the four stationary cylindrical electrodes (Fig. 5.23, right). The rotating field adds a radial component v_r to the ion velocity, based on the time spent in the RF region. From here, the ions coast through the field-free region and hit the detector plane at (R, ϕ_d). The angle ϕ_d tags the start time for the TOF across the field-free region and the radial deflection R the time spent in the RF region. Because all ions are singly charged, the combination of the acceleration voltage with R and ϕ_d yields m of the atoms. From start to detection, the particles touched no surface!

For calibration, gas entered STROFIO at 2–3 km/s. The analysis confirmed a sensitivity of 0.14 counts/s for a density of $1\,\mathrm{cm}^{-3}$ and

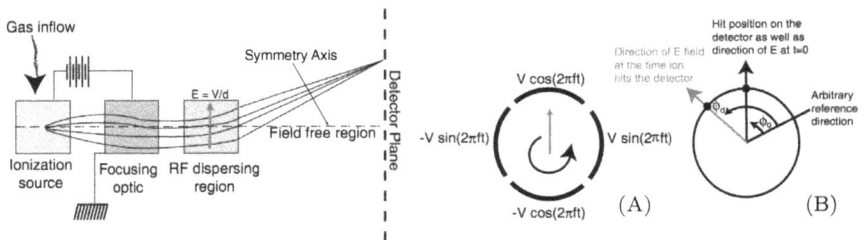

Fig. 5.23. Schematic view of STROFIO's functional elements (courtesy S. Livi). Left: Radial cut with the symmetry axis running through the center. Right: (A) The four alternatingly biased electrodes provide a rotating $\vec{E}$-field about the axis in the RF dispersion region. (B) The resulting phase relation ϕ_d and the radial detection location R determine the TOF timing and v_0 [50].

a mass resolution of $A/\Delta A = 84$ for $A = 1\text{--}64$ [61]. (According to S. Livi, the best achievable mass resolution is $A/\Delta A = 130$, but at the price of reduced sensitivity as in all mass spectrographs.) This is a remarkable jump in sensitivity from the gas mass-spectrograph in Sec. 5.1.1, given the two ambient pressures: $10^{-2}\text{--}10^{-8}$ mbar for Venus (Fig. 5.2) and 10^{-10} mbar for Mercury. TOF provides sensitive measurements of individual particles.

Here, we end the parade of selected ENA sensors — its start and end are two distinct ambient-gas analyzers, four decades and two planets apart. Chaps. 6 and 7 will highlight some results from observations in various regions of space, being mindful of the challenges of ENA diagnostics.

References

1. Von Zahn, U., & Mauersberger, K. (1978). Small mass spectrometer with extended measurement capabilities at high pressures, *Rev. Sci. Instr.*, **49**(11), 1539–1542.

2. Von Zahn, U., Krankowsky, D., Mauersberger, K., Nier, A. O., & Hunten, D.M. (1979). Venus Thermosphere: In situ composition Measurements, the temperature profile and the homopause altitude, *Science*, **203**(4382), 768–770.

3. Heikkila, W. J., Smith, J. B., Tarstrup, J., & Winningham, J. D. (1970). The soft particle spectrometer in the ISIS-I satellite, *Rev. Sci. Instr.*, **41**(10), 1393–1402.

4. Bernstein, W., Inouye, G. T., Sanders, N. L., & Wax, R. L. (1969). Measurements of precipitated 1–20 keV protons and electrons during a breakup aurora, *J. Geophys. Res.*, **74**(14), 3,601–3,608.

5. Witte, M., Rosenbauer, H., Keppler, E. *et al.* (1992). The interstellar neutral-gas experiment on Ulysses, *Astron. Astrophys. Suppl.*, **92**, 333–348.

6. Fuselier, S. A., Bochsler, P., Chornay, D. *et al.* (2009). IBEX-Lo sensor, *Space Sci. Rev.*, **146**, 117–147. Doi: 10.1007/s11214-009-9495-8.

7. Lallement, R., Raymond, J. C., Vallerga, J., Lemoine, M., Dalaudier, F., & Bertaux, J. L. (2004). Modeling the interstellar-interplanetary helium 58.4 nm resonance glow: Towards a reconciliation with particle measurements, *Astron. Astrophys.*, **426**, 875–884.

8. Möbius, E., Ruciński, D., Hovestadt, D., & Klecker, B. (1995). The Helium Parameters of the Very Local Interstellar Medium as Derived

from the Distribution of He$^+$ Pickup Ions in the Solar Wind, *Astron. Astrophys.*, **304**, 505–519.

9. Wenzel, K.-P., Marsden, R. G., Page, D. E., & Smith, E. J. (1992). The ULYSSES Mission, *Astron. Astrophys. Suppl.*, **92**, 207–219.

10. Johnson-Freese, J. (1987). Cancelling the US Solar Polar Spacecraft — Implications for International Cooperation in Space, *Space Policy*, **3**(1), 24–37.

11. Rosenbauer, H., & Fahr, H.-J. (1977). Direct Measurement of the Fluid Parameters of the Nearby Interstellar Gas Using Helium as Tracer, Experiment Proposal for the Out-of-Ecliptic Mission, *Internal Report*, Max-Planck-Institut für Aeronomie, Katlenburg-Lindau, Germany.

12. Raphelt, M. (1982). Untersuchungen zur Möglichkeit des Nachweises von interstellarem Neutralgas im interplanetaren Raum mit Hilfe der Sekundärionenemission, Ph.D. Thesis, Univ. Göttingen, Germany.

13. Bleszynski, S. (1985). Measurements of secondary ion, electron and photon yields of Pb-glass and LiF surfaces upon impact of He-atoms in the energy range of 23 to 3000 eV, *Internal Report MPAE-T-77-85-26*, Max-Planck-Institut für Aeronomie, Katlenburg-Lindau, Germany.

14. Banaszkiewicz, M., Witte, M., & Rosenbauer, H. (1996). Determination of Interstellar Helium Parameters from the ULYSSES-NEUTRAL GAS Experiment: Method of Data Analysis, *Astron. Astrophys. Suppl. Ser.*, **120**, 587–602.

15. McEntire, R. W., & Mitchell, D. G. (1989). Instrumentation for global magnetospheric imaging via energetic neutral atoms, in: Solar System plasma physics, *Geophys. Monogr., Ser.*, **54**, eds. J. H. Waite Jr., J. L. Burch, and T. E. Moore, pp. 69–80, AGU, Washington, D. C.

16. Barabash, S., C.:Son Brandt, P., Norberg, O. *et al.* (1997). Energetic Neutral Atom Imaging by the Astrid Microsatellite, *Adv. Space Res.*, **20**(4-5), 1055–1060.

17. Barabash, S., Lundin, R., Andersson, H. *et al.* (2006). The analyzer of space plasma and energetic atoms (ASPERA-3) for the Mars Express Mission, *Space Sci. Rev.*, **126**, 113–164.

18. McKenna-Lawlor, S., Balaz, J., Strharsky, I., *et al.* (2005). An Overview of the Scientific Objectives and Technical Configuration of the NeUtral Atom Detector Unit (NUADU) for the Chinese Double Star Mission, *Planet. Space Sci.*, **53**(1–3), 335–348.

19. Voss, H. D., Hertzberg, E., Ghielmetti, A. G. *et al.* (1992). Medium energy ion mass and neutral atom spectrometer, *J. Spacecraft & Rockets*, **29**(4), 566–569.

20. Voss, H. D. (1982). Energy and primary mass determination using multiple solid-state detectors, *IEEE Trans. Nucl. Sci.*, **NS-29**, 178–181.

21. Mason, G. M., Gold, R. E., Krimigis, S. M. *et al.* (1998). The Ultra-Low-Energy Isotope Spectrometer (ULEIS) for the ACE spacecraft, *Space Sci. Rev.*, **86**(1–4), 409–448. Doi: 10.1023/A:1005079930780.

22. Hovestadt, D., Hilchenbach, M., Bürgi, A. *et al.* (1995). CELIAS-Charge, Element and Isotope Analysis System for SOHO, *Sol. Phys.*, **162**(1–2), 441–281. Doi: 10.1007/BF00733436.

23. Domingo, V., B. Vleck., & Poland, A. I. (1995). The SOHO Mission: An Overview, *Sol. Phys.*, **162**(1–2), 1–37. Doi: 10.1007/BF00733425.

24. Hsieh, K. C. (2015). Detection of energetic neutral atoms in and out of the heliosphere, *Chin. J. Space Sci.*, **35**(3), 253–292. (In Chinese).

25. Hilchenbach, M., Hsieh, K. C., Hovestadt, D. *et al.* (1998). Detection of 55–80 keV hydrogen atoms of heliospheric origin by CELIAS/HSTOF on SOHO, *Astrophys. J.*, **503**(2), 916–922. Doi: 10.1086/306022.

26. Mitchell, D. G., Krimigis, S. M., Cheng, A. *et al.* (1996). Imaging-neutral camera (INCA) for the NASA Cassini mission to Saturn and Titan, *Cassini/Huygens: A Mission to the Saturn System*, ed. Horn, L., *Proc. SPIE*, **2803**, 154–161.

27. Mitchell, D. G., Krimigis, S. M., Cheng, A. *et al.* (1998). The imaging neutral camera for the Cassini mission to Saturn and Titan, in: *Measurement techniques in Space Plasms: Particles*, R.F. Pfaff, J.E. Borovsky and D. T. Young, eds., *AGU Monograph* **103**, 281.

28. Krimigis, S. M., Mitchell, D. G., Hamilton, D. C. *et al.* (2004). Magnetosphere imaging instrument (MIMI) on the Cassini mission to Saturn/Titan, *Space Sci. Rev.*, **114**(1), 233–329. Doi: 10.1007/s11214-004-1410-8.

29. Burch, J. L. (2000). IMAGE Mission Overview, *Space Sci. Rev.*, **91**, 1–14.

30. Mitchell, D. G., Jaskulek, S. E., Schlemm, C. E. *et al.* (2000). The high-energy neutral atom (HENA) imager for the IMAGE mission, *Space Sci. Rev.*, **91**(1–2), 67–112. Doi: 10.1023/A:1005207308094.

31. Mitchell, D. G., Paranicas, C. P., Mauk, B. H., Roelof, E. C., & Krimigis, S. M. (2004). Energetic neutral atoms from Jupiter measured with the Cassini magnetospheric imaging instrument: time dependence and composition, *J. Geophys. Res.*, **109** (A9), A09S11. Doi: 10.1029/2003JA010120

32. Pollock, C. J., Asamura, K., Baldonado, J. *et al.* (2000). Medium Energy Neutral Atom (MENA) Imager for the IMAGE Mission, *Space Sci. Rev.*, **91**, 113–154. Doi: 10.1023/A:1005259324933.

33. McComas, D. J., Allegrini, F., Baldonado, J. *et al.* (2009a). The two wide-angle imaging neutral-atom spectrometers (TWINS) NASA mission-of-opportunity, *Space Sci. Rev.*, **142**(1–4), 157–231. Doi: 10.1007/s11214-008-9467-4.

34. McComas, D. J., Funsten, H. O., & Scime, E. E. (1998). Advances in low energy neutral atom imaging, in: *Measurement Techniques in Space Plasmas: Particles*, R.F. Pfaff, J.E. Borovsky, D. T. Young, eds., *AGU Monograph* **103**, 275.

35. McComas, D. J., Allegrini, F., Pollock, C., & Funsten, H. O. (2004). Ultrathin (∼10 nm) carbon foils in space instrumentation, *Rev. Sci. Instr.*, **75**(11), 4863–4870. Doi: 10.1063/1.1809265.

36. Mitchell, D. G., Brandt, P. C., Westlake, J. H. *et al.* (2016). Energetic particle imaging: The evolution of techniques in imaging high-energy neutral atom emissions, *J. Geophys. Res.*, **121**(9), 8804–8820. Doi: 10.1002/ 2016JA022586.

37. Westlake, J. H., Mitchell, D. G., Brandt, P. C.-Son *et al.* (2016). The Low-Energy Neutral Imager (LENI), *J. Geophys. Res.*, **121**(9), 8228–8236. Doi: 10.1002/ 2016JA022547.

38. McComas, D. J., Allegrini, F., Bochsler, P. *et al.* (2009b). IBEX — Interstellar Boundary Explorer, *Space Sci. Rev.*, **146**, 11. Doi: 10.1007/s11214-009-9499-4.

39. McComas, D. J., Carrico, J. P., Hautamaki, B. *et al.* (2011). A new class of long-term stable lunar resonance orbits: Space weather applications and the Interstellar Boundary Explorer, *Space Weath.*, **9**, S11002. Doi: 10.1029/2011SW000704.

40. Funsten, H. O., Allegrini, F., Bochsler, P. *et al.* (2009). The interstellar boundary explorer high energy (IBEX-Hi) neutral atom imager, *Space Sci. Rev.*, **146**(1), 75–103. Doi: 10.1007/s11214-009-9504-y.

41. Scherrer, J., Carrico, J., Crock, J. *et al.* (2009). The IBEX Flight segment, *Space Sci. Rev.*, **146**, 35–73. Doi: 10.1007/s11214-009-9514-9.

42. Wurz, P., Fuselier, S. A., Möbius, E. *et al.* (2009). IBEX backgrounds and signal to noise ratio, *Space Sci. Rev.*, **146**, 173–206. Doi: 10.1007/s11214-009-9515-8.

43. Gruntman, M. A. (1993). A new technique for *in situ* measurement of the composition of neutral gas in interplanetary space, *Planet. Space Sci.* **41**(4), 307–319.

44. Wurz, P., Schletti, R., & Aellig, M.R. (1997). Hydrogen and oxygen negative ion production by surface ionization using diamond surfaces, *Surf. Sci.* **373**(1), 56–66.

45. Parks, G. K. (2015). Magnetosphere, in: *Encyclopedia of Atmospheric Sciences*, G. R. North, J. Pyle and F. Zhang, pp. 309–315, Elsevier B. V.

46. Kivelson, M. G., & Bagenal, F. (2014). Planetary Magnetospheres, in: *Encyclopedia of the Solar System*, pp. 137–157, Elsevier B. V.

47. Möbius, E., Kucharek, H., Clark, G. *et al.* (2009). Diagnosing the Neutral Interstellar Gas Flow at 1 AU with IBEX-Lo, *Space Sci. Rev.*, **146** (1–4), 149–172. Doi: 10.1007/s11214-009-9498-5.

48. Bochsler, P., Petersen, L., Möbius, E., *et al.* (2012). Estimation of the neon/oxygen abundance ratio at the heliospheric termination shock and in the local interstellar medium from IBEX observations, *Astrophys. J. Suppl.*, **198**(2), art. id. 13. Doi: 10.1088/0067-0049/198/2/13.

49. Bhardwaj, A., Wieser, M., Dhanya *et al.* (2009). Investigation of the Solar Wind–Moon Interaction Onboard Chandrayaan-1 Mission with the SARA Experiment, *Curr. Sci.*, **96**, 526–532.

50. Orsini, S., Livi, S., Tokar, K. *et al.* (2009a). SERENA: A novel instrument package on board BepiColombo-MPO to study neutral and ionized particles in the Hermean environment, *Future Perspectives of Space Plasma and Particle Instrumentation and International Collaborations*, *AIP Conf. Proc.*, **1144**, 76–90.

51. Orsini, S., Di Lellis, A. M., Milillo, A. *et al.* (2009b). Low energy high angular resolution neutral atom detection by means of microshuttering techniques: the BepiColombo SERENA/ELENA sensor, *Future Perspectives of Space Plasma and Particle Instrumentation and International Collaborations*, *AIP Conf. Proc.*, **1144**, 91–101.

52. Wieser, M., Wurz, P., Bochsler, P., Möbius, E. *et al.* (2005). NICE: an Instrument for Direct Mass spectrometric Measurement of Interstellar Neutral Gas, *Meas. Sci. & Technol.*, **16** (8), 1667–1676.

53. Möbius, E., P. Bochsler, M. Bzowski *et al.* (2012). Interstellar Gas Flow Parameters Derived from IBEX-Lo Observations in 2009 and 2010 — Analytical Analysis, *Astrophys. J. Suppl.*, **198**, art ID 11. Doi: 10.1088/0067-0049/198/2/11.

54. Rème, H., Bosqued, J. M., Sauvaud, J. A. *et al.* (1997). The CLUSTER ion spectrometry (CIS) experiment, *Space Sci. Rev.*, **79**, 303–350. Doi: 10.1023/A:1004929816409.

55. Möbius, E., Fuselier, S., Granoff, M. *et al.* (2008). Time-of-flight detector system of the IBEX-Lo sensor with low background performance for heliospheric ENA detection, *Proc. of the 30th Int. Cosmic Ray Conf.*, **1**, 841–844, Universidad Nacional Autónoma de México, Mexico City, Mexico.

56. Möbius, E., M. Bzowski, P. C. Frisch *et al.* (2015). Interstellar Flow and Temperature Determination with IBEX: Robustness and Sensitivity to Systematic Effects, *Astrophys. J. Suppl.*, **220**, art ID 24. Doi: 10.1088/0067-0049/220/2/24.

57. Moore, T. E., Chornay, D. J., Collier, M. R. *et al.* (2000). The low-energy neutral atom (LENA) imager, *Space Sci. Rev.*, **91**(1–2), 155–195. Doi: 10.1023/A:1005211509003.

58. Walton, D. M., James, A. M., & Bowles, J. A. (1998). 'High Speed 2-D Imaging for Plasma Analyzers Using Wedge-and-Strip Anodes', *Measurement Techniques Space Plasmas: Particles', Geophysical Monograph # 102*, AGU, Washington, DC, p. 295.

59. Wieser, M., Barabash, S., Wang, X.-D. *et al.* (2020). The advance small analyzer for neutrals (ASANA) on the Chang'E-4 rover Yutu-2, *Space Sci. Rev.*, **216**(4), article id.73

60. Wieser, M. & Barabash, S. (2016). A family of miniature, easily configurable particle sensors for space plasma measurements, *J. Geophys. Res. Space Physics*, **121**(12), 11,588-11,604. Doi: 10.1002/ 2016JA022799.

61. Orsini, S., Livi, S., Lichtenegger, H., Barabash, S. *et al.* (2021). SERENA: Particle instrument suite for determining the Sun-Mercury interaction from BepiColombo, *Space Sci. Rev.*, **217**:11.

62. Mattioli, F., Cibella, S., Leoni, R., Orsini, S. *et al.* (2011). A nanotechnology application for low energy neutral atom detection with high angular resolution for the BepiColombo mission to Mercury, *Mircroelectr. Eng.*, **88**(8), 2330–2333.

Chapter 6

ENAs from Magnetospheres and Small Bodies in the Solar System

"People only see what they are prepared to see."

Ralph Waldo Emerson, 1803–1882

Chaps. 4 and 5 described the means to detect and sort ENAs emanating from space plasmas beyond the reach of the S/C along with *in-situ* low-*E* ANAs. This chapter and the next will illustrate how ENA data can uncover essential information on the large-scale structure and dynamics of important regions throughout the heliosphere. While focusing on the methods, we will raise more questions rather than providing conclusive answers. This chapter will examine the environments of orbiting bodies in our solar system, *i.e.*, planets, moons, and small bodies. Section 6.1 will revisit the basic structure of Earth's magnetosphere (Sec. 1.2) directly tied to early ENA observations. We will leave definitions of and transformations between magnetospheric coordinate systems to textbooks [1, 2] and provide background information on other bodies as needed. In Chap. 7, we will explore the heliosphere and the *local* ISM.

Often ENA images reveal the more prominent features of the object of interest qualitatively prior to methodical extraction. Going beyond this first step, ENA diagnostics of Earth's magnetosphere, especially its inner region (Sec. 6.1.2), will serve as an excellent vehicle to discuss the critical and challenging process of extracting quantitative information on space plasmas from ENA observations (Secs. 6.1.2.1 and 6.1.2.2). In essence, obtaining spatial distributions of ions from ENAs is like finding the exospheric H distribution $n_H(r)$ from geocoronal data (Sec. 1.2.1).

6.1. Earth's Magnetosphere

The interaction between the SW including the embedded IMF and Earth's dipolar $\vec{B}$ including its trapped plasmas shapes Earth's magnetosphere, with all its distinct regions of interest (Fig. 1.3, right). Starting with *Explorer 1* and *Sputnik 3*, various Earth-orbiters have studied these regions for over six decades with *in-situ* particle and field instruments and remote-sensing EM-wave imagers. These observations have been interpreted and examined in the theoretical framework laid by Chapman & Ferraro [3], Dessler & Parker [4], Dungey [5], and Axford & Hines [6], with the modern extension of simulations [7, 8]. The coupling between the changing SW and the extended current systems, which connect the outer and inner magnetosphere to the ionosphere, is too complex to be evaluated when solely relying on *in-situ* observations. Global ENA imaging, preferably from multiple vantage points, is necessary. ENA imaging of the magnetosphere began with *ASTRID* PIPPI (Secs. 2.4 and 5.2.2) at the end of the 20th Century [9, 10]. In this Century, ENA diagnostics of Earth's magnetosphere, especially during substorms, received broader coverage by *IMAGE* [11] and *TWINS* [12]. We will feature recent results and work in progress.

We will treat the *outer* and *inner* magnetosphere separately: the *outer magnetosphere,* including the *magnetopause* (MP), *magnetosheath* (MS), polar cusps, and *magnetotail* (MT) in Sec. 6.1.1, and the *inner magnetosphere,* embracing the auroral zone and RC in Sec. 6.1.2. The geocentric radius r is the natural independent variable to distinguish regions of the magnetosphere. Roughly speaking, the *inner magnetosphere* lies inside 10 R_E and the *outer magnetosphere* beyond. These two distinct regions require ENA observations from different vantage points and different analytical methods. *IMAGE* and *TWINS* (Sec. 4.1.2) in polar orbits and with apogees at ≈ 7 R_E altitude, targeted the *inner magnetosphere*. In a highly elliptic near-equatorial orbit with ≈ 50 R_E apogee, *IBEX*, although optimized for observing the heliospheric boundaries and the ISM, is well-suited for viewing the *outer magnetosphere*. Key to these ENA diagnostics is the spatial distribution of neutral gases in Earth's *exosphere* (Sec. 1.2.1).

6.1.1.   *The outer magnetosphere*

Sections 6.1.1.1, 6.1.1.2, and 6.1.1.3 will touch upon the MS, polar cusps, and MT respectively. Given *IMAGE's* vantage point, we will primarily draw from the *IBEX* results, even though this is not the mission's prime goal. The plasmas in these regions are more dilute and less complex than those in the inner magnetosphere. Thus, it is less arduous to extract information from the ENA data on these regions.

6.1.1.1.   *The magnetosheath*

The MS is embedded in the exosphere and thus emits ENAs, almost exclusively H [13, 14]. Highly charged heavy ions of the subsonic SW emit X-rays after electron capture from exospheric H atoms [15]. The same holds for the cusp regions (Sec. 6.1.1.2). Because at these distances, n_H is low, the MS can be treated as "optically thin", and the extinction of the ENA fluxes can be ignored, setting $S(E, \vec{l}) = 1$. Using Eq. (3.18b) the differential H ENA flux in pixel (j, k) of the image is the line integral across the entire MS, taken from the front to the rear intersection of l with the BS (l_{fBS}, l_{rBS}) within a solid angle $\Delta\Omega$ along the LOS $\vec{l}$ in direction (θ_j, ϕ_k):

$$\frac{dJ(E, \theta_j, \phi_k)}{dEd\Omega} = \sigma_{H+H}(v)\frac{v^2}{m_p} \cdot \int_{l_{fBS}}^{l_{rBS}} n_H(l) \cdot f_p(l, v, \theta_j, \phi_k)dl \quad (6.1)$$

The CX cross-section $\sigma_{s+s}(v)$ for the species s under consideration connects the ion distribution $f_p(l, v)$ and exospheric density $n_H(l)$. $n_H(l)$ should preferably be observed concurrently or deduced from physics-based models of the region, possibly aided by inherent symmetries, to extract $f_p(l, v)$ from the ENA image. While ENA images also contain energy spectra of the H$^+$ distribution (Eq. (6.1)),

X-ray images often provide higher angular resolution and energy-integrated information on SW heavy ions, making these observations complementary.

The *IMAGE* LENA camera was the first to detect these ENAs [16], but it lacked the E resolution. Also, with *IMAGE*'s viewing of the magnetospheric boundary regions inside out and the relative phasing of its orbit to that of *Cluster*, there are no suitable conjunctions for simultaneous measurements of the ion distribution in the MS. However, with a correlation between the ENA fluxes and the dynamic SW pressure [13], *IMAGE* confirmed that the SW pressure affects the MP's location [17].

Figure 6.1 shows the ENA emission from the MS projected into the x_{GSE}-z_{GSE} plane taken with *IBEX*. *IBEX*-Hi obtained the image

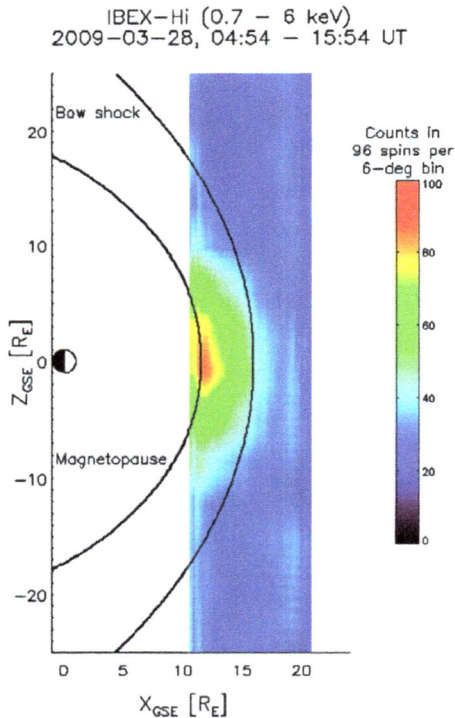

Fig. 6.1.   ENA image (0.7–6 keV) of the MS taken with *IBEX*-Hi [14]. (© AGU. Reproduced with permission).

scanning south to north during each spin and from smaller to larger geocentric distances during an ascending arc of the *IBEX* orbit. The emission is most intense right outside the subsolar MP. It extends to the BS and ≈ 10 R$_E$ to the north and south with decreasing intensity. The ENA intensity reflects the high proton density in the MS, and its variation indicates the decrease of $n_H(\vec{r})$ with r (Fig. 1.4). The assumptions $n_H(r) \propto r^{-3}$ and f_i independent of location facilitate the extraction or deconvolution of the ion distribution from this ENA image [14]. Adding another vantage point for ENA observation, *e.g.*, with *TWINS* [18], provides stereoscopic viewing with the possibility to unravel 3D inhomogeneities. This addition relaxes the dependence on symmetries or spatially homogeneous distributions in the deconvolution.

Figure 6.2 shows the ENA spectrum obtained with *IBEX*-Lo and Hi during the time interval shown in Fig. 6.1 and the H$^+$ spectrum observed concurrently with *Cluster 3* [14]. The ENA-derived H$^+$ spectrum, assuming a spherically symmetric exosphere with $n_H = 8\,\mathrm{cm}^{-3}$ at $10\,\mathrm{R}_E$ (the *Cluster* location), agrees well with

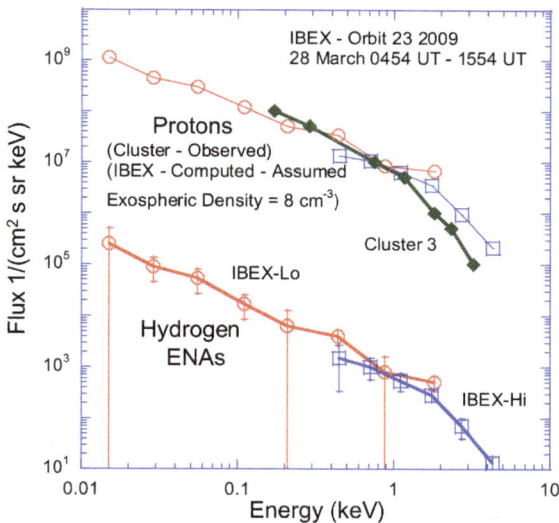

Fig. 6.2. Comparing the *Cluster 3* H$^+$ spectrum (green) with that deduced from concurrent *IBEX*-Lo H ENA observations (red) and Hi (blue) modeled for the *Cluster 3* location [14]. (© AGU. Reproduced with permission).

the observed H^+ spectrum below 1 keV. However, the value of n_H at 10 R_E is considerably lower than shown in Fig. 1.4. Raising n_H would lower the estimated proton flux, thus improving the agreement of the H^+ spectra at the three highest ENA energies measured by *IBEX*-Hi, while worsening it at the lower *IBEX*-Lo energies. The method used here is similar to the IHS thickness determination that combines the ENA flux with the local H^+ spectrum and the H density from other measurements, described in Sec. 7.5.2.

Like *Cluster*, also *THEMIS* provided *in-situ* data in the MS for the comparison with *IBEX* observations [19]. Broadening the database also allowed monitoring the temporal evolution with ENAs, particularly the IMF dependence of the MS ion populations [20]. Near the dayside MP (the inner MS boundary), *IMAGE* LENA detected neutral SW inflow [21] as another important ENA component and, in the process, identified cusp-related structures in the MS [22] which we turn to next.

6.1.1.2. *The polar cusp*

The global view facilitated by ENA imaging also enables diagnostics of the plasma's spatial distribution and temporal variation in the polar cusp regions, first monitored with *IMAGE* LENA [23]. Concurrent H^+ flux observations with *POLAR* allowed a quantitative comparison, resulting in a similar ENA-to-H^+ ratio as found later in the subsolar MS [14].

Figure 6.3 shows two ENA images of the cusp region during the southern summer at moderate SW flux ($n_p \cdot v_{SW} \approx 2 \cdot 10^8 \, \text{cm}^{-2}\text{s}^{-1}$) (left) and during the northern summer at strong SW flux ($n_p \cdot v_{SW} \approx 4 \cdot 10^8 \, \text{cm}^{-2}\text{s}^{-1}$) (right). Clearly, the SW flux modulates the emission of the cusp region, like that of the entire MS. In addition, the tilt of the magnetosphere determines which cusp (north or south) is illuminated more prominently. The mechanism at play for this variation becomes evident when comparing the cusp emissions for different IMF orientations. At southward IMF, the part of the cusp closer to the nose lights up, while the illumination shifts poleward for northward IMF. This behavior implicates reconnection [5], which occurs at the front side of the magnetosphere for southward IMF

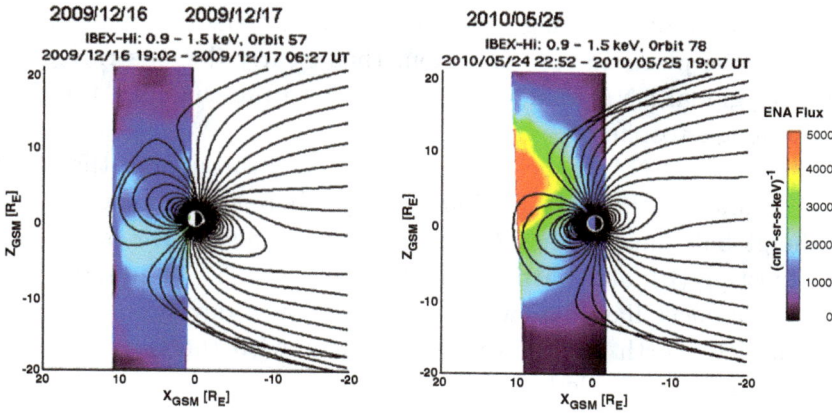

Fig. 6.3.   ENA images of the cusp regions projected into the x_{GSE}-z_{GSE} plane taken with *IBEX*-Hi (0.9–1.5 keV). Left: Southern summer at moderate SW flux. Right: Northern summer at strong SW flux [24]. (© AGU. Reproduced with permission, adapted).

and poleward of the cusp for northward IMF [24]. Apparently, reconnection amplifies the MS plasma flow into the cusp and increases the plasma density on the side of the cusp located next to the reconnection site [25].

6.1.1.3. *The magnetotail*

Driven magnetic reconnection in the Earth's MT leads to changes in the extension, plasma density, and ion temperature of the plasma sheet and disconnection of large parts as it reacts dynamically to SW pressure and IMF orientation changes [5, 26, 27]. *In-situ* observations over a range of distances into the MT, including multi-S/C measurements [28, 29], supported by simulations [8], have provided a general understanding of the physical processes. Nevertheless, there are numerous open questions because of the complexity of the interactions and the extensive range of spatial scales involved, which only become accessible with complementary imaging at a distance.

The highly elliptical orbit of *IBEX* with vantage points in the ecliptic on the dawn and dusk sides of the magnetosphere enabled its ENA cameras to image Earth's MT globally for the first time [30].

The sequence of two images potentially caught the disconnection of parts of the plasma sheet in action, thus clearly demonstrating the value of global ENA imaging for insight into the changing topology of the magnetosphere.

The many ENA images of the tail obtained throughout the *IBEX* mission allowed the investigation of the predicted dynamic variations of the global shape of the near-Earth plasma sheet, such as flapping (a), twisting (b), and warping (c) (Fig. 6.4, top) [31]. The signatures of plasma sheet flapping (wave-like north-south deviation, warping, or complete north and south displacement with the tilt of Earth's dipole) are identifiable based on the center z-position of the plasma sheet as a function of distance x from the Earth (Fig. 6.4, bottom).

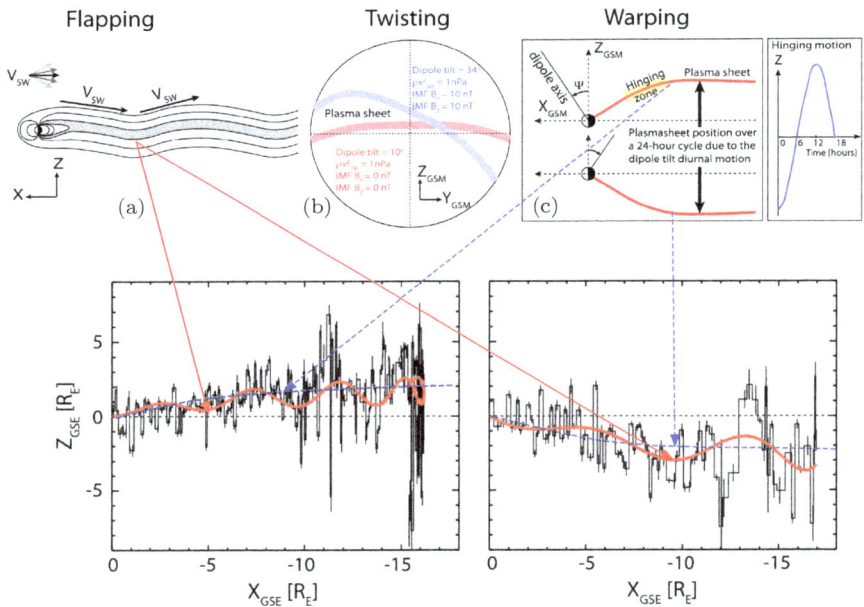

Fig. 6.4.　Observations of MT motion. Top: Illustration of plasma sheet flapping (a), twisting (b), and warping (c), in response to variations in the IMF and orientation of the Earth's dipole axis. Bottom: Observation of flapping and warping in *IBEX* ENA images. The center location of the plasma sheet in z_{GSE} is shown as a function of x_{GSE}, deduced from the distribution of ENA intensities [31]. (© AGU. Reproduced with permission, adapted).

However, it is impossible to obtain the plasma sheet twisting about the equatorial plane directly from *IBEX* images, which contain only east-west LOS integrals and thus are transparent to twisting. But a statistical study of all images sorted by IMF orientation likely provides indications of this effect [31].

One specific example of the plasma sheet evolution that ENA observations have shed some light on is the change of its thickness in the north-south direction in response to the extended presence of a northward IMF [32]. There have been reports that the plasma density increased under this condition and the ion temperature dropped over several hours [33]. This situation results in a cold, dense plasma sheet that extends considerably down the tail and is filled mainly with SW ions, likely from the MS [34].

Two-lobe reconnection at high latitudes has emerged as a leading candidate for the explanation because it naturally leads to the dominance of MS ions in the plasma sheet [35, 36]. This process would also likely inflate the plasma sheet volume with its thickening. Figure 6.5 shows two ENA images and flux distributions in the north-south direction at the beginning and end of an extended time with northward IMF.

The north-south extension of the plasma sheet grew from ≈ 11 R_E to ≈ 18.5 R_E, and the ENA flux increased, all consistent with the ion density increase observed earlier with *in-situ* measurements. The evolution of the ENA energy spectrum also revealed that the final plasma sheet is composed of a two-component population, with the cold portion below 1 keV strongly enhanced [32]. In a nutshell, features of the outer magnetosphere have been captured by ENA imaging from appropriate vantage points.

6.1.2. *The inner magnetosphere*

The inner magnetosphere is enclosed between the MP, on average at geocentric $r \approx 10$ R_E at the sub-solar point, and the atmosphere at altitude $\approx 10^2$ km (Fig. 1.3, right). This region is shaped by the interplay between the variable SW pressure and Earth's dipolar $\vec{B}$, which exhibits a seasonal tilt relative to the ecliptic, along with day-night and north-south asymmetries. The density and composition of

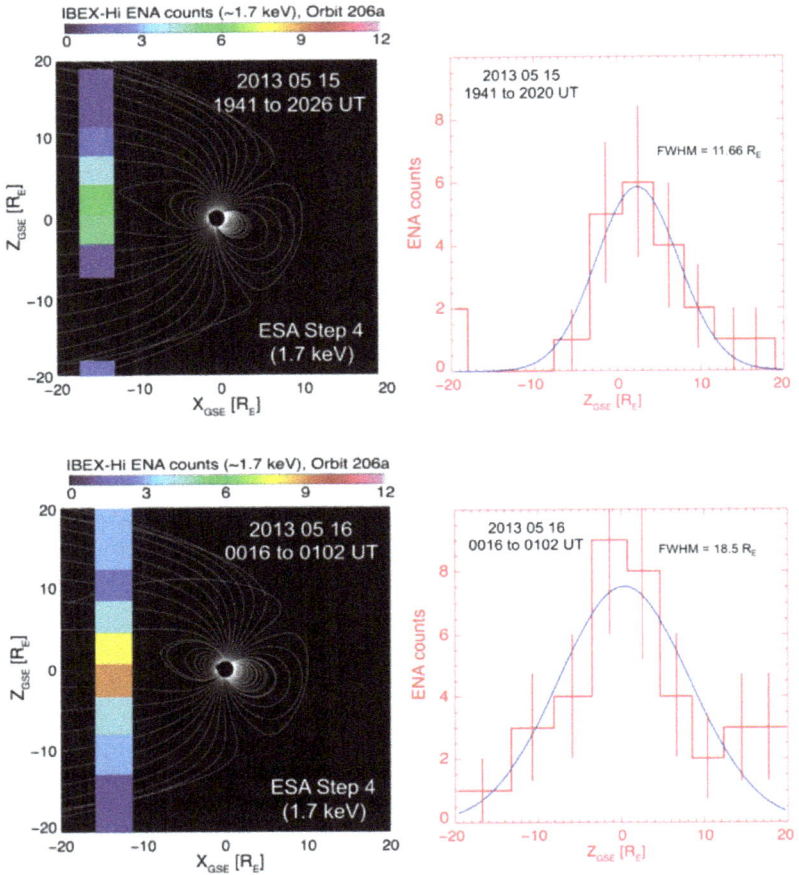

Fig. 6.5.   The plasma sheet thickens during northward IMF. Left: Two ENA images at 1.7 keV were accumulated for 46 min during the time shown on the plot. Right: The corresponding ENA count distributions in the north-south direction were obtained at the beginning (top) and end (bottom) of an approximately 6-hour long period [32]. (© AGU. Reproduced with permission).

the ionosphere depend on altitude, latitude, and LMT but also vary daily and seasonally. Variations in SW pressure, especially during solar transients, dictate the geometry and dynamics of Earth's $\vec{B}$, along with its trapped plasma. Currents coupling the ionosphere to the outer magnetosphere, especially the tail, make ENA diagnostics of the inner magnetosphere non-trivial. In turn, it offers an excellent opportunity to discuss how to extract information on spatial and temporal variations of energy and *pitch-angle* (α) distributions of ions from ENA data. Section 6.1.2.2 will outline the methods to do so after identifying the challenges in Sec. 6.1.2.1.

6.1.2.1. *ENA images and their intrinsic challenges*

We use one of the first ENA images taken with *IMAGE* HENA [37] (Fig. 6.6) to illustrate the ENA data challenges.

The ENA image (Fig. 6.6, left) is taken during a magnetic storm nearly over the magnetic north pole and accumulated over one S/C spin (2 min) in each of the 20 × 20 pixels for the TOF window of 39–60 keV H, including He and O at higher energies. Overlaid is the contour of the Earth, with four pairs of $\vec{B}$-lines separated by 6 hours. The pixel-smoothed image (Fig. 6.6, right) appears to be dominated by ENAs from *low-altitude emissions* (LAE) near the auroral zone on the dusk side. In contrast, *high-altitude emissions* (HAE) of the RC in the magnetic equator plane are fainter, especially in the pre- and post-dawn sectors. A clear separation of these emissions is not evident from the image.

Figure 6.7 shows six pixel-smoothed images of 50–60 keV ENAs of the same storm shown in Fig. 6.6, taken from six locations over one orbit. The first and sixth images are from the same location,

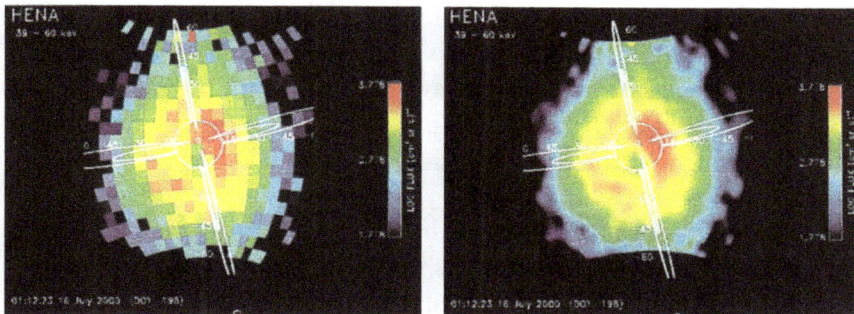

Fig. 6.6. ENA image taken with *IMAGE* HENA at 0112 UT, Jul 16, 2000. Left: flux based on raw counts per pixel (the bright pixel at the top is an artifact). Right: Pixel-smoothed image. An overlay of the Earth limb and four pairs of the dipole $\vec{B}$-lines for $L = 4$ and 8 at LMT 00 (mid-night, pointing up), 06, 12 (noon), and 18 guides the eye [37]. (© AGU. Reproduced with permission).

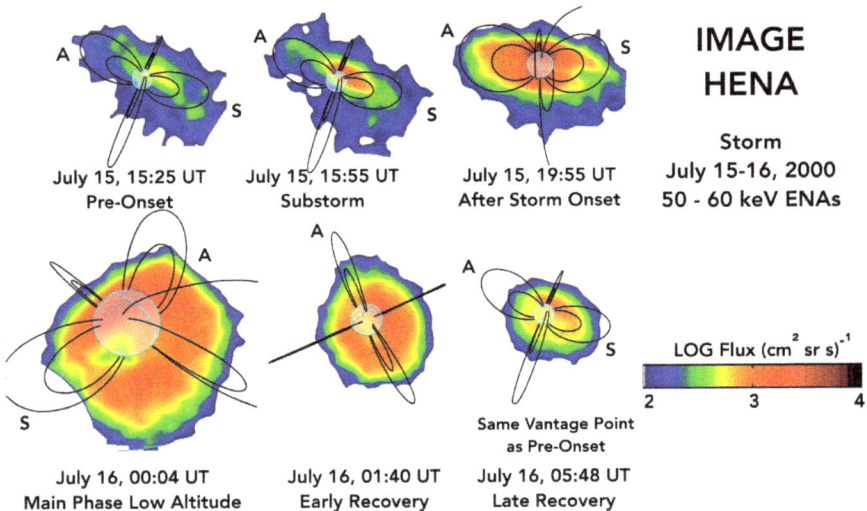

Fig. 6.7. ENA images taken with *IMAGE* HENA from six vantage points over an orbit between 1525 UT on July 15 and 0548 UT on July 16, 2000, throughout a magnetic storm. The Earth limb and the same dipole lines are overlaid as in Fig. 6.6. "A" denotes anti-sunward (00 LMT) and "S" sunward (12 LMT) [37]. (© AGU. Reproduced with permission, adapted).

one orbital period apart. The relative sizes of Earth indicate the imager's distances. The fifth image is taken 28 minutes after the one in Fig. 6.6, when the plasma has already evolved in the storm. Based on the overlaid dipole $\vec{B}$-lines, one can identify the brighter emissions shown inside the Earth's limb as LAE at high latitude. The bright emissions farther away from Earth appear to be **HAE** from the equatorial RC. The features in these images highlight three challenges for quantitative analysis:

- *Viewing limitations*: ENAs with nadir velocity components, such as those from precipitating ions, are very hard to observe because there is a lower limit for stable satellite orbits ($\approx 200\,\text{km}$) and interference from the upper atmosphere. Additionally, we know that only ENAs from ions with pitch angles that connect the source region with the imager along the LOS are visible. Thus, images in Fig. 6.7 show particular ranges of α.

- *Viewing resolution and contrast*: On any image, the brightest source on a LOS outshines and masks dimmer sources on that LOS. With limited angular resolution due to a wide FOV or scattering within the imager (Sec. 5.3.1), a bright source could bleed into adjacent pixels, thus complicating the separation of adjacent source regions and displacing their location. These effects, especially evident from higher latitudes due to the proximity of regions with different ENA emission rates, may be reduced by the precise point-spread function in the modeling of the observations [37].

- *Analytic dilemma*: As discussed in Chap. 3 and illustrated in Eq. (6.1), the number of ENAs accumulated in a pixel over a time interval is a LOS integral, which, for ENAs, is a convolution of the source ion and ANA distributions, with no unique solution, *i.e.*, more than one set of values can produce the same integral. Thus, imposing additional constraints on the extraction process, *e.g.*, based on physical symmetries, will be necessary.

6.1.2.2. *Deconvolution of ENA images*

Ignoring extinction (*cf.* Eq. (3.18a)), Eq. (6.1) or its variants link the ENA-producing ions to the detected ENAs. The ENA flux on the LHS is the observable, and the ion flux on the RHS is the quantity of pursuit, assuming all other terms are known. To find the ion distribution that best matches the observed ENA image is a mathematical challenge. When physical symmetries, *e.g.*, based on conservation laws, can be invoked, along with the description of the geometrical boundaries of the source region, mathematical *inversion* of Eq. (6.1) is possible. If there are no obvious symmetries, *forward modeling* of the expected ENA fluxes, based on the physical processes operating in the source region and their comparison with the observations with a χ^2-minimization, is the method of choice. The latter can also be applied to the source distribution obtained by *inversion* for further optimization in an iterative process.

- *Inversion*: To invert the observed ENA flux $dJ(E, \theta, \phi)/dEd\Omega$ along the LOS to the local ion phase-space density $f(E, \theta, \phi, l)$,

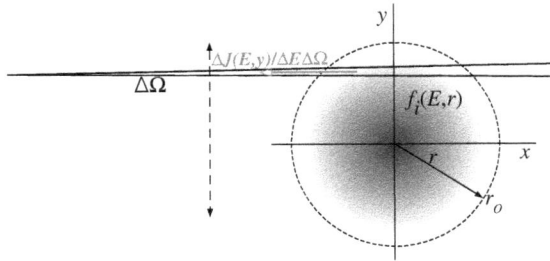

Fig. 6.8.   Scanning a cylindrically symmetric plasma column of radius r_o in ENAs along y allows obtaining the radial phase-space density $f_i(E,r)$.

with the ANA density $n_a(l)$ from different sources (Eq. 3.30), it is immediately evident that some information is missing because the sought-after quantity f depends on four variables, while the observable J only on three. Here is where symmetries or, in more complex situations, models come to the rescue. For the basics of inversion, we refer to the classic by S. Twomey [38]. Let us introduce the concept with a toy model shown in Fig. 6.8.

Consider the scan of a cylindrically symmetrical plasma column with radius r_o for ENA fluxes along y. For a constant ANA density n_a and isotropic ion velocity distribution f_i, the ENA flux of a specific E band with width ΔE and within the FOV $\Delta\Omega$ turns Eq. (6.1) into:

$$\frac{\Delta J(E,y)}{\Delta E \Delta\Omega} = \frac{2E}{m^2} \cdot \sigma_{aiCX}(E) \cdot n_a \cdot 2 \int_0^{x_{Max}} f_i(E,r) dx \qquad (6.2)$$

The integration along the LOS x is carried out symmetrically from 0 to the outer radius r_o, with $x_{Max} = \sqrt{r_o^2 - y^2}$. With $r = \sqrt{x^2 + y^2}$ and $dr/dx = x/r$, the integral $I(y) = (\Delta J(E,y)/\Delta E \Delta\Omega) \cdot (m^2/2E\sigma_{aiCX}(E)n_a)$ in Eq. (6.2) transforms as follows:

$$I(y) = 2 \int_0^{x_{Max}} f_i(E,r) dx = 2 \int_y^{r_o} \frac{f_i(E,r) \cdot r}{\sqrt{r^2 - y^2}} dr \qquad (6.3)$$

Eq. (6.3) is Abel's integral equation [39], which has the solution:

$$f_i(E,r) = -\frac{1}{\pi} \int_r^{r_o} \frac{dI(y)/dy}{\sqrt{y^2 - r^2}} dy = -\frac{1}{\pi r} \frac{d}{dr} \int_r^{r_o} \frac{I(y) \cdot y}{\sqrt{y^2 - r^2}} dy \quad (6.4a)$$

Due to the cylindrical symmetry, the ion PSD f_i turns into:

$$f_i(E, r) = -\frac{m^2}{2\pi E \cdot \sigma_{aiCX}(E) \cdot r} \frac{d}{dr} \int_r^{r_o} \frac{\Delta I(E, y)/\Delta E \Delta\Omega \cdot y}{\sqrt{y^2 - r^2}} dy$$

(6.4b)

For discrete ENA flux observations, a finite summation replaces the integral in Eq. (6.4b). Because of the rotational symmetry, the scan along y, which is equivalent to evaluating a line of pixels in an ENA image, provides the missing information for $f(r)$ inside the integral. It thereby reduces the dimensionality of the problem to one linear space dimension.

In analogy to this example, the 6D distribution (3D in $\vec{v}$ and configuration space) for ions of given m in the inner magnetosphere can reduce to the 3D information from ENA images, *i.e.*, E or v, and two-directional angles of the LOS [40]. The ion motion, constrained by the Earth's $\vec{B}$, is generally gyrotropic, which reduces the ion distribution instantly to 5D. Whether as an ideal dipole or empirically modified, *e.g.*, Tsyganenko [41] or later versions, $\vec{B}$ is a function of *magnetic local time* (MLT) ϕ and McIlwain's shell parameter L. The invariant latitude λ of the field line directly relates to L, as shown in Fig. 6.9 with the ENA viewing geometry. Without strong scattering, the local field strength $|\vec{B}|$ in the *first adiabatic invariant* μ constrains the ion pitch angle α

$$\mu = \sin^2\alpha \cdot E/|\vec{B}| = \sin^2\alpha_o \cdot E/|\vec{B}_o|$$

(6.5)

$|\vec{B}_o|$ and α_o are the $|\vec{B}|$ and α at the equator. Finally, *Liouville's theorem* maps the ion flux $j(l, \alpha) = j_o(L, \phi, \alpha_o)$ into the equatorial plane, where l denotes location $P(r, \phi, \theta)$ along the LOS $\vec{l}$ in Fig. 6.9. These relations reduce the ion distribution to 4D *in the equatorial plane*.

The remaining step down to 3D requires techniques equivalent to that in our toy model above, where the dipolar $\vec{B}$ field replaces the cylindrical symmetry and the ENA image the linear scan. The task at hand is finding the ion distribution along the LOS $\vec{l}$ by *deconvolution*. For ENAs produced in and propagated through optically thin regions, *i.e.*, with negligible extinction, Eq. (6.1) turns into Eq. (6.6), which yields the number of expected H ENAs $\bar{C}_i$ accrued in pixel i as:

$$\bar{C}_i(E) = \Delta E \Delta t \int_0^{l_{max}} A_i \sigma_{H+H} n_H(l) j(E,l) dl \qquad (6.6)$$

A_i is the response function of the ith pixel or the entire single-pixel sensor (Appx. A) and n_H the exospheric H density. The sensor accumulates the counts within the energy window ΔE, time interval Δt, and along the LOS up to l_{Max}, based on the geometry of the plasma domain under study. The sensor FOV determines the signal granularity Δs and ΔL along and across the field line (*cf.* Fig. 6.9). The choice of increments $\Delta \phi$ in MLT for the integration along l is commensurate with the granularity in the other two dimensions. Instead of a point at P, a volume element of thickness Δl and solid angle $\Delta \Omega$ is observed in the respective pixel. Its corresponding element in the equatorial plane is an annular sector $L \cdot \Delta \phi \cdot \Delta L$, bounded by an inner and outer L-shell separated by ΔL and two meridians subtending the azimuthal element $\Delta \phi$. Due to the shape of $\vec{B}$, the volume element in the equatorial plane is larger than

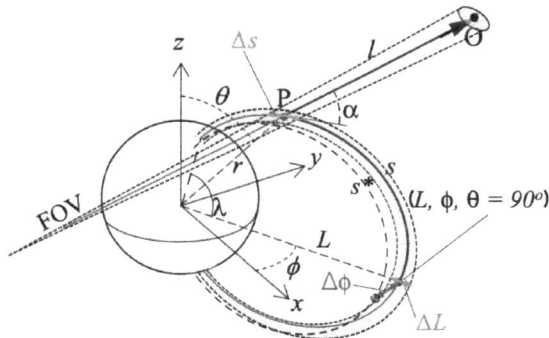

Fig. 6.9.   Viewing geometry for ENAs from RC ions at P, with the FOV pointing at pixel O. Visible are ions that gyrate about $\vec{B}$ with pitch angle α. The ENA accumulation occurs along the LOS, where increments Δl translate into a meridional angle increment $\Delta \phi$ at the magnetic equator. The FOV $\Delta \Omega$ is equivalent to an area $\Delta s \cdot \Delta L (r/L)$ around P. Also shown is the relation between shell parameter L and invariant latitude λ.

around P. Hence, angular uncertainties of the ENA image increase in its equatorial projection, thus limiting the precision of the inversion.

We use the multi-pixel image analogous to our toy model above to extract the remaining 4D ion distribution in the equatorial plane from the 3D ENA data. Instead of the single Eq. (6.2) that needs to be inverted to obtain f_i, a system of N equations of the type of Eq. (6.6), where N is the number of pixels with information on the inner magnetosphere, must be inverted simultaneously. If the governing equations are linearized or taken in linear approximation, it is a system of linear equations and, in principle, solvable by matrix inversion [42–44].

However, the finite resolution of the ENA image and the statistical and systematic uncertainties of the count rates in each pixel usually make it impossible to apply a direct inversion technique. Instead, approximate solutions for $j_o(L, \phi, \alpha_o)$ are obtained, parametrized in the three variables. The expected ENA counts in each pixel, based on Eq. (6.6), are compared with the observed counts C_i. Finally, a χ^2-minimization returns the optimized parameters [42–44]:

$$\chi^2 = \frac{1}{N} \sum_{i=1}^{N} \left(\frac{C_i - \bar{C}_i}{\sigma_i} \right)^2 \tag{6.7}$$

N is the total number of image pixels and $\sigma_i (= 1/\sqrt{C_i})$ the statistical uncertainty. Due to fluctuations from pixel to pixel and no *a priori* constraints along the LOS, often, the solution is not unique. This situation requires additional constraints on the trial set $\{j_o(L, \phi, \alpha_o)\}$, for example, "smoothness" [42] or "limits on the spatial gradient" of the ion distribution [43]. Favorably, the boundaries of the magnetospheric source regions are usually well-defined and provide the integration limits in Eq. (6.6).

As an example of matrix inversion [45], Fig. 6.10 shows an ENA image taken by *IMAGE* HENA (left) and the resulting ion intensity in the equatorial plane (*L*-LMT system) from the inversion (middle). Figure 6.10 (right) compares the deduced proton flux (*y*-axis) and the proton flux measured *in situ* by the *Cluster 4* ion spectrometer

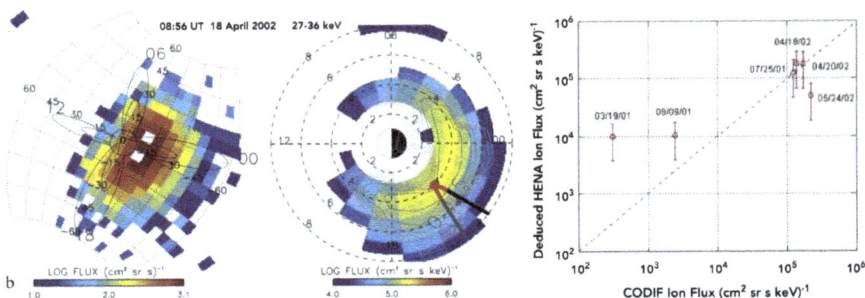

Fig. 6.10.   ENA image inversion and validation. Left: Image of 27–39 keV H ENAs taken by *IMAGE* HENA at 08:56 UT on Apr 18, 2007, overlaid with dipole lines of $L = 4$ and 8 at 00, 06, 12, and 18 MLT. Center: Equatorial H^+ flux distribution extracted by matrix inversion shown in annular sectors in (L, MLT), ignoring $L < 3\,R_E$ for exospheric "opacity". The contours represent relative inversion errors (inner contour 25% and outer contour 75%). The *Cluster* orbit is projected into the equatorial plane (red in the RC, black inbound, and gray outbound in the plasma sheet), with the red dot for the time of the ENA image. Right: Deduced ion flux *vs.* the *Cluster* observation at the red dot and for five similar cases along with statistical errors [45]. (© AGU. Reproduced with permission).

(x-axis) in the equatorial plasma sheet at the same time and for five similar cases.

The large error bars ($\pm 75\%$) in Fig. 6.10 (right) mainly reflect the low ENA count statistics. They are also due to uncertainties in the projection into the equatorial annular sectors and degradation in angular resolution from ENA scattering in the imager (Sec. 5.3.1). The limited range of accessible $\alpha(l)$ in any pixel forces the assumption of an isotropic *pitch-angle distribution*, while $j_o(\alpha_o)$ often peaks near $\alpha = 90°$. Additional contributors are simplifications in the $\vec{B}$-model and inaccuracies in the exospheric $n_H(l)$. For low H^+ fluxes on 03/19/01 and 08/09/01, the deduced H^+ fluxes appear overestimated, possibly due to unknown background. However, the agreement for 07/25/01, 04/18/02, and 04/20/02 lends support to the inversion method.

In most inversion examples, the complexity of the lower exosphere, *i.e.*, higher density and the presence of O in addition to H, are avoided, thus leaving altitudes $<2-3$ R_E blank, as in Fig. 6.10.

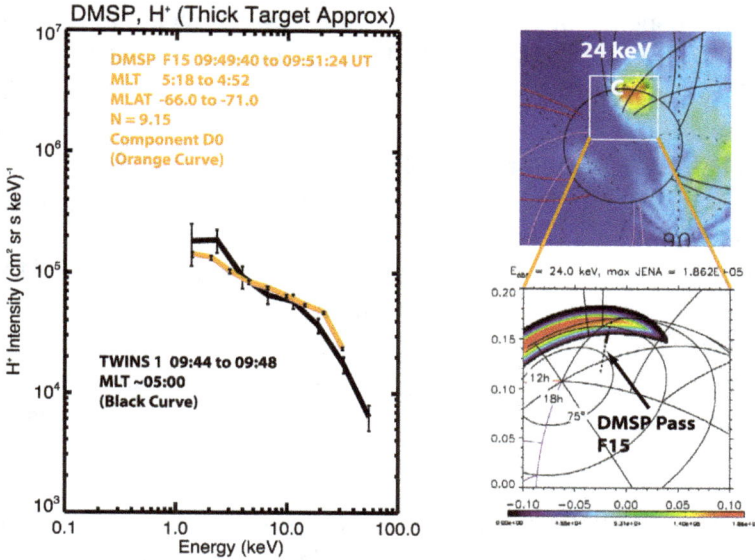

Fig. 6.11. Retrieval of LAE from the "optically opaque" polar region. Left: H^+ spectrum as derived from the *TWINS 1* ENA spectrum using Eq. (6.8) (black) compared with the one measured *in situ* by *DMSP* (orange). Right: ENA image of the LAE taken by *TWINS 1* (top) and modeled based on *DMSP* measurements using Eq. (6.8) (bottom) [46]. (© AGU. Reproduced with permission).

A simple approximation that reminds us of the first satellite-borne observation of ENA-generated protons (Sec. 2.3) makes this "optically opaque" region accessible to study (Fig. 6.11) [46]. The assumption of a "beam equilibrium" between ENA and ion production, *i.e.*, $\sigma_{iaCX} j_{ion} \approx \sigma_{aiCX} j_{ENA}$ (*cf.* Eq. (2.1)) allows an estimate of the total precipitating particle flux $j_{Tot} \approx j_{Ion} + j_{ENA}$ in a "thick target" at low altitude, including secondary ions and multiple generations of ENAs [46]. Thus, averaging j_{Tot} over the pixel FOV $\Delta\Omega$ yields:

$$\langle j_{Tot} \rangle = N \left(1 + \frac{\sigma_{aiCX}}{\sigma_{iaCX}} \right) \langle j_{ENA} \rangle \qquad (6.8)$$

The sequence *ai* in the subscript indicates ionization, *ia* neutralization, and N normalizes $\langle j_{Tot} \rangle$ to the *in-situ* observation. Figure 6.11 is an example for using Eq. (6.8) on the *TWINS1* ENA image with the concurrent *in-situ* ion flux measurement in the polar magnetosphere by *DMSP F15*. This clever and realistic approach sidesteps the complexity of the dense exosphere at <3 R$_E$ for ENA image inversion. However, without simultaneous *in-situ* observations and reliable cross-calibration of both E-dependent efficiencies, N remains undefined. Thus, concurrent remote ENA imaging and *in-situ* ion sampling are essential.

While the inversion methods for optically thin regions [42–44] address *RC emission*, the approximation in Eq. (6.8) yields the flux of precipitating ions from the LAE at high latitudes [46]. Together, they could adequately diagnose the inner magnetosphere with ENAs. However, they are regional and have limitations, *e.g.*, coarse resolution in the equatorial plane and the reliance on concurrent *in-situ* measurements to obtain N in Eq. (6.8). Finding the dynamic connections between the RC, precipitating and mirroring ions, ionospheric currents at high latitudes, and ion injection from the MT under varying solar conditions requires the alternate approach of *forward modeling*.

- *Forward modeling*: The goal of *forward modeling* is to simulate an ion distribution $f(m, E, \alpha, \vec{r})$ based on a comprehensive model of the region of interest and observations of the controlling conditions. These models *simulate* ENA images according to Eq. (6.6). Finally, χ^2-minimization optimizes the control parameters of the magnetospheric model in an iterative process (Eq. (6.7)) to best match the *observed* ENA image. The interpretation of the first fortuitous ENA image (Fig. 2.7) took advantage of forward modeling [47]. Before the *IMAGE* launch, algorithms were developed to minimize the variance between simulated and observed ENA images from space [42, 43]; *forward modeling* and *simulation* are inseparable. *Inversion* relies on closed magnetospheric $\vec{B}$-lines and ion motion that obeys the 1st *invariant* and *Liouville's theorem*. *Forward modeling* includes all important magnetospheric regions with particle sources and sinks, $\vec{E}$- and $\vec{B}$-fields, and relevant physical processes. G*lobal modeling* can harness the *global* potential of ENA imaging to complement *in-situ* ion measurements.

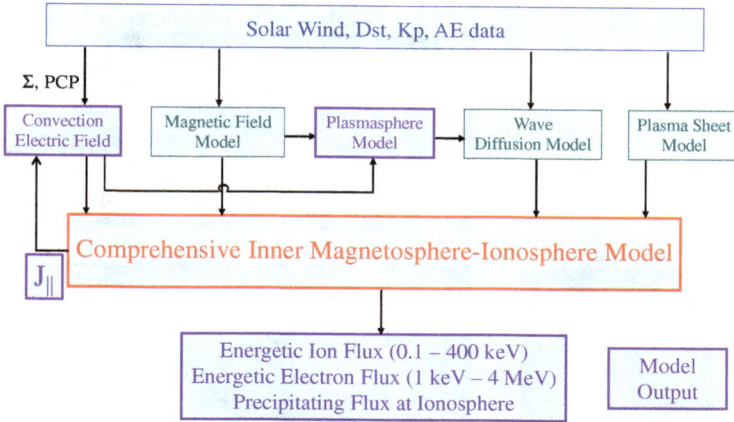

Fig. 6.12. Flow diagram of the CIMI model (courtesy M.-C. Fok). Σ stands for $\Sigma \leftrightarrow$, the ionospheric Hall-and-Pedersen conductance tensor, PCP for the polar cap potential, and $J_\parallel$ for the Region 2 current [48].

Figure 6.12 shows a flow diagram of the *Comprehensive Inner Magnetosphere-Ionosphere* (CIMI) model [48]. The CIMI model (in red box) allows the SW and geomagnetic conditions to affect the interactions between plasmas in different regions, focusing on the particles, fields, and waves. It computes the RC that completes the ionospheric-magnetospheric circuit [49]. CIMI optimizes the ion distribution and the related ENA emissions for comparison with the observations by tuning the input parameters.

Figure 6.13 compares simultaneous ENA images by *TWINS 1* and *2* (top, (a) and (b)) with two CIMI runs: Run 1 (center) assumes a self-consistent $\vec{E}$-field [50] and Run 2 (bottom) a Weimer $\vec{E}$-field [51]. Panels (c) and (f) show the equatorial distributions of 20 keV H^+, along with the simulated images on the right in panels (d, e) and (g, h), respectively. Run 1 returned a bright RC emission in the post-midnight sector, agreeing with the *TWINS* images, while it is shifted into the pre-midnight sector in Run 2. The observed eastward skewing of the RC emission may be due to the coupling between the RC and the ionosphere [49] as in Run 1. Overall, CIMI reproduced the RC emission well. An incomplete exobase model could have caused the absence of LAE [49].

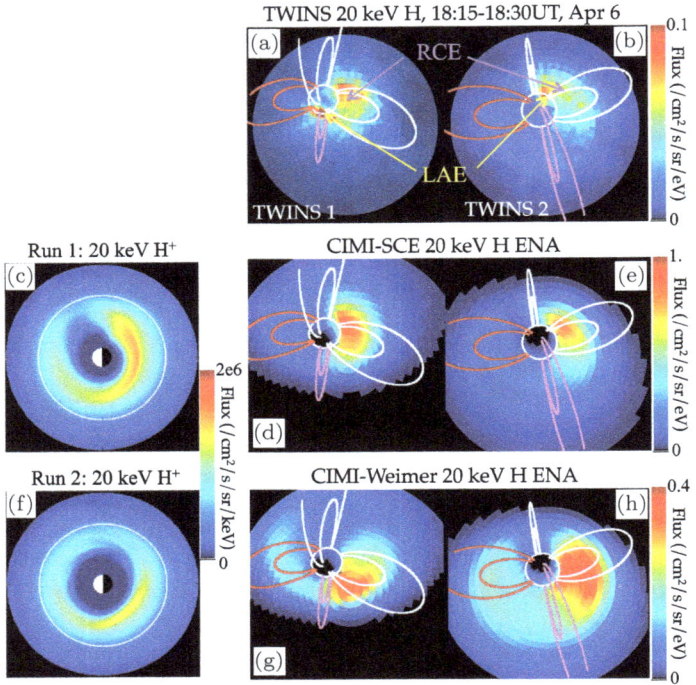

Fig. 6.13.   Observed H ENA images compared with CIMI simulations [48]. Top: 20-keV H images taken by *TWINS 1* and *2* at 18:15–18:30 UT on Apr 6, 2010. Center (Run 1): Equatorial 20-keV H^+ flux generated by CIMI using a self-consistent $\vec{E}$-field model (left) and simulated ENA H images (right). Bottom (Run 2): similar to the center row but using a Weimer $\vec{E}$-field. Earth's rim and four pairs of dipolar $\vec{B}$-lines ($L = 4$ and 8), 6 hours apart in MLT, are overlaid for spatial perspective (red lines at noon and purple at dusk). The simulations do not include the LAE. (© AGU. Reproduced with permission).

CIMI has enabled other examples of successful forward modeling [12, 52] and references therein. CIMI is still a work in progress, as some discrepancies between simulated and observed ENA fluxes might be traceable to missing magnetic coupling between the RC and other magnetospheric regions. A full MHD model of magnetospheric interactions may eliminate this deficiency [49].

Figure 6.14 summarizes the feedback loops involved in both *inversion* and *forward modeling*. Both methods can lead to results by themselves with significant uncertainties. However, they can also

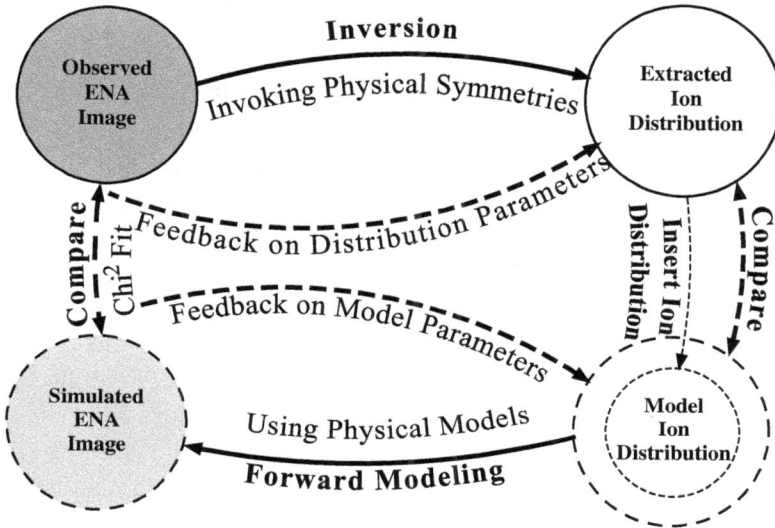

Fig. 6.14. An iterative process to extract the ion distributions. The observed ENA image yields a preliminary ion distribution through inversion compared with a model ion distribution that satisfies all known plasma conditions. "Viewing" this ion model through the same ENA imager by forward modeling produces a simulated ENA image to be compared with the observed ENA image. Iteratively revising the inputs for inversion and forward modeling minimizes the variance between simulated and observed ENA images.

reinforce each other iteratively, producing model ion distributions and simulated ENA images that closely match ion fluxes observed *in-situ* and ENA images obtained remotely.

Stressing the goal to establish a theoretical connection of the diverse magnetospheric phenomena to the principles of Newton and Maxwell, Parker [53] recognized the need for "the sometimes-arbitrary construction of idealized mathematical relations between various physical quantities as a guide to analyzing observational data." *Forward modeling* reinforced by *inversion* can provide a global view of the diverse phenomena for validation by ENA imaging and *in-situ* ion measurements. The techniques for extracting information on ion distributions from ENA images of Earth's magnetosphere, described above, are also useful for ENA diagnostics of other space plasmas, as we shall see below.

6.2. Magnetospheres and Ionospheres of Other Planets

Moving from Earth to Jupiter and Saturn, Mars and Venus, and Mercury is a natural extension of the ENA diagnostics to planetary magnetospheric plasmas. Before showing the ENA imaging highlights, let us sketch the geometries of these environments in the light of controlling parameters, which contextualize specific differences between the groups of planets.

Table 6.1 lists six planets in the order of their distances from the Sun with their spin axis, magnetic dipole orientation and strength, and the planet-centric distance of the sub-solar MP. Like at Earth, the $\vec{B}$-field and trapped plasma at Jupiter and Saturn interact with the SW and IMF to form a distinct magnetosphere. While the giant planets have the strongest dipole moments and largest magnetospheres, Venus and Mars have no intrinsic $\vec{B}$ but only SW-*induced magnetic boundaries* (IMBs) just above their atmosphere, a small fraction of the planet radius in altitude. In between is Mercury. We note that the spin axis of Venus and the dipoles of the giant planets point south, contrary to all others.

Figure 6.15 shows schematically two representative magnetospheres. The left is for planets without an intrinsic $\vec{B}$ (Mars and Venus) and is axially symmetric about the SW flow, similar to comets. The SW-induced $\vec{B}$ in the planet's ionosphere presents an

Table 6.1.   Planetary magnetic parameters.

Planet	Spin axis[1]	Dipole axis[2]	Dipole strength[3]	Magnetopause (Planet radii)
Mercury	$\approx 0°$	$14°$	5×10^{-4}	1.5
Venus	$178°$	—	0	1.1
Earth	$23°$	$11°$	1	10
Mars	$25°$	—	0	1.2
Jupiter	$3°$	$170°$	3.5×10^4	42
Saturn	$27°$	$>179.5°$	5.7×10^2	19

[1]From ecliptic north.
[2]From spin axis.
[3]Relative to Earth ($7.9 \times 10^{15}\,\mathrm{T\,m^3}$).

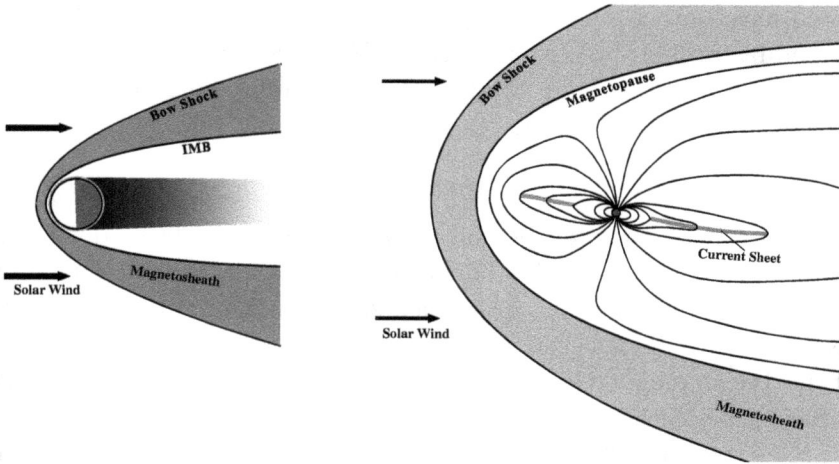

Fig. 6.15. Schematic representation of two types of magnetospheres, with a bow shock facing the supersonic SW. Left: SW-induced magnetosphere (Venus or Mars). The SW defines the axial symmetry and produces an IMB. The denser atmosphere below the exobase surrounds the planet as a thin shell. Sunlight casts a shadow behind the planet. Right: Magnetosphere of a giant planet (Jupiter or Saturn). Their rotation tilts and drags the dipolar $\vec{B}$ to form a current sheet. Mercury falls somewhere between the two types.

obstacle to the SW and IMF, forming a BS and IMB (analogous to MP). The upper atmosphere scatters some shocked SW ions back into the MS. Other ions penetrate the IMB and interact further with the lower atmosphere and the shadowed tail region. The right half of Fig. 6.15 shows a giant planet (Jupiter or Saturn) with its dipolar $\vec{B}$ and magnetosphere. The innermost plasma regions corotate with the planet, leading to a current sheet in the transition region. In addition, the giant magnetospheres enclose moons and dust rings, which make the interactions and production of ENAs even more complex. Mercury's magnetosphere notably straddles between these two types.

6.2.1. *Jupiter and Saturn*

When passing by these two planets, the *Voyager1* Low-Energy Charged-Particle instrument (LECP) detected energetic particles without any $\vec{B}$ connection to likely sources [54, 55]. These fluxes could have been ENAs or X-rays. Only 19 years later, when *Cassini* passed by Jupiter *en route* to Saturn, INCA (Sec. 5.3.1) verified that they were ENAs through the first and only ENA diagnostics of the Jovian magnetosphere to date.

6.2.1.1. *Jupiter seen in ENA images*

Cassini INCA observed Jovian ENAs for 80 days from distances 950 to 140 R_J [56]. INCA's boresight was close to that of *Cassini's* Optical Remote Sensing instruments, training on Jupiter. Thus, INCA had the Jovian system in full view. Figure 6.16 shows an image of the Jovian magnetosphere at 140 R_J. At such distances, INCA had sufficient resolution for meaningful spatially resolved studies.

The right panel of Fig. 6.16 compares Jovian H and O ENA spectra taken by *Cassini* INCA with those deduced from ion spectra taken by *Voyager1* LECP [54] at almost the same distance from Jupiter. The study showed an overall r^{-2} dependence in the ENA flux without noticeable temporal variation [56]. Further studies await *in-situ* measurements of the relative contribution of Io and Europa to the ion fluxes and UV sensing of the ANA distributions to quantitatively assess the ENA production.

Since Jul 5, 2016, *NASA's JUNO* has been gathering Jovian ion and UV data. It prepares the ENA diagnostics of the Jovian

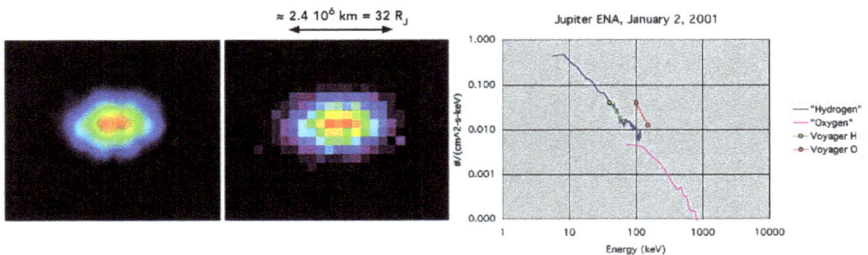

Fig. 6.16. ENAs from Jupiter. H ENA image of the Jovian magnetosphere taken by *Cassini* INCA at the closest approach of 0.07 AU (140 R_J), left: pixel-smoothed, center: raw image. Right panel: Energy spectra of Jovian H and O ENAs obtained with *Cassini* INCA in comparison with H (green) and O (red) spectra deduced from *Voyager* LECP data [56]. (© AGU. Reproduced with permission, adapted).

system by JENI (Sec. 5.3.2) after *ESA's* Jupiter Icy Moon Explorer (*JUICE*) arrives in 2030, promising breakthroughs like those with *Cassini* INCA at Saturn.

6.2.1.2. *Saturn seen in ENA images*

On Feb 20, 2004, at $\approx 10^3$ R$_S$ (0.43 AU) from Saturn, *Cassini* INCA was turned on and detected Kronian ENAs. Operations continued as planned from orbit insertion on Jul 1, 2004, to plunging into Saturn's atmosphere on Sep 15, 2017. Beyond a wealth of ENA results already published [57], further in-depth studies will likely enhance ENA diagnostics of giant magnetospheres on future missions.

Two early findings are: (1) The H ENA emission corotates with the planet at an ≈ 11-hour period and exhibits a day-night

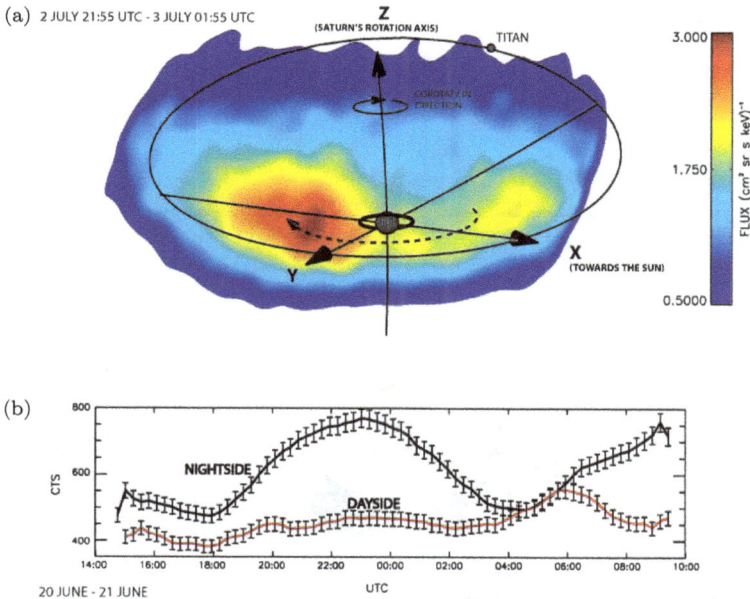

Fig. 6.17.   Saturn's corotating ENA source. (a) 20–50 keV H ENA image, taken before the orbit injection from below the ring (*x-y*) plane, shows a bright area that corotates with Saturn. For context, the figure includes Saturn, its A ring, its sense of rotation, and Titan's orbit. (b) Day-night asymmetry in the 3–80 keV H ENA intensities with an ≈ 11-hour period [58]. (© AAAS. Reproduced with permission).

asymmetry. (2) An additional inner radiation belt is seen. Figure 6.17 highlights the usefulness of ENA diagnostics to identify unique phenomena in a magnetosphere [58].

Cassini's in-situ observations of the $\vec{B}$-field [59], plasma ($E <$ 2 keV) [60], >2 keV electrons and ions [61], and *Saturn kilometric radiation* (SKR) [62], along with ENA imaging, corroborated the ≈ 11-hour period of a Saturnian day [63]. The co-rotation of the

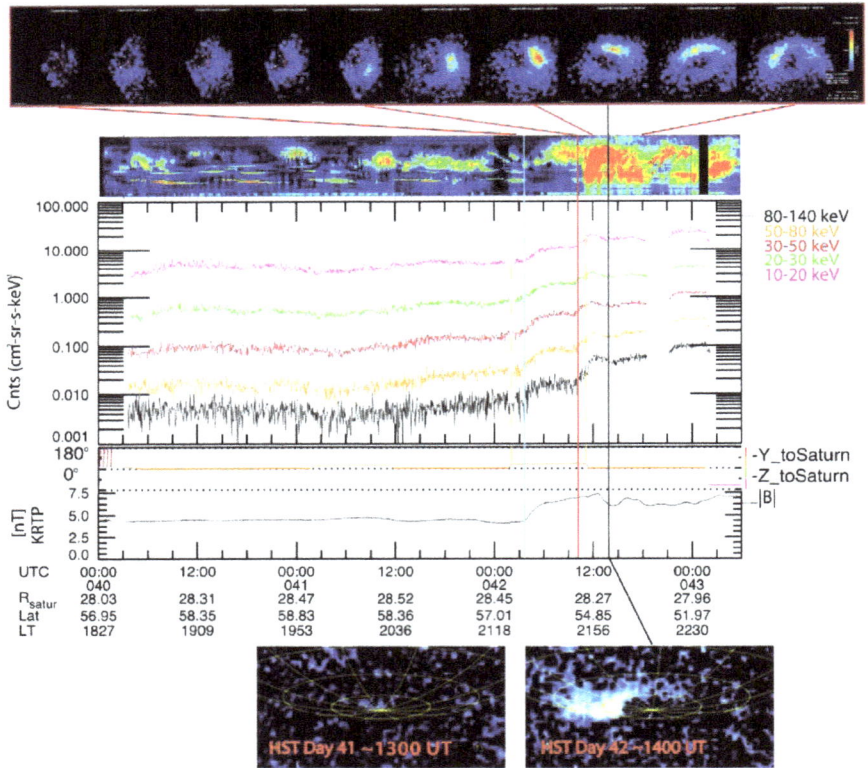

Fig. 6.18. Time sequence of correlated emissions on DOY 40–43, 2007, with location and time of observation at the x-axis. From top to bottom: 50–80 keV H images centered on Saturn, with the noon-midnight direction from the upper left to the lower right; SKR spectra from 1 kHz to 1 MHz; H ENA fluxes, colored according to their energies; $\vec{B}$ direction and strength. Straight lines mark the times of the ENA images. Bottom: Two auroral UV images, taken by the *Hubble Space Telescope* at the two indicated times [64]. (© Elsevier. Reproduced with permission)

Kronian $\vec{B}$-field, which drags along the trapped plasma generating ENAs and SKR, may explain these observations [64].

Figure 6.18 shows the timing during an enhanced SW compression. The timing revealed a close correlation among SKR, auroral UV, and ENA emissions in recurring events related to enhanced SW. Based on the ENA data (H and O images and spectra), the recurring events typically start with ion injection near or past local midnight, forming a co-rotating partial RC, which dissipates after the maximum emission around dawn. They coincide with SKR and the aurora once *per* Saturnian day.

In Fig. 6.19, Brandt *et al.* [65] showed that the ion pressure in the periodic partial RC could produce the current that induces the observed periodic oscillations in $|\vec{B}|$. They derived the ion pressure distribution in the equatorial plane from ENA images and spectra by inversion, converting it to a current with a previously developed method [47, 66, 67].

The example above demonstrates how ENA diagnostics can help extract a periodic partial RC that can cause observed variations in $|\vec{B}|$, which also manifests itself *in-situ* in SKR, plasma pressure, ion spectra, and through remote sensing of auroral UV. However, the question remains: How is the partial RC generated at the Saturnian rotational periodicity? A few plausible mechanisms come to mind [68], inspired by those in Earth's magnetosphere, still requiring further studies. On this note, we move on to Mars and Venus, planets without an intrinsic $\vec{B}$-field.

6.2.2. *SW-induced magnetospheres*

ESA's Mars Express (MEX) and *Venus Express (VEX)* were launched from Baikonur on Jun 2, 2003, and Nov 9, 2005, respectively [69]. While *MEX* is still working as of this writing, *VEX* ended its mission on Dec 16, 2014. Each S/C has an ASPERA suite, including two ENA sensors, NPI (Sec. 5.2.2) and NPD (Sec. 5.3.3). Both planets have SW-induced magnetospheres (Fig. 6.15, left), but differ in their distance from the Sun, atmospheric composition, and density. We highlight some of the *MEX* and *VEX* findings below.

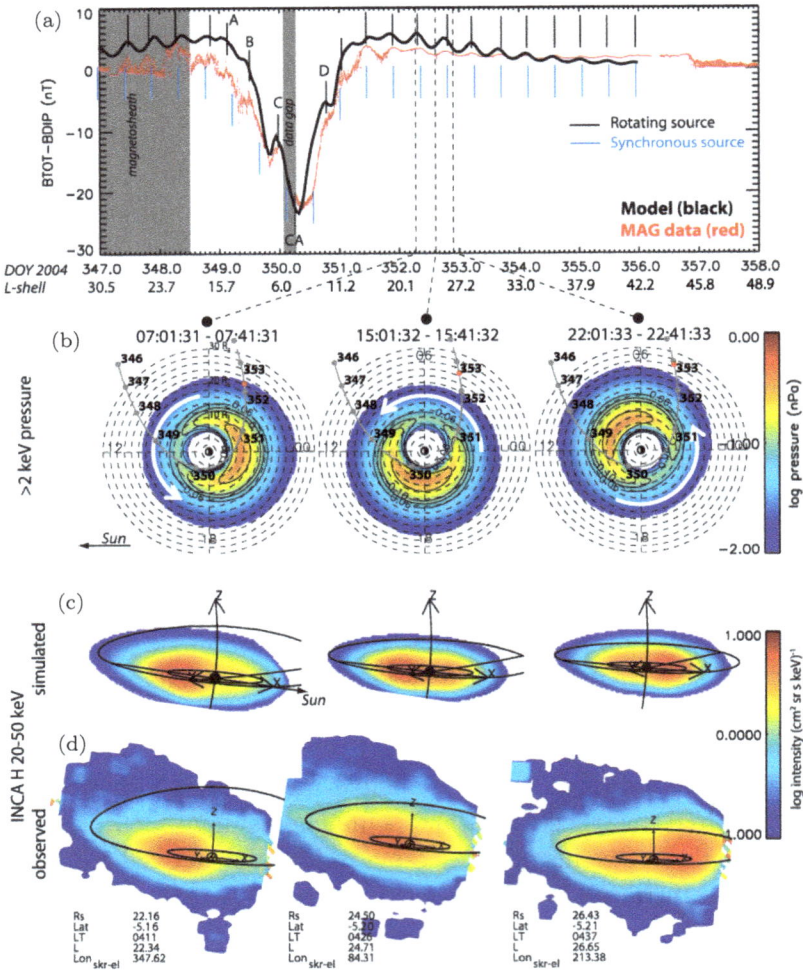

Fig. 6.19. Illustration of the steps from observed ENA images to the periodicity in $|\vec{B}|$. Bottom to top: (d) Observed 20–50 keV H ENA images; (c) simulated 24–55 keV H ENA images, obtained by forward modeling to match those in (d); (b) ion pressure distributions in the equatorial plane used to produce the images shown in (c); (a) measured (red) and modeled (black) $|\vec{B}|$ along the Cassini orbit at the end of 2004. Blue bars mark the peaks in the measured $|\vec{B}|$ and black ones the peaks obtained from the partial RC model [65]. (© AGU. Reproduced with permission).

6.2.2.1. *ENAs at Mars*

Based on ASPERA-3 results [70], the ENAs in the Martian environment come from SW ions after CX with the Martian exosphere, extending from the exobase at about 600 km altitude to beyond the BS. The pre-shock ENAs show a smaller angular spread than those from the hotter post-shock ions. Further scattering occurs in the upper atmosphere, but anti-sunward flow dominates, especially in the shadowy tail. Specific Martian ENA emissions comprise a jet from the subsolar IMB [71–73], ENAs from the upper Martian atmosphere [74, 75], the MS [73, 76] and the tail region [77–80]. In general, H ENAs from a non-magnetized planet carry the signature of scattering and anisotropy due to the convection $\vec{E}$-field produced by the SW bulk flow [73, 75, 81].

We highlight a unique study on the *crustal magnetic anomalies* (CMAs) [82, 83] and the ENA emissions in the Martian MS in Fig. 6.20. The CMAs may divert the shocked SW ions on the dayside towards the flanks of the IMB [84], guiding their ENAs into the NPD1 and 2 FOV, similar to the ENA jet from the subsolar IMB. Assuming the exospheric density drops off as r^{-2} and the shocked SW H$^+$ flows primarily anti-sunward, the ENA flux is highest at the IMB. It drops off as r^{-2}, similar to the observations of MS ENAs at Earth (Sec. 6.1.1.1).

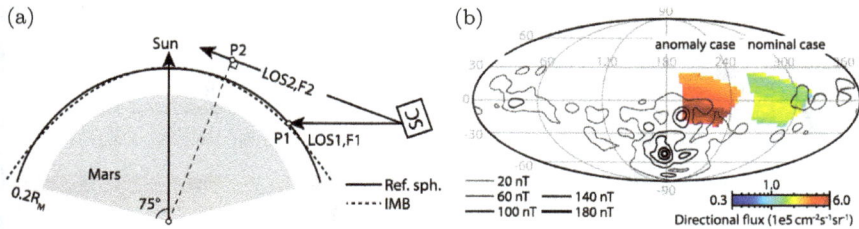

Fig. 6.20. Observing CMAs in ENAs. (a) Viewing geometry: The spherical shell of 1.2 R_M radius approximates the dayside IMB of Mars (dashed curve) within a 75°-cone. The LOS1 intersects the IMB at P1, and P2 is the point closest to the IMB for LOS2. (b) ENA flux images, scanned from LOS1 to LOS2, for two Martian regions shown in geographic coordinates superposed with *equi*-$|\vec{B}|$ contours at 400 km altitude [84]. (© AGU. Reproduced with permission).

Figure 6.20(a) shows the viewing geometry for two LOS. Along each LOS, the deduced ENA source is brightest close to the IMB (intersection P1 for LOS1 and closest approach P2 to the IMB for LOS2). The ENA source dims with distance from the IMB due to the decrease in the exospheric density. Based on these localizations, the 0.3–3.0-keV ENA directional fluxes from two dayside regions, taken at the same LT and SW pressures (<1.8 nPa), seem to differ, likely related to different CMA strengths at 400 m altitude (Fig. 6.20(b)) [84]. Using the same approach, Fig. 6.21 shows a statistical survey for the ±30° latitude region across Mars, for which *MEX* provides sufficient coverage.

Figure 6.21 shows the median values per 20° × 20° bin for: a) the directional ENA flux, b) the CMA strength at 400 m altitude, c) SW dynamic pressure (<1.8 nPa), and d) the number of orbits that provided the data, in the three latitude bands. There is a visible correlation, especially in the center ±10° latitude band. It is assumed that the influence of a stronger CMA field extends and overshadows the adjacent weaker fields.

Figure 6.22 (left) shows the maximum $|\vec{B}|$ along the three latitudinal bins in each longitudinal band from Fig. 6.21(b) as the

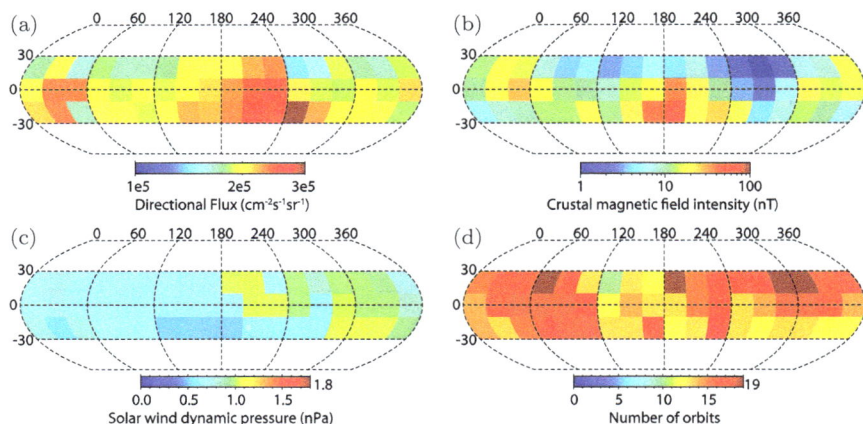

Fig. 6.21.   Maps in 20° × 20° bins between ±30° Martian geographic latitude: (a) 0.3–3-keV ENA flux, (b) CMA $|\vec{B}|$ at 400-km altitude, (c) SW dynamic pressure <1.8 nPa, (d) number of orbits viewed each bin to assess the uncertainties [84]. (© AGU. Reproduced with permission).

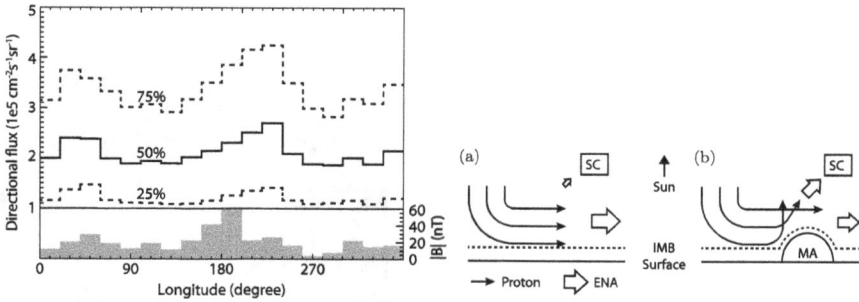

Fig. 6.22. ENA flux variations with CMA strength. Left: The top three histograms represent the first three quartile values of the directional ENA flux in the 10°S–10°N latitude band from Fig. 6.21(a). The shaded histogram shows the maximum in $|\vec{B}|$ for the three latitude bands at each longitude (Fig. 6.21(b)), which is used as reference $|\vec{B}|$. Right: The schematic drawings show how a CMA can affect the ion trajectories and, thus, the emerging ENAs [84]. (© AGU. Reproduced with permission).

reference $|\vec{B}|$ to accentuate this correlation. Comparing the ENA flux of the three latitude bands centered on −20°, 0°, and +20° with the "reference $|\vec{B}|$" resulted in 0.214, 0.527, and 0.787, respectively, for Pearson's correlation coefficient. The results lend credibility to a correlation with the 0° and +20° bands, but not −20°. Their conclusion appears somewhat counterintuitive, because $|\vec{B}|$ is often most intense in the −20° band (Fig. 6.21(b)). After a bin-to-bin correlation between panels (a) and (c) of Fig. 6.21, the authors attribute this lack of correlation in the southern band to a noticeable correlation between the SW pressure (<1.8 nPa) and ENA flux.

This tantalizing result is inconclusive. However, results and open questions in an earlier study of ion acceleration at a CMA exemplify the value of exploration with a new tool, such as ENA imaging [85].

6.2.2.2. *ENAs at Venus*

Like Mars, Venus also has a SW-induced magnetosphere but differs in the subsolar locations of the BS and IMB (1.4 and 1.1 R_V, compared to 1.6 and 1.2 R_M for Mars). The Venusian ENA sources observed by *VEX* are similar to those observed by *MEX* at Mars [86, 87].

Similar to a detailed study of Martian H ENAs [78], the best-fitting temperature and density of a Venusian model exosphere are

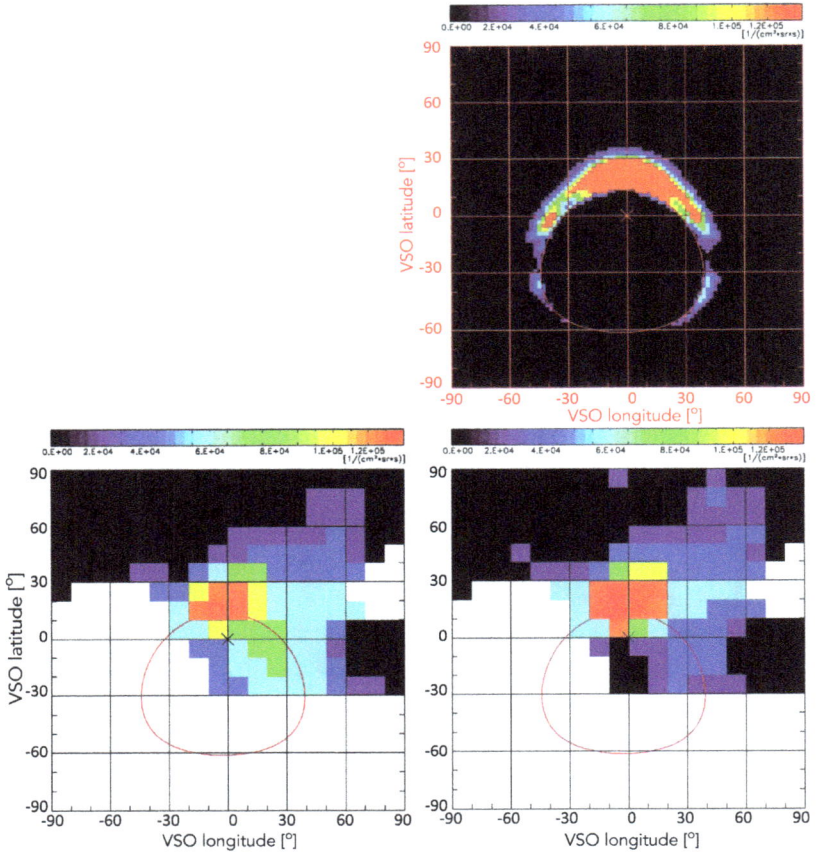

Fig. 6.23. Venusian exosphere in ENAs in Venus-Solar-Orbital coordinates [88]. Lower left: H ENA image taken by VEX NPD2 at 0.64 ± 0.20 R$_V$ altitude over 14 Earth days ($\approx$14 orbits), when Venus' center moved over $-3° \pm 12°$ in longitude and $-24° \pm 10°$ in latitude. The average FOV footprint covers $70° \times 45°$ on the surface. Top: Simulated image of tailward moving H ENA flux with the best-fitting Venusian exospheric H model. Lower right: Simulated image with the NPD2 resolution compared to the observation (lower left). The red line marks Venus' limb, and the X points toward the Sun. (© AGU. Reproduced with permission).

obtained, comparing the viewing geometry and spectrum of tailward flowing H ENAs with simulated images, shown in Fig. 6.23 [88]. The predominantly anti-solar ENA flow and NPD's UV-sensitivity limited the useful observations to the Venusian shadow, shown in Venus-Solar-Orbital coordinates [86].

Before we leave Venus, let us briefly show how combining ASPERA-4 IMA and NPI observations on the polar-orbiting *VEX* shed light on the SW-ionosphere interaction and the related energy and momentum balance in a SW-induced magnetosphere [89]. IMA distinguishes between 0.01–1.5 keV H^+ and 1–300 eV O^+, and NPI (Sec. 5.2.2) covers 0.01–1.0 keV ENAs without mass resolution. The observed ENA flow directions and the ionospheric O^+ fluxes at altitudes 200–600 and 600–1,200 km are consistent with the SW H^+ flow relative to Venus' orbital motion. Apparently, the SW drives the local O^+ ions and the ENAs from CX with the Venusian ANAs, as shown in Fig. 6.24.

The ENA flux increases with decreasing altitude as the ANA density in the thermosphere rises. The red curves are closer to reality due to the E-dependent response of NPI, inferred from calibration and the E-resolved IMA ion spectra because the ENAs inherit the momentum of the parent ions. The BS at $\approx$5,000 km and the IMB at $\approx$1,200 km altitude are evident. The ENA flux in the dawn sector is

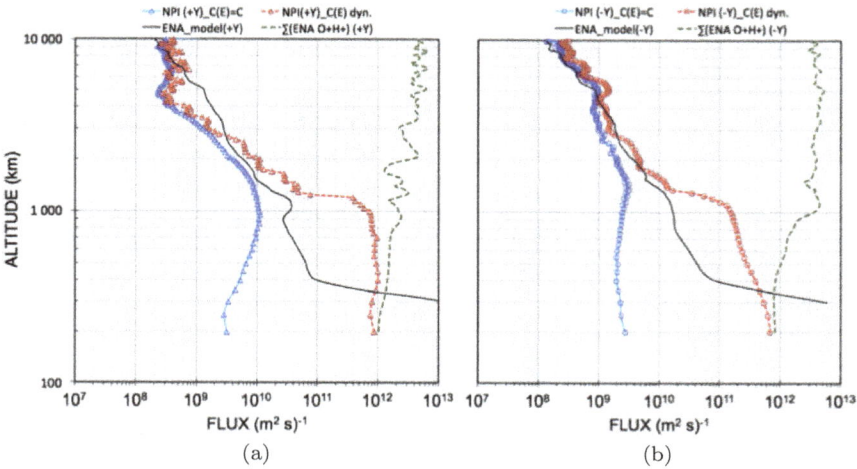

Fig. 6.24.   ENA fluxes *vs.* altitude over Venus' northern polar region in the dawn (a) and dusk (b) sectors. Shown are the observed ENA fluxes, obtained with a constant detector sensitivity (in blue), with an E-dependent sensitivity (in red), and the model ENA fluxes (in black). The green dashed line is the sum of ENA (red), O^+, and SW H^+ flux. The flow points from dusk to dawn with an anti-sunward component [89]. (© AGU. Reproduced with permission).

higher and better organized because the ENA production increases toward the $+y$ direction. The steady rise in ENA flux in the MS between the BS and IMB (Fig. 6.24(a)) and the almost constant level below the IMB should yield a better Venusian atmospheric model than the one used in the current ENA model (black line). The convergence of the red and green curves validates the conservation of momentum, energy, and the number of particles in the ENA production. Also, the near-total conversion of ions into ENAs at ≈ 200 km altitude is evident [89]. These results show the power of concurrent ion and ENA observations but also call for m and E resolution and Doppler-shift observations in Ly-α for H and at 1304 Å for O ENAs.

6.2.3. *Mercury*

At this writing, we await the safe arrival of *ESA's BepiColombo* at Mercury in 2025 and the first Hermean ENA signals by ELENA and STROFIO (Sec. 5.3.3) [90]. The outcome will be something very new because Mercury is unlike any of the planets we have surveyed (Table 6.1). It even has an intrinsic $\vec{B}$-field, albeit very weak. Being closest to the Sun, the SW pressure shapes Mercury's MP more substantially than that of Mars and Venus (Fig. 6.15, left).

Mercury has a tenuous atmosphere consisting of H, He, and O from neutralized SW ions and Na, Mg, Al, Si, S, K, Ca, Mn, and Fe from SW and magnetospheric ions impacting the surface [91–93]. Although a planet, Mercury's peculiar condition makes it akin to our Moon and small bodies, such as asteroids and meteorites, which we shall discuss next.

6.3. Small Bodies in the Solar System

We define *small bodies* as solid objects in the solar system without an atmosphere and intrinsic $\vec{B}$, such as the Moon, asteroids, and even dust grains. Their surfaces are directly exposed to the SW and some magnetospheric ions, *e.g.*, when the Moon is in Earth's magnetotail and Mercury under its compressed magnetosphere. In the ion-solid interactions, the impacting ions lead to backscattering and sputtering of ions and ENAs. Cosmic rays are left out; they are much more energetic and produce different kinds of interactions. The sputtered ENAs form another category different from heliospheric or magnetospheric ENAs in their spatial, directional, energetic, and compositional characteristics [94, 95]. They enable the remote analysis or *in-situ* sampling from landers or rovers (Sec. 5.3.3). We will discuss lunar ENAs in some detail and then end with some remarks on even smaller bodies.

6.3.1. *ENAs from the Moon*

Sputtered particles from the Moon were first detected as O^+ and Mg^+ PUIs when *AMPTE* IRM mission was in the SW upstream of Earth's BS during the new Moon phase [96]. An early *IBEX* observation reported lunar ENAs, even before completing the first all-sky map [97]. This unexpected happy discovery right after the *IBEX* payload commissioning validated the ENA sensitivity of the instrumentation early in the mission. It also provided evidence for the SW as the generator of ENAs on the lunar surface.

As shown in Fig. 6.25 (top), lunar ENAs produce a substantial signal in *IBEX*-Hi that stands out at the appropriate spin angle when *IBEX* scans across the Moon. In this configuration, the Moon was about 30 R_E from the S/C. The ENA flux correlates with the SW flux observed by *ACE* SWEPAM time-shifted for the *ACE*-Moon distance (Fig. 6.26), which implicated SW ions as the source of backscattered ENAs from the Moon.

In comparison with SRIM modeling of the energy distribution of backscattered ENAs, the observed ENA to SW flux ratio, integrated over the entire lunar surface, yields a ≈10% albedo for H ENAs [97]. This value is consistent with the approximate H^+ albedo of ≈0.1–1% found earlier [98] because the expected ionized fraction of backscattered particles is a few %. Using *IBEX*-Lo observations, which covered the entire energy range of backscattered ENAs to 10 eV, the integrated lunar H ENA albedo was refined to 0.09 ± 0.005 [99] and subsequently to 0.11 ± 0.06, including modeled scattering functions [100]. These values are also consistent with the local albedo of 0.16–0.2 found with *Chandrayan* CENA (Sec. 5.3.3)

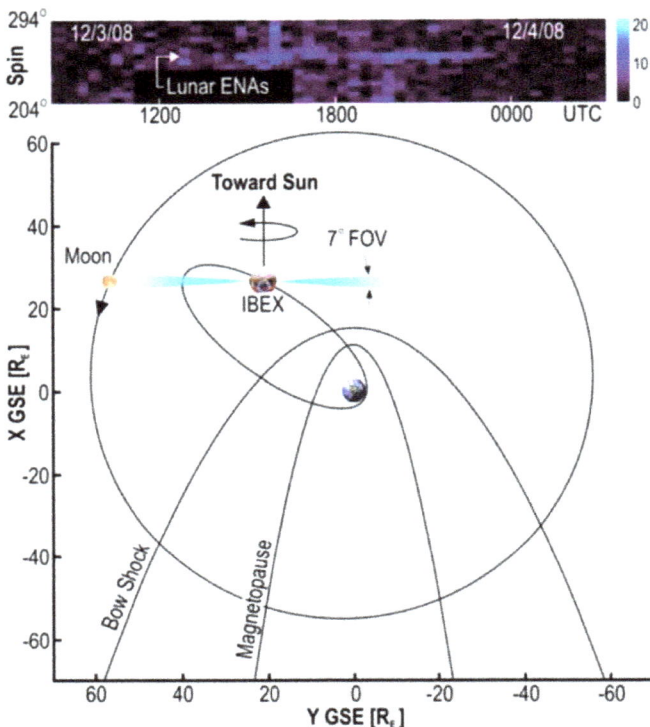

Fig. 6.25. Detection of lunar ENAs. Bottom: Lunar ENAs show up when the Moon is at a right angle relative to the *IBEX* spin axis and the matching spin angle for the *IBEX*-Hi FOV. Top: Color-coded ENA count rate as a function of time and spin angle [97]. (© AGU. Reproduced with permission).

for the lunar equatorial region where the SW impacts the surface at $\approx 90°$ [101]. A detailed study of ten close lunar encounters with *IBEX* showed that the albedo decreases approximately linearly with the SW energy [102]. As is evident from these examples, ENA diagnostics provides essential information about backscattering and sputtering of ions from solid surfaces, which can provide remote diagnostics of the surface structure, composition, erosion rate, and history [95].

6.3.2. *Smaller bodies and dust*

In our discussion of the Moon and Mercury, we have shown that airless and $\vec{B}$-less objects in the solar system are subject

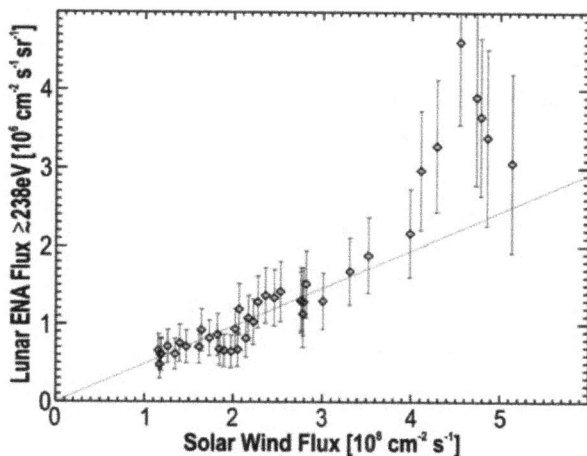

Fig. 6.26. Total lunar ENA flux as a function of SW flux, as observed with *ACE* SWEPAM, time-shifted for the distance from *ACE* at L1 to the Moon [97]. (© AGU. Reproduced with permission).

to continuous SW or magnetospheric ion bombardment, yielding two categories of ENAs: backscattered SW and sputtered surface atoms. This inference can be readily extended to asteroids and dust grains from interplanetary and interstellar sources, thus making ENA observations potentially potent diagnostics for this debris in the solar system. The charged particles emerge as PUIs, which, for dust in the SW, were identified as *inner-source* PUIs [103, 104].

The interaction between dust and SW was implicated early on as the origin of this ion population. However, essential questions remained, such as the abundance of these ions relative to the amount of observed dust [105] and the dominant processes by which the SW releases ions and neutral atoms in its interaction with dust grains [106]. The combination of observations with comprehensive modeling of likely interaction processes and subsequent PUI evolution has recently provided new constraints, but critical questions remain unanswered [107]. With its capability to determine the angular distribution of the neutral atoms that emerge from their source location, lost in the PUI distributions, ENA diagnostics may become a complementary contributor of information to resolve these questions. Since various interaction processes predict different Q ratios among the particles [108], combined ENA and PUI observations would likely provide decisive insight into these processes.

This chapter has presented the ongoing efforts with ENA diagnostics on a large variety of sources in our solar system, all intimately related to the effects of the dynamic SW. We have covered the intrinsic magnetospheres of Earth and the giant planets, induced magnetospheres of smaller planets, airless and $\vec{B}$-field free bodies like the Moon, and micron-size IP dust. We move now to the ENA diagnostics at the final frontier of the SW, *i.e.*, its interaction with the ISM.

References

1. Russell, C. T. (1971). Geophysical coordinate transformations, *Cosm. Eletrodyn.*, **2**, 184–196.
2. Laundal, K. M., & Richmond, A. D. (2017). Magnetic coordinate systems, *Space Sci. Rev.*, **206**, 27–59. Doi: 10.1007/s11214-016-0275-y.
3. Chapman, S., & Ferraro, V. C. A. (1940). The theory of the first phase of a geomagnetic storm, *Terrest. Mag. Atmo. Elect.*, **45**(3), 245–268.
4. Dessler, A. J., & Parker, E. N. (1959). Hydromagnetic theory of geomagnetic storms, *J. Geophys. Res.*, **64**(12), 2230–2252.
5. Dungey, J. W. (1961). Interplanetary magnetic field and the auroral zones, *Phys. Rev. Lett.*, **6**(2), 47–48.
6. Axford, W. I., & Hines, C. O. (1961). A unifying theory of high-latitude geophysical phenomena and geomagnetic storms, *Canad. J. Geophys.*, **39**(10), 1433–1464.
7. Raeder, J. (2003). Global magnetohydrodynamics — a tutorial, in: Space plasma simulation, J. Büchner, C. Dum, M. Scholer eds., *Lecture Notes in Physics*, **615**, 212–246.
8. Hesse, M., Aunai, N., Birn, J. *et al.* (2016). Theory and modeling for the magnetospheric multiscale mission, *Space Sci. Rev.*, **199**, 577–630. Doi: 10.1007/s11214-014-0078-y.
9. Barabash, S., C:son Brandt, P., Norberg, O. *et al.* (1997). Energetic atom imaging by the Astrid micro satellite, *Adv. Space Res.*, **20**(4/5), 1055–1060.
10. C:son Brandt, P., Barabash, S., Norberg, O. *et al.* (1997). ENA imaging from the Swedish micro satellite Astrid during the magnetic storm of 8 February 1995, *Adv. Space Res.*, **20** (4–5), 1,061–1,066.

11. Pollock, C. J., C:Son-Brandt, P. C., Burch, J. l. *et al.* (2003). The role and contributions of energetic neutral atom (ENA) imaging in magnetospheric substorm research, *Space Sci. Rev.*, **109**(1), 155–182.

12. Goldstein, J., & McComas, D. J. (2018). The big picture: Imaging of the global geospace by the TWINS mission, *Rev. Geophys.*, **56**(1), 251–277.

13. Collier, M. R., Moore, T. E., Fok, M.-C. *et al.* (2005). Low-energy neutral atom signatures of magnetopause motion in response to southward B_z, *J. Geophys. Res.*, **110**, A02102. Doi: 10.1029/2004JA010626.

14. Fuselier, S. A., Funsten, H. O., Heirtzler, D. *et al.* (2010). Energetic neutral atoms from the Earth's subsolar magnetopause, *Geophys. Res. Lett.*, **37**(13), L1310. Doi: 10.1029/2010GL044140.

15. Connor, H. K., & Carter, J. A. (2019). Exospheric neutral hydrogen density at the nominal 10 RE subsolar point deduced from XMM-Newton x-ray observations, *J. Geophys. Res.: Space Phys.*, **124**(3), 1612–1624.

16. Moore, T. E., Chornay, D. J., Collier, M. R. *et al.* (2000). The low-energy neutral atom imager for IMAGE, *Space Sci. Rev.*, **91**, 155–195.

17. Hosokawa, K., Taguchi, S., Suzuki, S. *et al.* (2008). Estimation of magnetopause motion from low-energy neutral atom emission, *J. Geophys. Res.*, **113**, A10205. Doi: 10.1029/2008JA013124.

18. McComas, D. J., Buzulukova, N., Connors, M. G. *et al.* (2012). Two wide-angle imaging neutral-atom spectrometer and interstellar boundary explorer energetic neutral atom imaging of the April 5 2010 substorm, *J. Geophys. Res.*, **117**, A03225.

19. Ogasawara, K., Angelopoulos, V., Dayeh, M. A. *et al.* (2013). Characterizing the dayside magnetosheath using energetic neutral atoms: IBEX and THEMIS observations, *J. Geophys. Res.: Space Physics*, **118**(6), 3126–3137. Doi: 10.1002/jgra.50353.

20. Ogasawara, K., Dayeh, M. A., Funsten, H. O. *et al.* (2015). Interplanetary magnetic field dependence of the suprathermal energetic neutral atoms originated in subsolar magnetopause, *J. Geophys. Res.: Space Physics*, **120**(2), 964–972.

21. Collier, M. R., Moore, T. E., Ogilvie, K. W. *et al.* (2001a). Observations of neutral atoms from the solar wind, *J. Geophys. Res.*, **106**(A11), 24,893–24,906.

22. Moore, T. E., Collier, M. R., Fok. M.-C. *et al.* (2003). Heliosphere-geosphere interactions using low energy neutral atom imaging, *Space Sci. Rev.*, **109**(1), 351–371.

23. Taguchi, S., Chen, S.-H., Collier, M. R. *et al.* (2005). Monitoring the high-altitude cusp with the Lo Energy Neutral Atom imager: Simultaneous observations, *J. Geophys. Res.*, **110**(A12), A12204. Doi: 10.1029/2005JA011075.

24. Petrinec, S. M., Dayeh, M. A., Funsten, H. O. *et al.* (2011). Neutral atom imaging of the magnetospheric cusps, *J. Geophys. Res.*, **116**(A7), A07203. Doi: 10.1029/2010JA016357.

25. Trattner, K. J., Mulcock, J. S., Petrinec, S. M., & Fuselier, S. A. (2007). Probing the boundary between antiparallel and component reconnection during southward interplanetary magnetic field conditions, *J. Geophys. Res.*, **112**(A8), A08210. Doi: 10.1029/2007JA012270.

26. Hones, E. W. Jr., Asbridge, J. R., Bame, S. J., & Strong, I. B. (1967). Outward flow of plasma in the magnetotail following geomagnetic substorms, *J. Geophys. Res.*, **72**(23), 5879–5892. Doi: 10.1029/JZ072i023p05879.

27. Fairfield, D. H., & Ness, N. F. (1970). Configuration of the geomagnetic tail during substorms, *J. Geophys. Res.*, **75**(34), 7032–7047. Doi: 10.1029/JA075i034p07032.

28. Escoubet, C. P., Schmidt, R., & Goldstein, M. L. (1997). CLUSTER-science and mission overview, *Space Sci. Rev.*, **79**, 11–32. Doi: 10.1023/A:1004923124586.

29. Burch, J. L., Moore, T. E., Torbert, R. B., & Giles, B. L. (2016a). Magnetospheric multiscale overview and science objectives, *Space Sci. Rev.*, **199**(1–4), 5–21. Doi: 10.1007/s11214-015-0164-9.

30. McComas, D. J., Dayeh, M. A., Funsten, H. O. *et al.* (2011). First IBEX observations of the terrestrial plasma sheet and a possible disconnection event, *J. Geophys. Res.*, **116**, A02211. Doi: 10.1029/2010JA016138.

31. Dayeh, M. A., Fuselier, S. A., Funsten, H. O. *et al.* (2015). Shape of the terrestrial plasma sheet in the near-Earth magnetospheric tail as imaged by the Interstellar Boundary Explorer (IBEX), *Geophys. Res. Lett.*, **42**(7), 2115–2122.

32. Fuselier S. A., Dayeh, M. A., Livadiotis, G. *et al.* (2015). Imaging the development of the cold dense plasma sheet, *Geophys. Res. Lett.*, **42**(19), 7867–7873.

33. Terasawa, T., Fujimoto, M., Mukai, T. *et al.* (1997). Solar wind control of density and temperature in the near-Earth plasma sheet WIND/GEOTAIL collaboration, *Geophys. Res. Lett.*, **24**(8), 935–938.

34. Lennartson, O. W., & Shelley, E. G. (1986). Survey of 0.1- to 16-keV/e plasma sheet ion composition, *J. Geophys. Res.*, **91**(A3), 3061–3076. Doi: 10.1029/JA091iA03p03061.

35. Øieroset, M., Raeder, J., Phan, T. D. *et al.* (2005). Global cooling and densification of the plasma sheet during an extended period of purely northward IMF on October 22–24, 2003, *Geophys. Res. Lett.*, **32**, L12S07. Doi: 10.1029/2004GL021523.

36. Fuselier, S. A., Petrinec, S. M., Trattner, K. J., & Lavraud, B. (2014). Magnetic field topology for northward IMF reconnection: Ion observations, *J. Geophys. Res.: Space Physics*, **119**(11), 9051–9071. Doi: 10.1002/2014JA020351.

37. Mitchell, D. G., Hsieh, K. C., Curtis, C. C., Hamilton, D. C., & Voss, H. D. (2001). Imaging two geomagnetic storms in energetic neutral atoms, *Geophys. Res. Lett.*, **28**(6), 1151–1154.

38. Twomey, S. (1977). *Introduction to the Mathematics of Inversion in Remote Sensing and Indirect Measurements* (Elsevier Sci., New York).

39. Wazwaz, A. M. (2011). Abels' integral equation and singular integral equations. In: *Linear and Nonlinear Integral Equations* (Springer, Berlin, Heidelberg). https://doi.org/10.1007/978-3-642-21449-3_7.

40. Roelof, E. C. (1997). Energetic neutral atom imaging of magnetospheric ions from high and low-altitude spacecraft, *Adv. Space Res.*, **20**(3), 341–350. Doi: 10.1016/S0273-1177(97)00689-3.

41. Tsyganenko, N. A. (1989). A magnetospheric magnetic field model with a warped tail current sheet, *Planet. Space Sci.*, **37**(1), 5–20. Doi: 10.1016/0032-0633(89)90066-4.

42. Perez, J. D., Fok, M.-C., & Moore, T. E. (2000). Deconvolution of energetic neutral atom images of the Earth's magnetosphere, *Space Sci. Rev.*, **91**, 421–436. Doi: 10.1023/A:1005277307611.

43. Roelof, E. C., & Skinner, A. J. (2000). Extraction of distributions from magnetospheric ENA and EUV images, *Space Sci. Rev.*, **91**, 437–459. Doi: 10.1007/978-94-011-4233-5_15.

44. DeMajistre, R., Roelof, E. C., C:son Brandt, P., & Mitchell, D. G. (2004). Retrieval of global magnetospheric ion distributions from high-energy neutral atom measurements made by the IMAGE/HENA instrument, *J. Geophys. Res.*, **109**(A4), A04214. Doi: 10.1029/2003JA010322.

45. Vallat, C., Dandouras, I., C:son Brandt, P. *et al.*, (2004). First comparisons of ion measurements in the inner magnetosphere with energetic atom magnetospheric image inversions: Cluster-CIS and IMAGE-HENA observations, *J. Geophys. Res.*, **109**, A04213. Doi: 10.1029/2003JA010224.

46. Bazell, D., Roelof, E. C., Sotirelis, T. *et al.* (2010). Comparison of TWINS images of low-altitude emission of energetic neutral atoms with DMSP precipitating ion fluxes, *J. Geophys. Res.*, **115**(A10), A10204. Doi: 10.1029/2010JA015644.

47. Roelof, E. C. (1987). Energetic neutral atom image of a storm-time ring current, *Geophys. Res. Lett.*, **14**(6), 652–655.

48. Fok, M. C., Buzulukova, N. Y., Chen, S.-H. *et al.* (2014). The comprehensive inner magnetosphere-ionosphere model, *J. Geophys. Res., Space Phys.*, **119**(9), 7522–7540. Doi: 10.1002/2014JA020239.

49. Fok, M.-C., Wolf, R. A., Spiro, R. W., & Moore, T. E. (2001b). Comprehensive computational model of Earth's ring current, *J. Geophys. Res.*, **106**(A5), 8417–8424.

50. Vasyliunas, V. M. (1970). Mathematical models of magnetospheric convection and its coupling to the ionosphere, in *Particles and Fields in the Magnetosphere*, ed. McCormac, B., pp. 60–71 (D. Reidel, Norwell, Mass.)

51. Weimer, D. R. (2001). An improved model of ionospheric electric potentials including substorm perturbations and applications to the geospacer environment modeling November 24, 1996, event, *J. Geophys. Res.*, **106** (A1), 407–416. Doi: 10.1029/2000JA000604.

52. Perez, J. D., Edmond, J., Hill, S. *et al.* (2018). Dynamics of a geomagnetic storm on 7–10 September 2015 as observed by TWINS and simulated CIMI, *Ann. Geophys.*, **36**, 1439–1456. Doi: 10.5194/angeo-36-1439-2018.

53. Parker, E. N. (2000). Newton, Maxwell, and magnetospheric physics, *Magnetospheric Current Systems*, eds.: Ohtani, S.-i., Fujii, R., Hesse, M., & Lysak, R. L., *AGU Monograph Series* **118**, 1–10.

54. Kirsch, E., Krimigis, S. M., Kohl, J. W., & Keath, E. P. (1981a). Upper limits for x-ray and energetic neutral particle emission from Jupiter: Voyager 1 results, *Geophys. Res. Lett.*, **8**(2), 169–172.

55. Kirsch, E., Krimigis, S. M., Ip, W.-H., & Gloeckler, G. (1981b). X-ray and energetic neutral particle emission from Saturn's magnetosphere, *Nature*, **292**, 718– 721.

56. Mitchell, D. G., Paranicas C. P., Mauk, B. H., Roelof, E. C., & Krimigis, S. M. (2004). Energetic neutral atoms from Jupiter measured with the Cassini magnetospheric imaging instrument: Time dependence and composition, *J. Geophys. Res.*, **109**, A09S11. Doi: 10.1029/2003JA010120.

57. Special issue (2009). *Saturn from Cassini-Huygens*, eds. Dougherty, M., Esposito, L. and Krimigis, S., (Springer Verlag, Berlin).

58. Krimigis, S. M., Mitchell, D. G., Hamilton, D. C. *et al.* (2005). Dynamics of Saturn's magnetosphere from MIMI during Cassini's orbit insertion, *Science*, **307**(5713), 1270–1273. Doi: 10.1126/science.1105978.

59. Giampieri, G., Dougherty, M. K., Smith, E. J., & Russell, C. T. (2006). A regular period for Saturn's magnetic field that may track its internal rotation, *Nature*, **441**, 62–64.

60. Gurnett, D. A., Persoon, A. M., Kurth, J. B. *et al.* (2007). The variable rotation period of the inner regions of Saturn's plasma disk, *Science*, **316**(5823), 442–445. Doi: 10.1126/science.1138562.

61. Carbary, J. F., Mitchell, D. G., Krimigis, S. M., Hamilton, D. C., & Krupp, N. (2007). Charged particle periodicities in Saturn's outer magnetosphere, *J. Geophys. Res.*, **112**(A11), 6246. Doi: 10.1029/2007JA012351.

62. Kurth, W. S., Averkamp, T. F., Gurnett, D. A., Groene, J. B., & Lecareux, A. (2008). An update to a Saturnian longitude system based on kilometric radio emission, *J. Geophys. Res.*, **113**(A5), A05222. Doi:10.1029/2007JA012861.

63. Carbary, J. F., Mitchell, D. G., Brandt, P., Parannicas, C., & Krimigis, S. M. (2008). ENA periodicity at Saturn, *Geophys. Res. Lett.*, **35**, L07102. Doi: 10.1029/2008GL033230.

64. Mitchell, D. G., Krimigis, S. M., Paranicas, C. *et al.* (2009a). Recurrent energization of plasma in the midnight-to-dawn quadrant of Saturn's magnetosphere, and its relationship to auroral UV and radio emissions, *Planet. Space Sci.*, **57**, 1732–1742.

65. Brandt, P. C., Khurana, K. K., Mitchell, D. G. *et al.* (2010). Saturn's periodic magnetic field perturbations are caused by a rotting partial ring current, *Geophys. Res. Lett.*, **37**, L22103. Doi: 10.1029/2010GL045285.

66. Vasyliunas, V. M. (1984). Fundamentals of current descriptions, *Magnetospheric Currents*, ed: Potemra, T. A., *AGU Monograph Series* **28**, 63–66.

67. Roelof, E. C., Brandt, P. C., & Mitchell, D. G. (2004). Derivation of currents and diamagnetic effects from global plasma pressure distributions obtained by IMAGE/HENA, *Adv. Space Res.*, **33**(5), 747–752. Doi: 10.1016/S0273-1177(03)00638-0.

68. Mitchell, D. G., Carbary, J. F., Cowley, S. W. H., Hill, T. W., & Zarka, P. (2009b). The dynamics of Saturn's magnetosphere, Chap. 10 in *Saturn from Cassini-Huygens*, eds. Dougherty, M., Esposito, L. & Krimigis, S., Springer Verlag, Berlin.

69. Mars Express/Venus Express (2008). *Planet and Space Science* **56**(6), 779–880.

70. Special issue on MEX results (2006). *ICARUS*, **182**(2).

71. Futaana, Y., Barabash, S., Grigoriev, A. *et al.* (2006a). First ENA observations at Mars: Subsolar ENA jet, *ICARUS*, **182**(2), 413–423.

72. Grigoriev, A., Futaana, Y., Barabash, S., & Fedorov, A. (2006). Observations of the Martian subsolar ENA jet oscillations, *Space Sci. Rev.*, **126**(1–4), 299–313.

73. Wang, X.-D., Alho, M., Jarvinen, R., Kallio, E., Barabash, S., & Futaana, Y. (2016). Emission of hydrogen energetic neutral atoms from the Martian subsolar magnetosheath, *J. Geophys. Res. Space Phys.*, **121**(1), 190–204. Doi: 10.1002/2015JA021653.

74. Futaana, Y., Barabash, S., Grigoriev, A. *et al.* (2006b). First ENA observations at Mars: ENA emission from the Martian upper atmosphere, *ICARUS*, **182**(2), 424–430.

75. Wang, X.-D., Alho, M., Jarvinen, R., Kallio, E., Barabash, S., & Futaana, Y. (2018). Precipitation of hydrogen energetic neutral atoms at the upper atmosphere of Mars, *J. Geophys. Res. Space Phys.*, **123**(10), 8730–8748. Doi: 10.1029/2018JA025188.

76. Gunell, H., Brinkfeldt, K., Holmström, M. *et al.* (2006). First ENA observations at Mars: Charge exchange ENAs produced in the magnetosheath, *ICARUS*, **182**(2), 431–438.

77. Brinkfeldt, K., Gunell, H., C:son Brandt, P. *et al.* (2006). Frist ENA observations at Mars: Solar-wind ENAs on the nightside, *ICARUS*, **182**(2), 439–447.

78. Galli, A., Wurz, P., Barabash, S. *et al.* (2006). Energetic hydrogen and oxygen atoms observed on the nightside of Mars, *Space Sci. Rev.*, **126**(1–4), 267–297.

79. Galli, A., Wurz, P., Kallio, E. *et al.* (2008a). Tailward flow of energetic neutral atoms observed at Mars, *J. Geophys. Res.*, **113**(E12), E12012. Doi: 10.1029/2008JE003139.

80. Milillo, A., Mura, A., Orsini, S. *et al.* (2009). Statistical analysis of the observations of the MEX/ASPERA-3 NP1 in the shadow, *Planet. Space Sci.*, **57**(8–9), 1000–1007.

81. Kallio, E., Barabash, S., Brinkfeldt, K. *et al.* (2006). Energetic neutral atoms (ENA) at Mars: Properties of the hydrogen atoms produced upstream of the martian bow shock and implications for ENA sounding technique around non-magnetized planets, *ICARUS*, **182**(2), 448–463.

82. Acuña, M. H. (1998). Magnetic and plasma observations at Mars: Initial results of the Mars Global Surveyor mission, *Science*, **279**(5357), 1679–1680.

83. Lillil, R. J., Purucker, M. E., Halekas, J. S., Louzada, K. L. *et al.* (2010). Study of impact demagnetization at Mars using Monte Carlo modeling and multiple altitude data, *J. Geophys. Res: Plantes*, **115** (E7), 2401–2428. Doi: 10.1029/2009JE003556.

84. Wang, X.-D., Barabash, S., Futaana, y., Grigoriev, A., & Wurz, P. (2014). Influence of Martian crustal magnetic anomalies on the emission of energetic neutral hydrogen atoms, *J. Geophys. Res., Space Phys.*, **119**, 8600–8609. Doi: 10.1002/2014JA020307.

85. Lundin, R., Winningham, D., Barabash, S. *et al.* (2006). Plasma acceleration above Martian magnetic anomalies, *Science*, **311**(5763), 980–983.

86. Special issue on VEX results (2008). *Planet. Space Sci.*, **56**(6).

87. Galli, A., Wurz, P., Bochsler, P., Barabash, S., & Grigoriev, A. (2008b). First observation of energetic neutral atoms in the Venus environment, *Planet. Space Sci.*, **56**(6), 807–811. Doi: 10.1016/j.pss. 2007.12.011.

88. Galli, A., Fok, M.-C., Wurz, P. *et al.* (2008c). Tailward flow of energetic neutral atoms observed at Venus, *J. Geophys. Res.*, **113**(E9), E00B15. Doi: 10.1029/2008JE003096.

89. Lundin, R., Barabash, S., Futaana, Y., Holmström, M., Sauvaud, J.-A., & Fedorov, A. (2014). Solar wind-driven thermospheric winds over the Venus north polar region, *Geophys. Res. Lett.*, **41**(13), 4413–4419. Doi: 10.1002/2014GL060605.

90. Orsini, S., Livi, S., Lichtenegger, H. *et al.* (2021). SERENA: Particle instrument suite for determining the Sun-Mercury interaction from BepiColombo, *Space Sci. Rev.*, **217**(11).

91. Mangano, V., Massetti, S., Milillo, A., Mura, A., Orsini, S., & Leblanc, F. (2013). Dynamical evolution of sodium anisotropy in the exosphere of Mercury, *Planet. Space Sci.*, **82–83**, 1–10. Doi: 10.1016/j.pss.2013.03.002.

92. Vervack, Jr., R. J., Killen, R. M., McClintock, W. E. *et al.* (2016). New discoveries from MESSENGER and insights into Mercury's exosphere, *Geophys. Res. Lett.*, **43**(22), 11, 555–11551. Doi: 10.1002/2016GL071284.

93. Massetti, S., Mangano, V., Milillo, A., Mura, A., Orsini, S., & Plainaki, C. (2017). Short-term observations of double-peaked Na emission from Mercury's exosphere, *Geophys. Res. Lett.*, **44**(7), 2970–2977. Doi: 10.1002/2017GL073090.

94. Wurz, P., & Lammer, H. (2003). Monte-Carlo simulation of Mercury's exosphere, *ICARUS*, **164**(1), 1–13.

95. Milillo, A., Orsini, S., Hsieh, K. C. *et al.* (2011). Observing planets and small bodies in sputtered high energy atom (SHEA) fluxes, *J. Geoph. Res.*, **116**, A07229. Doi: 10.1029/2011JA016530.

96. Hilchenbach, M., Hovestadt, D., Klecker, B., & Möbius, E. (1992). Detection of singly ionized energetic lunar pick-up ions upstream

of the earth's bow shock, *Solar Wind VII*, eds.: E. Marsch and R. Schwenn, COSPAR Colloquia Series, **3**, 349–356.

97. McComas, D. J., Allegrini, F., Bochsler, P. *et al.* (2009). Lunar backscatter and neutralization of the solar wind: First observations of neutral atoms from the Moon, *Geophys. Res. Lett.*, **36**(12), L12104. Doi: 10.1029/2009GL038794.

98. Saito, Y., Yokota, S., Tanaka, T. *et al.* (2008). Solar wind proton reflection at the lunar surface: Low energy ion measurement by MAP-PACE onboard SELENE (KAGUYA), *Geophys. Res. Lett.*, **35**, L24205. Doi: 10.1029/2008GL036077.

99. Rodriguez, M. D. F., Saul, L., Wurz, P. *et al.* (2012). IBEX-Lo observations of energetic neutral hydrogen atoms originating from the lunar surface, *Planet. Space Sci.*, **60**, 297–303.

100. Saul, L., Wurz, P., Vorburger, A. *et al.* (2013). Solar wind reflection from the lunar surface: The view from far and near, *Planet. Space Sci.*, **84**, 1–4.

101. Wieser, M., Barabash, S., Futaana, Y. *et al.* (2009). Extremely high reflection of solar wind protons as neutral hydrogen atoms from regolith in space, *Planet. Space Sci.*, **57**(14–15), 2132–2134. Doi: 10.1016/j.pss.2009.09.012.

102. Funsten, H. O., Allegrini, F., Bochser, P. A. *et al.* (2013). Reflection of solar wind hydrogen from the lunar surface, *J. Geophys. Res.*, **118**(2) 292–305. Doi.org/10.1002/jgre.20055.

103. Geiss, J., Gloeckler, G., Mall, U. *et al.* (1994). Interstellar oxygen, nitrogen and neon in the heliosphere, *Astron. & Astrophys.*, **282**(3), 924–933.

104. Geiss, J., Gloeckler, G., Fisk, L. A., & von Steiger, R. (1995). C^+ pickup ions in the heliosphere and their origin, *J. Geophys. Res.*, **100**(A12), 23373–23377. Doi: 10.1029/95JA03051.

105. Schwadron, N. A., Geiss, J., Fisk, L. A. *et al.* (2000). Inner source distributions: Theoretical interpretation, implications, and evidence for inner source protons, *J. Geophys. Res.*, **105**(A4), 7465–7472. Doi: 10.1029/1999JA000225.

106. Allegrini, F., Schwadron, N. A., McComas, D. J., Gloeckler, G., & Geiss, J. (2005). Stability of the inner source pickup ions over the solar cycle, *J. Geophys. Res.*, **110**(A5), A05105. Doi: 10.1029/2004JA010847.

107. Quinn, P. R., Schwadron, N. A., Möbius, E., Taut, A., & Berger, L. (2018). Inner source C^+/O^+ pickup ions produced by solar wind recycling, neutralization, backscattering, sputtering, and sputtering-induced recycling, *Astrophys. J.*, **861**(2), 98–109.

108. Bochsler, P., Möbius, E., & Wimmer-Schweingruber, R. (2007). Inner source pickupions: Sputtering of small dust particles and charge exchange of solar wind ions, in *The Second Solar Orbiter Workshop*, 16–20 October 2006, Athens, Greece. eds. Marsch, E., Tsinganos, K., Marsden, R., and Conroy, L. (ESA SP-**641**, Noordwijk.)

Chapter 7

ENA Diagnostics of the Heliosphere-Interstellar Medium Interaction

"Discovery follows discovery, each both raising and answering questions, each ending a long search, and each providing the new instruments for a new search."

J. Robert Oppenheimer, 1904–1967

This chapter will push neutral-atom imaging diagnostics to their limits exploring the boundary regions of the heliosphere and the ISM just outside. Because the ISM in our neighborhood is a partially ionized plasma, the combined investigation of ANAs (ISN flow through the heliosphere) and ENAs (products of CX with the SW and the IS plasma) provides a global perspective of the heliosphere. It also probes the neutral component of the ISM, along with the physical processes involved in the interaction. We will focus on the arguably most important examples of using the same technique in ENA and ANA diagnostics in the heliosphere.

Section 7.1 will continue our brief description in Sec. 1.3 of the global structure of the heliosphere and its interactions with the ISM, schematically depicted in Fig. 7.1 along with the four distinct ANA and ENA populations. The methods described in this book make them accessible. Next, we will take on the *in-situ* diagnostics of the ISN flow through the solar system in Sec. 7.2, emphasizing direct imaging of the neutrals and extracting the parameters in the local IS environment. Section 7.3 will highlight the recent discovery of the

ENA Ribbon, which provides us with information about the ISMF. Section 7.4 will describe the *secondary ISN* population generated from the IS plasma flow in the OHS, and Sec. 7.5 tackles the GDF of ENAs and their sources in the IHS. Each section starts with observations, followed by proposed interpretations and diagnostic methods, and concludes with challenges and opportunities.

As Fig. 7.1 illustrates, the largely unimpeded access to the ISN flow, the Ribbon, the secondary IS neutral atoms, and the GDF of ENAs at 1 AU enables unique synergistic diagnostic opportunities for the heliospheric topology and plasma flows in the boundary regions. The analysis of the pristine ISN flow and the Ribbon ENAs provide a local window on the ISM conditions in the solar neighborhood, including the ISMF. In concert with global heliospheric modeling, the detailed analysis of these messengers provides multiple quantitative constraints on the shape and structure of the heliosphere and the physical processes involved in the interactions.

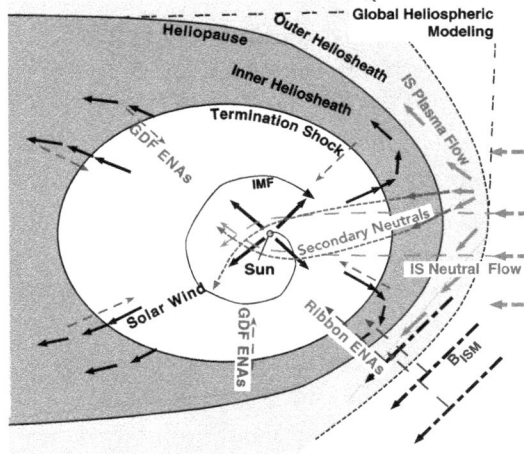

Fig. 7.1.   Schematic cut through the heliosphere in the plane subtended by the ISN flow and $\vec{B}_{ISM}$ vector, along with the associated IS (grey) and SW (black) plasma flows. Grey dashed lines of various styles indicate different neutral atom populations, such as the ISN flow (ANAs), Ribbon ENAs, secondary IS neutral atoms, and the GDF of ENAs visible with neutral imaging instruments at 1 AU. These populations originate in the ISM (ISN flow), ISM and OHS (Ribbon ENAs), OHS (secondary ENAs), and IHS (GDF of ENAs).

7.1. Getting Acquainted with the Heliosphere

The local galactic environment of the Sun consists of a warm, dilute, partially ionized, and quite structured IS gas cloud [1, 2]. Apparently, the Sun is now near the boundary of the *local interstellar cloud* (LIC), possibly with a significant gradient in the He ionization fraction [3–6]. The environment and structure of the LIC, including column densities and relative velocities, have been studied on scales of several parsecs through the absorption of stellar UV lines by the surrounding medium [7–11]. The LIC is just one of many warm IS clouds moving relative to the Sun [12], all embedded in a very hot and dilute plasma, the Local Bubble [13, 14].

A recent compilation showed that the conditions in the surrounding ISM have dramatically changed over time, with consequences for the history of the solar system [15]. Within the hot and dilute plasma of the Local Bubble, the heliosphere moves through a succession of warm clouds similar to the one it currently traverses. In fact, it may be at the edge of one cloud and in transition to another at this point [12, 16].

Two questions arise: 1) Are there local opportunities to diagnose the surrounding IS conditions, especially with ANAs and ENAs? 2) How do the IS conditions influence the state of the heliosphere?

Had one asked the astronomers before the 1960s whether one could "see" the influence of the ISM inside the solar system, the answer would have been, "No, we are likely living inside a Strömgren Sphere." [17] It is a void of the ISN gas ionized by the Sun's UV up to a radius of ≈ 700 AU and swept away by the SW. In 1968, however, H.-J. Fahr realized that the Sun is moving too fast through the ISM to allow settling to the required ionization equilibrium in $< 10^6$ years [18], during which time the Sun would have moved by ≈ 5 million AU. Instead of a static equilibrium, the ISN flow establishes a dynamic balance, with H depleted to $1/e$ at around 3 AU and He at 0.3 AU during solar minimum [19]. The combined effect of the Sun's motion, ionization, and gravitation (for H, also radiation pressure, Sec. 1.2.1) leads to a characteristic spatial distribution and pattern of the ISN species in the inner solar system, which provides a unique and crucial sample for detailed diagnostics of the LIC conditions.

Deep inside the heliosphere, the ISN flow has other consequences, *e.g.*, the generation of PUIs [20, 21] and anomalous cosmic rays [22, 23] along with the slowdown of the SW [24, 25]. Evidently, the inventory of ISN gas, its spatial distribution, and its products in the inner heliosphere change substantially with external conditions [26], as does the filtering at the interface [27, 28].

Due to the Sun's motion, the pervading presence of ISN gas throughout the heliosphere and its CX with the ions generate ENAs and PUIs, both important daughter populations. Consequently, CX intimately links the ISN flow (ANAs that surround the heliosphere) to the IS plasma forced to flow around the heliosphere with the resulting PUIs and ENAs.

7.2. Interstellar Neutral Wind through the Heliosphere

The Sun's motion relative to the surrounding ISM ushers the ISN gas into the inner heliosphere as an *interstellar wind*. Solar gravitation and ionization shape the ISN flow patterns and density structures in the inner heliosphere for each species. The resulting spatial distributions include a cavity close to the Sun and, except for H due to radiation pressure (Sec. 1.2.1), a gravitational focusing cone (Fig. 4.1) on the downwind side of the flow.

Modeling the ISN flow to obtain the velocity distribution of the atoms at the observer and their spatial distribution in the inner heliosphere largely follows an approach used to describe the accretion of gaseous matter onto a star [29]. The trajectories of atoms in the Sun's gravitational field start as a Maxwell-Boltzmann distribution with the characteristic speed $v_T = \sqrt{2k_B T_{ISN}}$ and bulk velocity $\vec{v}_{ISN\infty}$ relative to the Sun outside the heliosphere. Without ionization loss, their phase space density f^* is conserved according to Liouville's Theorem [30–33].

$$f^*(\vec{v}(\vec{r})) = f_\infty(\vec{v}_\infty) = n_{0\infty}(\sqrt{\pi}v_T)^{-3}\exp\left[-\frac{(\vec{v}_\infty - \vec{v}_{ISN\infty})^2}{v_T^2}\right] \quad (7.1)$$

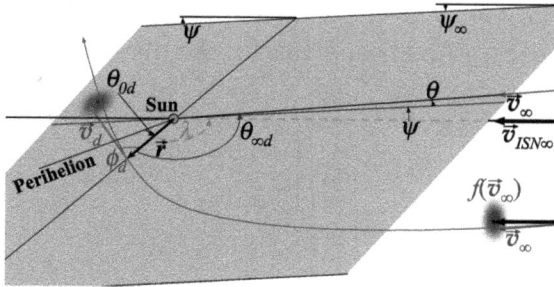

Fig. 7.2. Schematic view of the ISN flow velocity distribution $f(\vec{v}_\infty)$ outside the heliosphere, mapped to the observer location $\vec{r}$ along a sample trajectory, with velocity vectors and angles based on Lee *et al.* [2012] [33]. The sample trajectory starts with a velocity offset relative to $\vec{v}_{ISN\infty}$ in angle θ within and ψ out of the trajectory plane (grey-shaded).

We use a polar coordinate system aligned with $\vec{v}_{ISN\infty}$ and the vector $\vec{r}$ from the Sun to the observer location (Fig. 7.2) [33].

Keplerian trajectories in the Sun's gravitational field map any velocity $\vec{v}_\infty(v_\infty, \phi_\infty, \psi_\infty)$ of the ISN distribution outside the heliosphere to the observer at $\vec{r}$ with the corresponding velocity $\vec{v}(\vec{r})$ given by (v, ϕ, ψ). ϕ is the polar and ψ the azimuthal angle, referenced to the ISN flow direction. We obtain $\psi_\infty = \psi$ because ψ is the tilt of the plane with the observed trajectories, relative to the $\vec{v}_{ISN\infty} \times \vec{r}$ plane. Shown is a direct trajectory d, with the atom still heading toward its perihelion along angle $-\theta_{0d}$, which has swept out the angle $\theta_{\infty d}$ from infinity to $\vec{r}$. We can largely treat the ISN flow neglecting collisions because the typical mean free paths are of the order of or larger than the size of the heliosphere, turning this description into a Vlasov problem according to Eqs. (3.21a), (3.21b), and (3.22). The eccentricity ε of the resulting hyperbolic Keplerian trajectory to the observer at $\vec{r}$ is given by:

$$\varepsilon^2 = 1 + \frac{2L^2 E}{mk^2} > 1 \text{ and angle } \theta_\infty \text{ by } \theta_\infty = -\theta_0 + a\cos\left(\frac{-1}{\varepsilon}\right)$$

$$(7.2)$$

With vectors $\vec{v}_{ISN\infty}$, $\vec{v}_\infty$, and $\vec{v}$ in polar coordinates as defined in Fig. 7.2 and Eqs. (3.20) and (3.21), the distribution $f^*(\vec{v})$ at the observer location $\vec{r}$ is:

$$f^*(v, \phi, \psi) = n_{0\infty}(\sqrt{\pi}v_T)^{-3} \exp\left\{ \frac{-1}{v_T^2}[v_\infty^2 + v_{ISN\infty}^2 \right.$$

$$\left. - 2v_{ISN\infty}v_\infty(\cos(\theta_\infty)\cos(\lambda) - \sin(\theta_\infty)[\cos(\psi)\sin(\lambda)])] \right\}$$

(7.3)

We evaluate the equations in the trajectory plane, neglecting the arrival of the ISN flow at a small negative angle $\beta_{ISN\infty} \approx -5°$ relative to the ecliptic plane. We will come back to this issue in Sec. 7.2.2 in connection with a simplified treatment of the ISN flow observations by *IBEX*.

The total ionization rate is treated as a loss process along the trajectories and described by a survival probability S in Eqs. (3.24a) and (3.24b), invoking a total loss rate ν_{Ion}. Combining Eq. (7.3) with Eq. (3.24a) or (3.24b) yields the ISN flow distribution at the location $\vec{r}$:

$$f(\vec{r}, \vec{v}, t) = f^*(\vec{r}, \vec{v}) \cdot S(\vec{r}, \vec{v}, t) \qquad (7.4)$$

Equations (3.21a), (3.21b), (3.22), and (7.2)–(7.4) are the foundation for the analysis using any *in-situ* observation method to obtain information on the ISN flow through the inner heliosphere. It is straightforward to obtain either differential fluxes $\vec{j}_s(\vec{r}, \vec{v}, t)$ or the density $n_s(\vec{r})$, s indicating the species, for comparison with *in-situ* observations at any location $\vec{r}$ in the heliosphere.

7.2.1. *Overview of in-situ diagnostic methods*

Taking advantage of the theoretical framework laid out above, the *in-situ* observation of the ISN flow through the inner solar system started with backscattered solar Ly-α intensity sky maps [34, 35], which provided access to ISN H around the solar system. The first reasonable H bulk speed and temperature values were obtained using high-resolution Ly-α line profiles with *Copernicus* [36]. A substantial improvement came with H absorption cells [37]. (Sec. 1.3.1.)

Using the Sun's gravitational focusing of ISN He downwind in maps of the backscattered solar He I line at 58.4 nm [38] enables the determination of the bulk flow velocity vector, density, and temperature of He. The local density distributions $n_0(\vec{r})$, $\vec{v}$-integrals of $f(\vec{r}, \vec{v})$ in Eq. (7.4), result from comparing computed and observed scattered solar line intensities [31, 32].

The discovery of IS He PUIs at 1 AU [20] introduced the first *in-situ* particle observation of ISN gas. The local ionization rate $\nu_{Ion}(\vec{r})$, which controls the production of PUIs, connects the observed PUI flux with the spatial distribution of the ISN density $n_0(\vec{r})$, or the PUI parent population [39]. Observations out to 5.4 AU and access to H$^+$ and He^{2+} PUIs with *Ulysses* SWICS provided a more precise determination of the H (0.089 ± 0.022 cm^{-3}), and He (0.015 ± 0.0015 cm^{-3}) densities, referenced to the TS [40, 41]. PUIs are also the tool of choice to determine the abundance of minor species in the neutral component of the ISM, such as N, O and Ne in the LIC [42].

Direct observations of the ISN velocity distribution with ENA imaging instruments have become available first for He [43, 44]. *IBEX* expanded this ability to include the He, H, O, and Ne ISN [45–52]. As discussed in Sec. 5.2.1 and 5.3.3, the energy resolution of the ENA sensors used for ISN flow imaging is poor or lost almost entirely in the detection of He and Ne atoms *via* sputtering. However, the use of the Sun as a gravitational velocity spectrograph provides accurate observations of the angular distribution, referenced to fixed stars, which can lead to reliable ISN flow direction, speed, and temperature (Eqs. (3.21a), (3.21b), (3.22), and (7.4)).

Comparing the pros and cons of the *in-situ* ISN observation techniques, PUIs appear to provide the most accurate neutral densities and abundance ratios when extrapolated to the TS. Using He^{2+} PUIs, generated through resonant double charge exchange with SW He^{2+}, leads to a He density that only hinges on the CX cross-section and the PUI/SW He^{2+} ratio, measured with the same sensor [42]. The observation of H PUIs [41] and the SW slowdown due to mass loading with implanted IS H PUIs [25] provide complementary results for this more challenging case. However, obtaining the ISN flow vector and temperature from PUIs faces severe challenges from PUI transport [53, 54] and stark temporal variations [55, 56]. The flow longitude by itself can be determined more accurately from the symmetry axis of the PUI cut-off speed variation with ecliptic longitude, which depends on the vector sum of the local ISN flow and SW, with minimum PUI transport effects [57–59]. The He cone can also be an ENA source in the inner heliosphere and thus be diagnosed when it interacts with solar energetic ions, *e.g.*, those in a passing corotating interaction region [60].

UV backscattering observations provide consistency checks, along with an absolute speed determination *via* Doppler shift of the H and He resonance lines. Most importantly, they make up the longest historical record because they opened the first glimpse of the ISN.

In turn, neutral atom imaging provides the most direct access to the local ISN flow velocity distribution and thus the kinetic parameters of the ISM. Obtaining abundances and densities relies on the more challenging absolute calibration of ENA sensors, with substantially larger error bars.

7.2.2. *Connecting ISN flow observations with the ISM*

The following demonstrates how to extract the ISN flow parameters outside the heliosphere from neutral-atom images taken at observation points near the Sun. It is instructive to discuss this observation with a FOV pointing precisely at $90°$ from the Sun, which selects the portion of the ISN flow distribution at its perihelion, thus greatly simplifying the analysis. This situation occurs when the *IBEX* spin axis points precisely at the Sun [61] (Fig. 7.3). Along its orbit around the Sun, *IBEX* scans different parts of the distribution. This configuration occurs twice a year: in early February, ramming into the ISN flow, and in October, receding from it. Because the instrument sensitivity depends on E, the ram direction is highly favored (Fig. 4.2, right). Combining Eqs. (3.21a), (3.21b), and (3.22) relates the angle $\theta_{\infty 0}$ swept out from infinity to the observer at perihelion along the hyperbolic ISN flow trajectories to the ISN bulk

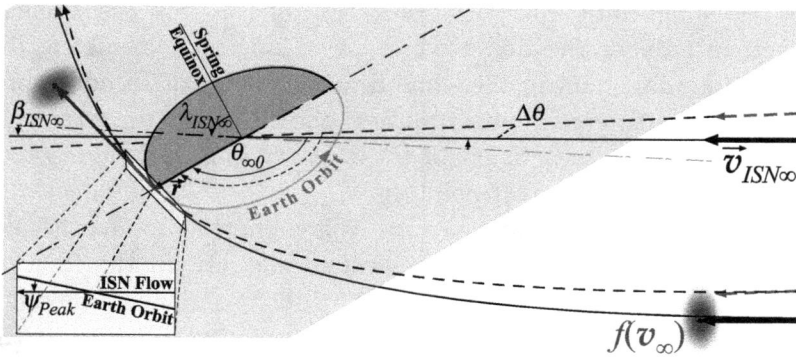

Fig. 7.3. An Earth orbiter observes two sample ISN bulk flow trajectories (solid and dashed) offset at infinity by $\Delta\theta$ in the trajectory plane at their perihelion. Both trajectories with different true anomalies $\theta_{\infty 0}$ or arrival directions $\lambda_{ISN\infty}$ and speeds at infinity $v_{ISN\infty}$ satisfy Eqs. (7.5) and (7.6). Moving with the Earth exactly into their local flow direction, the sensor observes them at 90° relative to the Sun (location A in Fig. 4.2). The inset shows the inclination angle ψ_{Peak} of the ISN flow at the observer. ψ_{Peak} depends on the flow latitude $\beta_{ISN\infty}$ relative to the ecliptic plane (darkened above the trajectory plane) and $\theta_{\infty 0}$ [65]. (© AAS. Reproduced with permission, adapted).

flow speed $v_{ISN\infty}$.

$$\frac{-1}{\cos\theta_{\infty 0}} = 1 + \frac{r \cdot v_{ISN\infty}^2}{GM_S} = 1 + \frac{v_{ISN\infty}^2}{v_E^2} \qquad (7.5)$$

v_E is the average orbital speed of the Earth. The longitude λ_{Peak} where the S/C intercepts the ISN bulk flow is identified by the peak of the flow distribution. $\theta_{\infty 0}$ relates the ISN flow direction $\lambda_{ISN\infty}$ and λ_{Peak} [33, 47, 62]:

$$\theta_{\infty 0} = \lambda_{ISN\infty} + 180° - \lambda_{Peak} \qquad (7.6)$$

Equations (7.5) and (7.6) establish a relation between $\lambda_{ISN\infty}$ and $v_{ISN\infty}$ that is solely controlled by the Sun's gravitation and known with the measurement uncertainty of λ_{Peak}.

Observationally, the maximum neutral-atom flux $f(\vec{v}) \cdot \vec{v}$ identifies the ISN bulk flow, while the bulk velocity vector marks the maximum ISN phase space density $f(\vec{v})$. Because the speed of the part of the distribution seen at perihelion increases with the observer longitude (trajectories with decreasing θ_∞), the maximum flow shifts to larger longitudes. Likewise, ionization of the ISN gas along its trajectory reduces the flux toward lower longitude, resulting in a shift of the maximum in the same direction [47, 63]. In addition to statistical measurement uncertainties, the resulting uncertainty in λ_{Peak} depends on the fidelity of these corrections and the absolute precision of the instrument pointing.

Because the ISN flow pattern is symmetric about the flow direction at infinity, it would be ideal to also obtain the ISN flow maximum at perihelion in the anti-ram direction, making maximum use of the symmetry. However, the energy of the ISN flow at that location falls below the *IBEX*-Lo detection threshold for both ISN He and O [64].

Equation (7.5) is exact for all angles in the ISN trajectory plane, but the ISN flow arrives with an inclination $\beta_{ISN\infty} \approx -5°$ relative to the ecliptic plane. *IBEX* determines this angle through the spin angle ψ. Per geometry, the angle ψ_{Peak}, at which the ISN flow arrives relative to the ecliptic plane at *IBEX* in the inertial frame, is related to $\beta_{ISN\infty}$ as:

$$\tan \psi_{Peak} = \frac{\tan \beta_{ISN\infty}}{|\sin \theta_{\infty 0}|} \tag{7.7}$$

While Eqs. (7.5) and (7.6) are frame-independent because the observations occur precisely in the ram direction, ψ_{Peak} must be transformed into the observer frame with Earth's orbital speed v_E [33]:

$$\frac{\sin \psi'_{Peak}}{\sqrt{v_{ISN\infty}^2 + v_E^2}} = \frac{\sin(\psi_{Peak} - \psi'_{Peak})}{v_E} \tag{7.8}$$

Because the transformation depends on $v_{ISN\infty}$, $\beta_{ISN\infty}$ also depends on the relation between $\lambda_{ISN\infty}$ and $v_{ISN\infty}$ in Eq. (7.5).

The fourth parameter deduced from the ISN flow distribution is the temperature T_{ISN} of the ISN gas surrounding the Sun. Figure 7.4 shows the flow distribution recorded as an angular distribution in the sky maps. The left panel shows that the angular width θ'_M is that of the Mach cone for the local flow distribution in the observer frame.

The right panel shows the latitudinal distribution, *i.e.*, perpendicular to the ecliptic plane, where the characteristic thermal speed

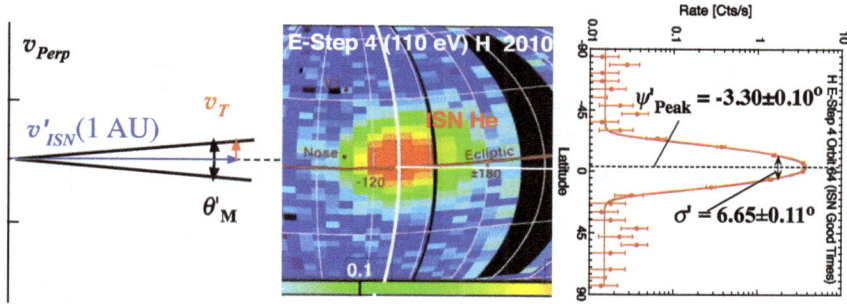

Fig. 7.4. Skymap image (center) of the He ISN flow intercepted by *IBEX* at 1 AU, along with a latitude scan of the observed count rates (right) at the peak longitude. The σ-width of a fit to a Gaussian is adopted as the angular width θ'_M of the Mach cone (left) in the observer frame of the ISN flow that arrives with the flow speed v'_{ISN} (1 AU) and the characteristic thermal speed v_T at the S/C. In line with Eq. (7.7) and after transformation into the observer frame, the peak for $\beta_{ISN\infty} \approx -5°$ is at about $-4°$ latitude.

v_T remains constant from the undisturbed ISM to the observer at 1 AU. Thus, the width of the Mach cone yields:

$$\theta'_M = \frac{v_T}{v'_{ISN}(1AU)}$$

with $\quad v_T = \sqrt{\dfrac{2k_B T_{ISN}}{m}} \quad$ and $\quad v'_{ISN}(1AU) = \sqrt{v^2_{ISN\infty} + 2v^2_E} + v_E$

$$(7.9)$$

where k_B is the Boltzmann constant and m the atomic mass. v_T is the characteristic thermal speed that corresponds to σ of the Maxwellian distribution with temperature T_{ISN}, not the mean thermal speed, which is:

$$v_{Th} = \sqrt{\frac{8k_B T_{ISN}}{\pi m}} \qquad (7.10)$$

Based on Eq. (7.9), it is evident that the deduced T_{ISN} depends on $v_{ISN\infty}$ and thus its knowledge. Together with Eqs. (7.5), (7.6), and (7.8), this result leads to a tight constraint on the ISN parameters in the form of a narrow 4-D parameter tube with substantially larger uncertainties along the tube when obtained from neutral-atom

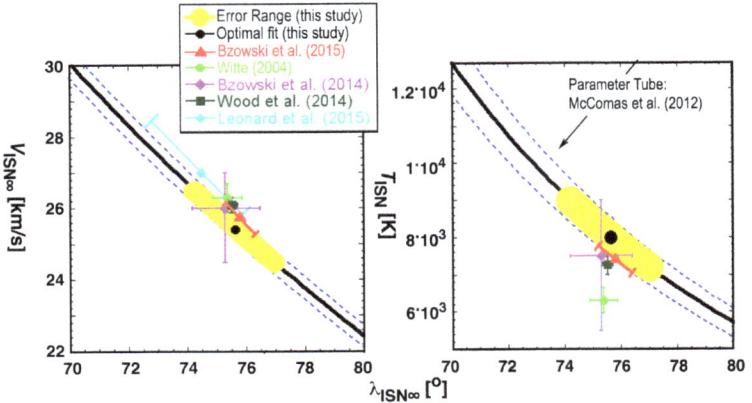

Fig. 7.5. ISN flow parameter tube for $v_{ISN\infty}(\lambda_{ISN\infty})$ and $T_{ISN}(\lambda_{ISN\infty})$, as obtained with *IBEX*, together with results from *Ulysses* GAS [67]. (© AAS. Reproduced with permission).

images with a limited range of observation points [47, 66]. This relation is shown in Fig. 7.5 for *IBEX* observations [67]. While a tight error bar is achieved perpendicular to the parameter tube, the uncertainties along the tube are substantially larger. If the *IBEX* observations solely returned λ_{Peak} (Eqs. (7.5) and (7.6)), ψ_{Peak} (Eqs. (7.7) and (7.8)), and θ'_M (Eq. (7.9)), they would only constrain the resulting parameters to the tube shown in Fig. 7.5, with no constraint along the tube at all, constituting a degeneracy for the ISN parameters.

However, *IBEX* scans the entire ISN flow distribution while moving with the Earth in ecliptic longitude. The sensor pointing drifts within 83° and 97° of the Earth-Sun line, providing access to parts of the ISN distribution with different arrival speeds and a distinct variation of ψ'_{peak} and θ'_M as a function of the observer longitude λ_{Obs}. The functional dependences $\psi'_{Peak}(\lambda_{Obs})$ and $\theta'_M(\lambda_{Obs})$ show variations with the ISN flow longitude $\lambda_{ISN\infty}$, which can constrain the result [63]. These constraints are also important drivers in the analysis of the complete *IBEX* ISN flow distributions using a χ^2-minimization in comparison with different models of the expected ISN flow at 1 AU [48, 67–70], typically with larger error bars along the tube.

The *Ulysses* GAS sensor (Sec. 5.2.1) has obtained images of the ISN flow over the three fast latitude scans of the S/C from almost above the south pole to the north pole of the Sun, where the relative speed of the S/C and the ISN flow exceeded the GAS detection threshold [44, 71]. The *Ulysses* orbit around the Sun is almost perpendicular to the ISN flow vector.

Therefore, GAS has observed the ISN flow for a wide range of trajectories whose planes intersect at the line of direct approach of the ISN to the Sun. The combination enables a method to reduce the uncertainty of the flow vector that is akin to triangulation [71]. Figure 7.6(a) shows the complement of ISN flow arrival directions in the S/C frame, as obtained from the maxima of the counts in the flow images, which forms a horseshoe-shaped arc in the sky. The remaining panels of Fig. 7.6 contain the model arc for the best-fit

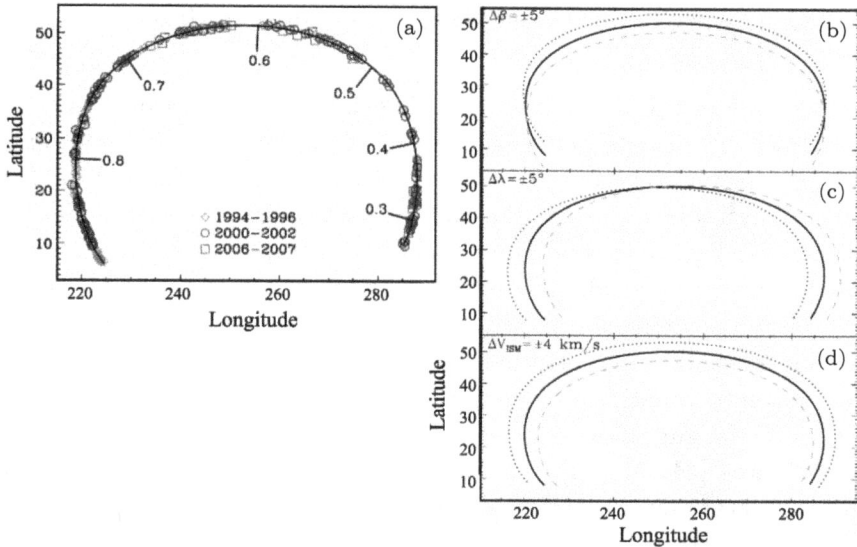

Fig. 7.6. Observed He ISN bulk flow locations in the sky with *Ulysses* GAS for three fast latitude scans (a), which arrange along a horseshoe-shaped arc. Best fit model arc, along with expected deviations for different ISN flow vectors by positive (dotted) or negative (dashed) increments, as shown in the upper left corner, in ecliptic latitude $\beta_{ISN\infty}$ (b), ecliptic longitude $\lambda_{ISN\infty}$ (c), or speed $v_{ISN\infty}$ (d) [71]. (© AAS. Reproduced with permission, adapted).

ISN flow vector along with the variation of a single parameter (corner $\beta_{ISN\infty}$ (b), $\lambda_{ISN\infty}$ (c), and $v_{ISN\infty}$ (d)) by either a positive (dashed) or negative (dotted) increment shown in the upper left. Clearly, a wider range of vantage points provides a stronger constraint on the correct ISN flow vector.

The range of observation points with *Ulysses*, spread in ecliptic latitude and longitude, has removed the degeneracy from the GAS observations of the ISN flow [71]. However, as can be seen from GAS data points in Fig. 7.5, the originally deduced temperature T_{ISN} [44] was substantially lower than that found with *IBEX* [72, 73] and there appears to remain a slight difference in T_{ISN} after the more recent analyses of *Ulysses* GAS and *IBEX* data. These issues point to observational challenges for a precision determination of the ISN flow parameters.

7.2.3. *Observational challenges and opportunities for the ISN flow*

To appreciate the uncertainties in determining the ISN flow parameters, let us discuss the key challenges in the *Ulysses* GAS and *IBEX* observations and their analysis. Involved are the knowledge and accuracy in pointing, the counting statistics, the sensitivity to various backgrounds, and the discovery of competing signal sources, which, in turn, provide new opportunities.

That the attitude stability and knowledge of the *Ulysses* S/C were less reliable than planned called for alternate means to establish the absolute pointing of the GAS sensor. Its natural sensitivity to stellar UV light allowed relating its pointing to the astronomical coordinates [44, 71, 74], using the very sensor that takes the ISN flow observations. Therefore, a small star sensor was implemented as part of the *IBEX*-Lo camera that observes the ISN flow. Mounted on the mechanical structure that establishes the boresight and pointing, mechanical and thermal tolerances are under the instrument control and thus minimized to within $\pm 0.05°$ [75]. In-flight verification revealed an overall *IBEX*-Lo pointing uncertainty relative to the stars and the spacecraft attitude of within $\pm 0.1°$ in longitude and latitude [76]. Including asymmetries in the *IBEX*-Lo point-spread function, the uncertainties are still within $\pm 0.15°$ [63], similar to *Ulysses* GAS.

The differing results in temperature are partially due to the low signal-to-noise (or background) ratios S/N $\approx$ 2–5 of GAS compared to S/N $\geq$ 1000 of *IBEX*-Lo in the ISN flow peak (Fig. 7.5). Suppose only the peak portion of the angular distribution, as GAS observed above the background, is treated as the entire thermal distribution. In that case, a noticeably lower temperature may be deduced, making the observations sensitive to the background levels. Adapted analyses have led to higher T_{ISN} values more recently [71, 74]. Also, the simultaneous presence of O and Ne in the peak may slightly affect the T_{ISN} determination with GAS, although a significant influence is ruled out [77]. *IBEX*-Lo (Secs. 5.3.3) uses a TOF spectrograph to distinguish between He and H as well as O and Ne. This new capability assures the T_{ISN} determination for a single species and, at the same time, opens the diagnostics for multiple ISN species.

Alternate ENA sources originating in different regions can influence the flow vector and temperature results of affected ISN species. Even He, thought to be least affected by the presence of the heliosphere, shows a noticeable secondary neutral component due to CX in the OHS in the *IBEX* data [48, 78, 79]. The blending of the primary ISN distribution with the broader and offset secondary distribution distorts the deduced ISN velocity vector and increases the temperature, if not fully taken into account [63, 69]. With its much larger S/N ratio, *IBEX*-Lo is more sensitive to the secondary neutrals and their effects, but in turn, it opens a new window on the shape of the heliosphere and the ISM charge state, as we will discuss in Sec. 7.4.

The substantial improvement in the S/N ratio will separate different sources in the analysis, thus providing a much deeper view into the ISN velocity distributions, extending into their tails. This part of the velocity distribution could show any non-thermal effects in the ISN gas if they exist, most likely evident in the form of a κ-distribution [80].

7.3. The ENA Ribbon, a "Compass" for the Interstellar Magnetic Field

Like the ISN flow through the heliosphere (Sec. 7.2), the ISMF plays a defining role in shaping the heliosphere, controlling its boundaries, and diverting the flow of the *IS plasma*. The *Voyager 1 & 2* TS passages provided indications about the orientation of $\vec{B}_{ISM}$ in our galactic neighborhood [81], in conjunction with global heliospheric modeling [82–84]. However, it was the surprising observation of the Ribbon in the first *IBEX* all-sky ENA maps [85] that handed us a precision tool to determine the ISMF orientation. The unanticipated bright Ribbon withstood all tests to rule out any artifact in the observation. Interestingly, its locations in the sky form an almost full circle that satisfies $\vec{B} \cdot \vec{r} = 0$ [86], where $\vec{B}$ is the field in the source region. On the heliospheric scale, $\vec{r}$ from the Sun is equivalent to the LOS vector $\vec{l}$ from Earth. This orientation of $\vec{B}$ coincided with the ISMF direction in global heliospheric models that satisfied the boundary conditions established by the *Voyager* encounters with the TS and the suspected $\vec{B}_{ISM} \times \vec{v}_{ISN}$ deflection plane for H based on Ly-α backscatter observations [87]. Even with the differences in the processes invoked in the models, several simulations arrived at similar ISMF orientations [83, 84, 88].

Besides placing solid constraints on the ISMF orientation, the Ribbon also suggested that the magnetic pressure with $B_{ISM} \approx 2.5\mu G$ [86] must be comparable to the ISM ram pressure, putting the heliosphere between the two limiting cases considered earlier (ISM flow or $\vec{B}_{ISM}$ dominating) [89]. This unanticipated discovery is an excellent example of a new observational technique offering completely new diagnostic opportunities.

7.3.1. *Ribbon overview*

The unexpected Ribbon discovery called for independent verification, for which the *IBEX* mission design had already built in a crucial test. Considerable overlap of *IBEX*-Hi and -Lo in the anticipated core ENA energy range (0.5–2 keV) (Secs. 5.3.2 and 5.3.3) with different detection methods leads to two independent observations. *IBEX*-Hi converts ENAs to positive and *IBEX*-Lo to negative ions, with different internal background and noise reduction techniques while covering the same FOV during each S/C spin. Both sensors see the same structure, albeit with substantially lower statistics in *IBEX-Lo*, which rules out any artifact. It was quickly evident that the Ribbon traces out locations in the sky that satisfy $\vec{B} \cdot \vec{r} = 0$, but the reason for the ENA enhancement in the Ribbon is still a matter of debate. Before we address this debate, we review the most telling observations with the help of Figs. 7.7 and 7.8.

As shown in Fig. 7.7, the Ribbon forms almost a perfect circle in the sky, centered on $(219°, 40°)$ in ecliptic longitude and latitude with an opening cone angle of $\approx 75°$ for $E = 0.7$–2.7 keV [90]. The Ribbon center moves southward with energy by $\approx 5°$ along the heliographic meridian (aligned with the Sun's axis) [90, 91].

The Ribbon energy spectra stand out with a 2–3 times higher flux over the GDF (Sec. 7.5) with a knee between 1 and 3 keV [92]. The Ribbon shows a clear peak in the ratio Ribbon/GDF around 1 keV at low latitudes and near 2–3 keV at high latitudes [93], mimicking the behavior of the SW during solar minimum [94]. Note, *IBEX*

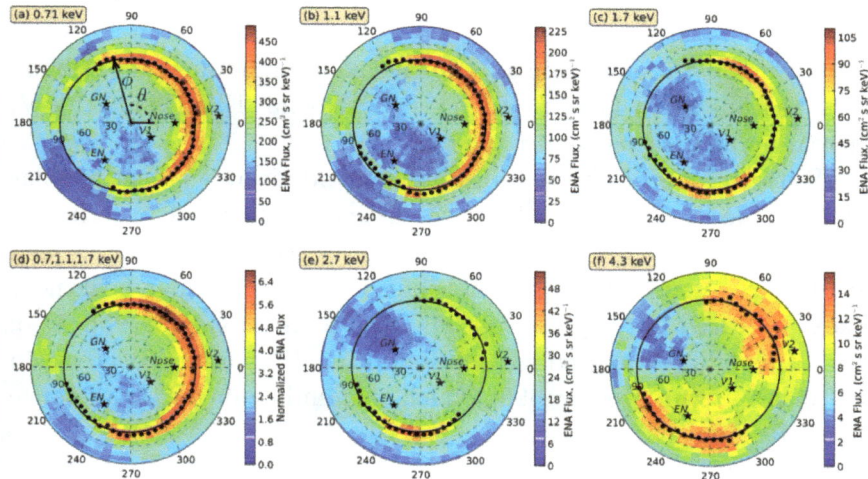

Fig. 7.7.   ENA maps in the six *IBEX*-Hi energy ranges centered on the circle fitted to the Ribbon, emphasizing its circularity and $\approx 75°$ opening cone angle [90]. (© AAS. Reproduced with permission).

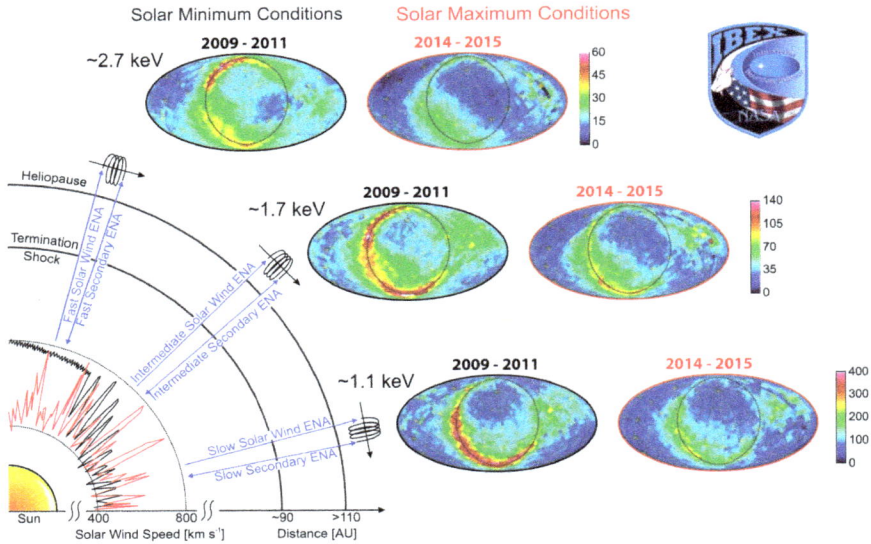

Fig. 7.8. *IBEX* ENA maps centered on the Ribbon for 1.1, 1.7, and 2.7 keV, for 2009–2011 and 2014–2015, respectively. It is suggestive that the Ribbon mirrors the latitudinal speed distribution of the SW. A cartoon model in the lower left shows how the SW and the ISMF may conspire to produce the Ribbon in the secondary ENA model [94]. (© AAS. Reproduced with permission).

started its operation during a prolonged solar minimum. The Ribbon persists from 0.2 to 4.3 keV and has a width of 23° FWHM at 1.1 keV, increasing at higher energies [93].

With the *IBEX* observations now extending over an entire solar cycle, it became evident that the distinct latitudinal ordering in energy all but disappeared after reaching the solar maximum, as seen in Fig. 7.8. A more isotropic distribution with a broader peak near 1 keV appeared in the years following the solar maximum [95]. Also, the Ribbon flux decreased with the expected delay after a decrease in SW flux and ram pressure [96], which is an indicator for the distance of the Ribbon source region.

An image by itself does not show the source distance, which requires either temporal tracking of the source and signal or a parallax analysis. The latter method provides a geometric way to obtain the distance of nearby stars, which the *Gaia* mission has recently pushed to cover a large fraction of the Milky Way [97]. Comparing Ribbon images taken six months apart returned an average Ribbon source distance $D = 140^{+84}_{-38}$ AU [99], battling the challenge that one image points in the S/C ram and the other in the anti-ram direction, with a vast difference in counting statistics.

We will now briefly comment on each model or scenario and how they relate to the key observations. Section 7.3.2 will expand on the model family that appears to hold the likely explanation. Table 7.1 organizes the key observations (in columns) and models (in rows), ordered by increasing source distance, as in an early review [98].

The model that puts the Ribbon source closest to the Sun calls upon specularly reflected SW ions at the perpendicular TS, which then gyrate around the IMF downwind of the shock and produce ENAs through CX with the ISN gas [100]. To form the Ribbon, the model assumes that the TS is pushed the farthest inward where the ISMF is perpendicular to the radial direction and thus exerts maximum pressure. The SW as the parent population for the ENAs explains the latitudinal and temporal organization with the SW.

Simulating PUIs and SW, processed in the IHS, and integrating the resulting ENA fluxes over the entire IHS for an axisymmetric heliosphere has produced a Ribbon-like structure with a comparable angular width and a flux that surpasses the GDF [101–103]. Another scenario, proposed early on [85, 86], which puts the Ribbon source at the outer edge of the IHS, draws on the preferential compression of the plasma in the IHS stagnation region where the ISMF is tangential to the HP to explain the $\vec{B} \cdot \vec{r} = 0$ ordering. However, so far there is no physical process for the needed pressure enhancement in the correct location. A final scenario in the IHS assumes a density wave (dubbed H-wave) whose front is oriented perpendicular to the ISMF and convected with the ISN flow [104]. This density structure would move freely through the IHS and thus produce an ENA enhancement in the Ribbon location. Depending on the wave speed in the ISM, such an H-wave would move at 1.5–5.5 AU/year ($0.85° - 3.2°$/year for the source distance) in the direction of the ISN flow, thus challenging this scenario directly since the Ribbon has not moved for at least ten years.

Table 7.1. Ribbon Observations and Models

Model Description [Ref.]	Source Location	Ribbon location along $\vec{B}\cdot\vec{r}=0$ [85, 86]	Circular Ribbon with ≈75° Cone [90]	E-Dependence of Ribbon Center [90]	$j = 2$–3×200 (cm² sr s keV)⁻¹ GDF [93]	Latitude Dependent Peak Energy [85, 92]	Solar Cycle Dependent Structure [95]	$D = 140$ +84/-38 AU Parallax [99]
TS reflected SW & PUIs [100]	**TS**	YA	NA	NA	Yes	Yes	Yes	<D
TS processed PUIs & SW [85, 101–103]	**IHS**	NA	Yes	NA	Yes	NA	NA	<D
IHS Stagnation Region [85, 86]	**IHS**	Yes	NA	NA	QL	No	No	≤D
H-Wave [104]	**IHS**	Yes	NA	NA	QL	No	No	≤D
Reconnection at HP [85]	**HP**	NA	NA	NA	NA	NA	NA	=D
K-H & R-T Inst. at HP [85, 105]	**HP**	NA	NA	NA	NA	NA	NA	=D
ISMF Compression & Mirror [85, 86]	**OHS**	Yes	Yes	NA	Yes	No	No	=D
Secondary ENAs [85, 106–110]	**OHS**	Yes	Yes	NA	Yes	Yes	Yes	=D
— Reduced instability [111, 112]								
— Spatial Retention [113, 114]								
— ISMF Trapping [115]								
— SW Latitude distribution [91]				Yes				
LIC-Local Bubble Boundary [116]	**ISM**	No	Yes	NA	YA	No	No	≫D

Yes: Model addresses observation   No: Model contradicts

YA: Yes, but specific assumptions   NA: Not addressed in the model

QL: Only addressed qualitatively

Note: References are given by their number in the citation list, for observations in the top row, and for models in the first column.

The list of scenarios contains two suggested sources at the HP, or close to the center-of-the-parallax distance range, that involve ion flux enhancements in regions with reconnection, or Kelvin-Helmholtz and Rayleigh-Taylor instabilities [85]. Reconnection could potentially explain the $\vec{B} \cdot \vec{r} = 0$ ordering and indeed occur at the HP, similar to a situation at Ganymede in the Jovian magnetosphere [105].

We leave the Secondary-ENA model, which likely produces the Ribbon ENAs in the OHS through a sequence of three CX processes, to Sec. 7.3.2. Instead, we jump to the Ribbon model that puts the source region far away from the heliosphere to the boundary of the LIC with the surrounding hot gas of the Local Bubble [116]. CX between evaporating neutral gas from the LIC and the hot ions of the Local Bubble is the suggested Ribbon source. With a spherical shell boundary structure and its curvature as a free parameter, this model reproduces the 75° Ribbon cone angle. However, it side-steps the organization with $\vec{B} \cdot \vec{r} = 0$, does not explain the latitude and temporal variation similar to the SW structure and requires a very stable Ribbon over many decades. The distance limitation to see localized structures due to the optical thickness of the ISM outside the heliosphere for ENAs also challenges this model (Sec 3.4.3).

7.3.2. The Secondary-ENA model as likely explanation

As shown in Fig. 7.9, the secondary-ENA Ribbon model uses a thus far underappreciated source of ENAs, albeit through a sequence of steps, the neutral SW (the primary ENAs in this sequence). On its way from the Sun to the TS, the SW interacts *via* CX with the ISN gas, increasing the amount of PUIs. Its fraction of H^+ reaches $\approx 20\%$ at the TS [108] and causes the SW to decelerate [25]. In the same CX process, a comparable fraction of neutral SW is generated that continues to flow radially outward unimpeded by the IMF in the IHS and the ISMF, which ends up "polluting" the Sun's neighborhood. In the next step, this neutral SW is subject to CX with IS H^+ ions with an estimated density of $\approx 0.07 \, \text{cm}^{-3}$ [2].

As indicated in Fig. 7.9, PUIs from this CX injected into the ISMF at 90° pitch angle, satisfying $\vec{B} \cdot \vec{r} = 0$ for an observer near the Sun, form a ring distribution that remains in place. PUIs injected at

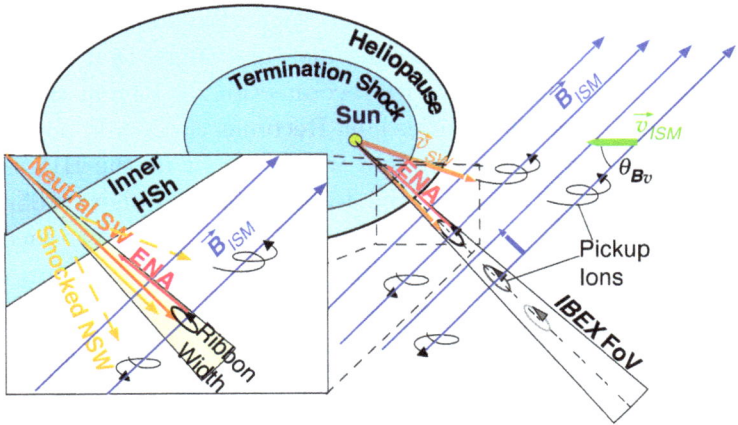

Fig. 7.9.   ENA generation in the Secondary-ENA Model for the Ribbon. Neutral SW expands radially outward beyond the HP, producing PUIs through CX with IS H^+. PUIs injected into the ISMF at a 90° pitch angle, satisfying $\vec{B} \cdot \vec{r} = 0$, pile up in a PUI ring distribution, while PUIs injected at different angles quickly escape along $\vec{B}_{ISM}$. Inset: Neutrals from the supersonic SW, the shocked SW of the IHS, and PUIs can form the Ribbon at their respective energies [108]. (© AAS. Reproduced with permission).

different angles quickly escape along $\vec{B}_{ISM}$. The ISMF sweeps the PUI ring across the heliosphere. The ions can undergo a third CX forming a second generation of ENAs (generally termed secondary ENAs) that move toward the Sun if created at the proper phase of the gyration.

The PUIs injected at $\vec{B} \cdot \vec{r} = 0$ stay around and concentrate the ENA fluxes that form the Ribbon. After CX into neutral atoms, the supersonic SW from inside the TS and those from the shocked SW in the IHS within the appropriate directions can contribute to the Ribbon, as do PUIs in the SW or the IHS, all at their original energy (Fig. 7.9, inset). However, ENA fluxes from the supersonic SW (≈1keV slow and 2–3 keV fast) stand out in the observed Ribbon energy distribution.

7.3.2.1.   *Simplified analytical model*

To show how the consecutive CX processes lead to continuous Ribbon ENA flux accumulation along the LOS, we take a simplified analytical model [108]. The TS is at $r_0 = 100$ AU from the Sun, the HP at $1.5r_0 = 150$ AU, and the neutral SW evolves in the OHS as:

$$n_{SW0}(r) = n_{SW0}(100 \text{ AU}) \frac{r_0^2}{r^2} e^{\frac{-(r-1.5r_0)}{\lambda_{CX}}} \tag{7.11}$$

We use $n_{SW0}(100 \text{ AU}) = 8.2 \cdot 10^{-5} \text{ cm}^{-3}$ or 16.5% of the total SW density, accumulated from the Sun to the TS at the CX rate $\sigma_{HpCX} = 1.71 \cdot 10^{-15} \text{ cm}^2$ at 1 keV [117]. The ISN density at the TS [41] $n_H = 0.08 \text{ cm}^{-3}$ is kept constant for simplicity. The neutral SW density decreases according to its expansion and extinction by CX with $\lambda_{CX} = 1/(\sigma_{CX} \cdot n_p)$ (species subscripts H and p dropped), where n_p is the IS proton density. Conversely, this neutral SW extinction produces PUIs in the OHS, whose density $n_{PUI}(r_1)$ as a function of distance r_1 from the Sun can be estimated based on Eq. (7.11):

$$n_{PUI}(r_1) = \frac{n_{SW0}(r_0) \cdot \sigma_{CX} \cdot n_p \cdot v_{SW}}{v_{ISN} \cdot \sin\theta_{BV}} \int_{r_1}^{\infty} \frac{r_0^2}{r^2} e^{\frac{-(r-1.5r_0)}{\lambda_{CX}}} \cdot e^{\frac{-(r-r_1)}{\lambda_{Eff}}} dr \tag{7.12}$$

$\sigma_{CX} \cdot n_p \cdot v_{SW}$ is the ionization rate or the PUI source rate at distance r. The last exponential describes the PUI extinction that produces ENAs with the effective mean free path $\lambda_{Eff} = v_{ISN}\sin\theta_{BV}/(v_{SW}\sigma_{CX}n_H)$, during the PUI advection toward the HP. While the PUIs move through the ISN gas at v_{SW}, they are convected with $\vec{B}_{ISM}$ at speed $v_{ISN}\sin\theta_{Bv}$. Note that we restrict the treatment to the $\vec{B}_{ISM} \times \vec{v}_{ISN}$-plane.

All ENAs that emerge from the PUI ring toward the Sun within the sensor FOV $\Delta\Omega$ along the LOS contribute to the observed Ribbon flux

$$j_{ENA}(r_2) = \frac{\Delta\psi}{2\pi} \frac{n_{SW0}(r_0) \cdot \sigma_{CX}^2 \cdot n_p \cdot n_H \cdot v_{SW}^2}{\Delta\Omega\Delta E \cdot v_{ISN} \cdot \sin\theta_{Bv}}$$
$$\times \int_{r_2}^{\infty} \left[\int_{r_1}^{\infty} \frac{r_0^2}{r^2} e^{\frac{-(r-1.5r_0)}{\lambda_{CX}}} \cdot e^{\frac{-(r-r_1)}{\lambda_{Eff}}} dr \right] \cdot e^{\frac{-(r_1-1.5r_0)}{\lambda_{CX}}} dr_1 \tag{7.13}$$

Equation (7.13) follows from Eq. (7.12) by including the ENA-producing CX rate $\sigma_{CX} \cdot n_H \cdot v_{SW}$ and integrating from infinity to r_2

in the OHS. $n_H = 0.16\,\mathrm{cm}^{-3}$ is the pristine ISN density [2]. The ENA extinction from their production at r_1 to the HP is similar to that of the neutral SW on its way out. Assuming that all second-generation ENAs of the supersonic neutral SW enter the sensor FOV $\Delta\Omega$ and energy window ΔE, Eq. (7.14) yields the differential flux. $\Delta\Psi/2\pi$ is the fraction of the PUI ring that falls within the sensor FOV, defined by its 1D cut $\Delta\Psi = 7°$ for *IBEX* [108].

Equation (7.13) describes the accumulation of the Ribbon ENA flux and its subsequent depletion along the LOS from infinity to the HP. Figure 7.10 shows the neutral SW density as a function of r, along with the $1/r^2$ expansion, in the top panel. In the bottom panel, the emission of the Ribbon ENA flux (solid line) and ENA

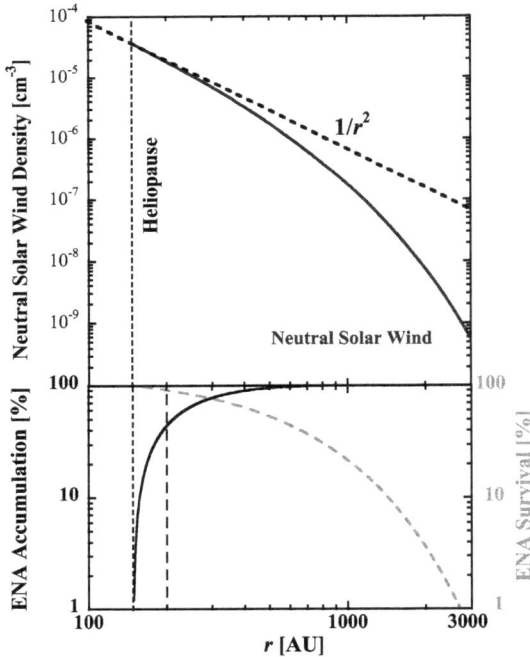

Fig. 7.10.   Top: Neutral SW density as a function of distance from the Sun r along with the $1/r^2$ expansion (dashed line). Bottom: Percentage accumulation of Ribbon ENA flux, starting from the HP (vertical dotted line) and survival of ENAs (dashed line) as a function of the source distance. The vertical dashed line marks 50 AU from the HP [108]. (© AAS. Reproduced with permission).

survival (dashed line) are shown as functions of the source distance r, starting at the HP. The decrease of the neutral SW with r (10^{-5} of the HP density at 3000 AU) and the extinction of the ENAs ensure that the lion share of the Ribbon ENAs stems from close to the HP, *i.e.*, $\approx$50% from within the first 50 AU (vertical dashed line). With a travel time of the neutral SW and ENAs of one year over 100 AU each, this emission length allows for time variations in the Ribbon flux over one year, as found, in response to SW changes [66, 94].

Adopting the working value $v_{ISN} = 26\,\text{km/s}$ [72], the upper value for the IS plasma density $n_p = 0.07\,\text{cm}^{-3}$ [2], and assuming a perfect PUI ring distribution, the Ribbon ENA flux at Earth seen in the heliosphere nose direction yields:

$$j_{ENA}(r_0) = \frac{\Delta\psi}{2\pi} \frac{n_{SW0}(r_0) \cdot \sigma_{CX}^2 \cdot n_p \cdot n_H \cdot v_{SW}^2}{\Delta\Omega\Delta E \cdot v_{ISN} \cdot \sin\theta_{Bv}} r_0^2$$

$$= 420\ (\text{cm}^2\,\text{s}\,\text{sr}\,\text{keV})^{-1} \tag{7.14}$$

The ENA flux estimated from this simple model does not include any draping of the ISMF and related compression of the IS plasma and gas in the OHS, which would lead to higher ENA fluxes. Yet, the estimate exceeds the observed Ribbon fluxes at $\approx 1\,\text{keV}$ by a factor of two [93, 94]. It thus allows for unaccounted losses, *e.g.*, due to scattering of the PUIs. However, scattering into an isotropic shell, which replaces $\Delta\Psi/2\pi = 0.019$ by $\Delta\Omega/4\pi = 9 \cdot 10^{-4}$ in Eq. (7.14), drops the model ENA flux to almost 1/10 of the observed flux. More importantly, through scattering toward isotropy, most PUIs would quickly escape along $\vec{B}_{ISM}$ (Fig. 7.9), which would wipe out almost any flux enhancement in the Ribbon region. A ring distribution is inherently unstable, leading to pitch angle scattering, the most severe challenge to the Secondary-ENA model.

The fast temporal changes of the Ribbon observed already two years into the *IBEX* mission [94], posed a second challenge for the Secondary-ENA model. This model seemed to require a long integration length along the LOS, averaging ENA fluxes with turnaround times different by several years. This long timescale for the Ribbon ENA accumulation would have heavily subdued any SW-related variations, the suggested cause for the observed temporal changes. However, the realization that $\approx 50\%$ of the ENA fluxes stem from the first 50 AU outside the HP for energies near 1 keV largely resolved this challenge [108], which is also evident in full simulations of the Ribbon [109].

7.3.2.2. *Tests of the Secondary-ENA model*

At this point, there is no consensus yet on a definitive model of the ENA Ribbon. However, the Secondary-ENA model has already passed many tests, as indicated by the number of "Yes" in Table 7.1, steadily increasing its support.

- *Solar Wind Structure*: One key observation pointed to the Secondary-ENA models early on, *i.e.*, the Ribbon energy spectra replicate the SW bulk speed distribution in latitude [93–95]. A quantitative comparison of the observed Ribbon centers as a function of ENA energy [90] with those obtained in an expanded analytical model [91] validated this correspondence. The model includes the observed latitudinal SW structure based on scintillation maps [118, 119].
- *Temporal Variation over Solar Cycle*: As an extension to the tests described above, the temporal evolution of the SW, with a time delay that allows its travel into the OHS the average time it takes to turn the PUIs in the source region into ENAs, and the return travel of the ENAs, should be reflected in the behavior of the Ribbon fluxes and spectra. The Ribbon discovery fell into an extended and deep solar minimum when it showed the distinct characteristics of SW under such conditions, with high-speed SW at high latitudes and slow wind at low latitudes [120], reflected in the latitudinal distribution of the Ribbon spectra [92, 94]. Toward solar maximum, the latitudinal dichotomy of the SW vanished [121], distributing the Ribbon spectra more evenly in heliolatitude. In addition, the SW flux and momentum flux decreased substantially during solar cycle 23, also reducing the Ribbon ENA flux [96].
- *Distance of the Ribbon Source Region*: The Secondary-ENA model puts the Ribbon into an extended region that starts just outside

the HP. The *Voyager* 1 HP crossing in 2012 [122, 123] places the expected distance of the Ribbon source region beyond 120 AU close to the *Voyager* 1 direction. There are two basic methods to obtain a distance estimate for the source region from ENA observations: 1) The time delay between changes in the SW speed and flux distributions and the related Ribbon ENA distributions invokes the SW and ENA travel times, 2) The parallax method uses Ribbon observations six months apart across a baseline of 2 AU.

For 1), the database is too short as of this writing. Yet, the temporal variations in the Ribbon over the first seven years appeared consistent with the SW travel time beyond the HP and the return of the ENAs [95].

For 2), a detailed χ^2 comparison between ENA maps taken six months apart makes use of a parallax-measurement baseline of 2 AU. A source distance of 100 AU results in an angular shift of 1.15° between the two observations, *i.e.*, $\approx 1/5$ of the *IBEX* FOV [99]. Five years of *IBEX* data yielded a parallax half-angle of $0.41 \pm 0.15°$, which translates into an estimated source distance of 140^{+84}_{-38} AU. The mean value of this result puts the source region right into the center of the expected range for the Secondary-ENA model, and it puts models that predict a source region at or just beyond the TS [100, 103] at the lower 1σ end of the uncertainty range and eliminates any model that puts the Ribbon source at the boundary of the LIC with the hot Local Bubble [116].

7.3.3. *Challenges and opportunities with the ENA Ribbon*

The Ribbon connection to the ISMF and the SW provides ample diagnostic opportunities for the interaction between the ISM and the heliosphere. Yet, it poses substantial challenges to extract this information quantitatively. Also, the leading model candidate poses challenges in terms of the efficacy of the three consecutive CX processes involved and the required stability of the intermediate ion population, which, in turn, provides opportunities to study the magnetic turbulence in the solar neighborhood.

7.3.3.1. Effectiveness of Ribbon generation and turbulence

At first sight, the Secondary-ENA model, with its three consecutive charge exchange processes, appears to be relatively ineffective in producing the observed flux enhancement within a reasonable distance along the LOS. However, with realistic IS gas and ion densities, about 50% of the Ribbon flux stem from the first 50 AU outside the HP. Yet, the final step in the Ribbon ENA production from the PUIs includes a storage time of $\approx$ 2–3 years. However, the ring distribution is notoriously unstable due to various plasma instabilities, notably the ion cyclotron instability [124–126].

First simulations, starting with the fully accumulated PUI density close to the HP, resulted in time scales of a few days for the PUI instability growth and scattering to isotropy [127] which would all but wipe out the Ribbon. Gradual injection of PUIs into the IS flow slows down the growth by a factor of ten, among others, due to an onset threshold, still leaving it orders of magnitude too fast [112]. A finite width of the PUI ring distribution, either due to preexisting turbulence [111] or a finite temperature [128, 129] may quench the instability growth.

Even if the initial PUI ring were stable, the $\approx$ 23°-wide Ribbon still needs further spatial retention [130]. At this angle, we find $v_\parallel = 0.11 v_{\mathrm{sw}}$, and the PUIs would escape along $\vec{B}_{ISM}$ from the region at a clip of 11 AU/year, or faster than the final CX could turn them into observed ENAs. Due to their wave generation, PUIs injected near 90° exhibit $v_\parallel < v_A$ and can thus be trapped (Fig. 7.11) because of a very slow spatial diffusion. Both a simulation [113] and a quasilinear analytic description of the instability and diffusion yield a narrow Ribbon structure [114]. This model combines PUI concentration in phase space near 90° pitch angle with spatial retention.

Alternatively, preexisting ISMF turbulence may lead to confinement of the Ribbon-generating PUIs close to $\vec{B} \cdot \vec{r} = 0$ [115]. The fluctuations $\delta|\vec{B}_{ISM}|$ act like local magnetic bottles that trap ions by mirroring, similar to the trapping in the Earth's ring current and

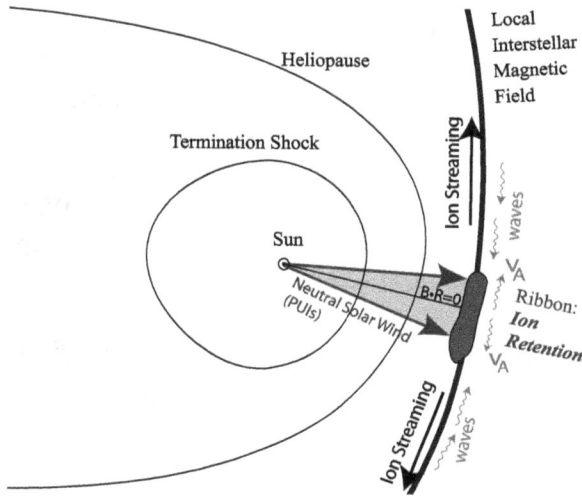

Fig. 7.11.   Trapping of PUIs from neutral SW origin near $\vec{B} \cdot \vec{r} = 0$ with waves and scattering [113]. (© AAS. Reproduced with permission).

radiation belts. The PUIs are trapped in these "bottles" if the cosine of their pitch angle μ_o satisfies:

$$|\mu_0| < \left(1 - \frac{B_{ISM}}{B_{ISM} + \delta B}\right)^{\frac{1}{2}} \approx \left(\frac{\delta B}{B_{ISM}}\right)^{1/2} \qquad (7.18)$$

With turbulence spectra from *Voyager* 1 outside the HP [131], numerical modeling returns ENA fluxes consistent with the Ribbon observations.

The specific predictions of any of these models involve the power spectra of the PUI-generated waves, resulting PUI velocity distributions, and related spatial structuring. A clear distinction between the processes will require observations of $\vec{B}_{ISM}$ and the PUI distributions inside the Ribbon source region with highly anticipated *Interstellar Probe* (*ISP*) [132]. In the light of the *IBEX* observations, the probe should travel into the Ribbon direction near the heliospheric nose. Then, the magnetic-field power spectra can be compared with those obtained beyond the HP on *Voyager* 1 and 2 [133], whose trajectories are clearly outside the Ribbon source region. That region may not align with the radial direction due to the draping of $\vec{B}_{ISM}$ around the heliosphere, as can be gleaned from radial cuts of the simulated Ribbon source region [109]. Therefore, *ISP* will likely exit the source region at some distance and reveal the spatial distribution of the Ribbon PUIs. At this point, a comparison of the energy-dependent angular Ribbon width with a model of the complete source distribution, including supersonic SW, subsonic SW from the IHS, and heliospheric PUIs, appears to support reduced diffusion due to self-generated waves [113].

7.3.3.2. *Determination of the interstellar magnetic field*

One of the most striking features of the Ribbon is its alignment normal to the ISMF, satisfying $\vec{B} \cdot \vec{r} = 0$, thus providing a compass for $\vec{B}_{ISM}$ just outside the heliosphere. To turn the Ribbon into a precision tool that would unravel the ISMF strength and direction, along with its draping around the HP, we must first understand the Ribbon-forming processes and the source distribution. As shown in Table 7.1, different Ribbon models invoke varied mechanisms and locations.

Therefore, let us start with the geometric characteristics of the Ribbon, independent of any model. The Ribbon seems almost precisely circular, with portions toward the heliotail hardly visible (Fig. 7.7) [90]. Whether the stronger magnetospheric ENA foreground obscures the Ribbon here or forms a much weaker signal due to neutral SW expansion to larger distances in these directions remains to be seen, likely with *IMAP* [134].

The Ribbon traces a conical circle at a $75.4 \pm 2.0°$ opening angle, with its center tilted toward the apex of the ISN flow, discussed in Sec. 7.2, consistent with tracing the draped OHS magnetic field, just outside the HP [135] instead of the pristine ISMF. The Ribbon center is still close to the pristine ISMF, which can be determined in global heliospheric models, assuming the Secondary-ENA model [109]. This direction is consistent with the anisotropy of TeV cosmic rays [136] and agrees with the average ISMF in the Sun's neighborhood from the polarization of starlight by IS dust [137].

However, the magnetic field orientation obtained by the *Voyagers* is substantially different from the Ribbon center [138]. More puzzling is that the magnetic field direction appeared to agree closely with the IMF inside the HP. Strong draping in the innermost layers of the OHS [135], coupling to the IMF [139], or that *Voyager 1* has not yet crossed the HP [140] may be potential explanations. *Voyager 1* saw a slow turning of the ISMF toward the Ribbon orientation [141], suggesting agreement by 2025. But solar disturbances that affect cosmic rays in the OHS [142] seemed to interrupt the turning, possibly due to dynamic draping around the HP [143]. At this writing, the jury on the differences in the ISMF is still out.

A more thorough understanding of the Ribbon formation will yield a substantial refinement of the local ISMF. Improved global heliospheric models and ENA diagnostics of the entire Ribbon will also enable probing the OHS topology.

7.3.3.3. *Reach of the Solar Wind into the ISM*

Suppose we adopt the Secondary-ENA model as the most likely explanation for the Ribbon. Then, the OHS is the ENA source region, and the Ribbon reflects the neutral SW as it penetrates the ISM. As observed *in-situ* in the ecliptic at 1 AU or *via* interplanetary scintillations for its 3D evolution, SW variations will show up in the Ribbon with the appropriate delay (SW travel time, PUI storage, and return to 1 AU at the ENA energies). With reasonable estimates, the temporal variation of the Ribbon at various energies is consistent with the SW variation over the current solar cycle [95]. Figure 7.12 shows an early recovery at the lowest ENA energies, indicating closer proximity of the respective source regions.

7.4. Secondary Interstellar Neutrals

The studies of the ENA populations discussed here and in Sec. 7.5 become possible only after appropriately accounting for the primary ISN flow in the case of the secondary IS neutrals and for the Ribbon in the case of the GDF of ENAs. The mean free path for CX between

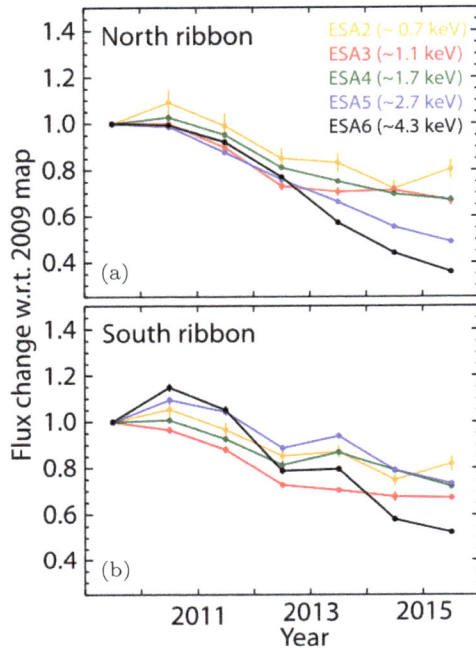

Fig. 7.12.   ENA fluxes from two Ribbon regions for five *IBEX*-Hi energies as a function of time, high latitude in the north (top) and low latitude in the south (bottom) [95]. (© AAS. Reproduced with permission).

ISN H and both IS H^+ and O^+ ions in the heliospheric boundary regions is comparable with the overall size of the heliosphere. Therefore, secondary neutrals that emanate from these regions were predicted [144–147]. For H, the signature of the ISN flow is a combination of the primary flow and a secondary component of comparable strength. Consequently, modern global heliospheric models contain CX and the secondary neutrals [27, 148–152].

The first *IBEX*-Lo ENA maps, dominated by the He, H, and O ISN flow signatures at $E = 10$ to $600\,eV$, also showed indications of a secondary O component [46]. Surprisingly, the first detailed analysis of the He ISN flow in the *IBEX* observations revealed a secondary He component, despite a longer mean free path for CX by about one order of magnitude [48]. Conversely, the secondary He component substantially affects the more precise determination of the ISN flow vector from *IBEX* observations [63, 68, 69], which required its detailed analysis.

7.4.1. *Extraction of secondary interstellar neutrals*

While *IBEX*-Lo resolves the neutral species (H, He, O, and Ne), incoming direction $\hat{l}$, and E (with wide passbands), it cannot unambiguously differentiate between the primary and secondary components. Therefore, an iterative process is used, starting with identifying and characterizing the dominant primary ISN flow. After subtracting the primary ISN with the best fitting parameters, the analysis of the secondary neutrals starts (Sec. 7.2). Their parameterization uses the same method. Accounting for this component improves the primary ISN flow parameters and, subsequently, the extraction of the secondary component. If this route is not possible, a combined fit could resolve the two components. However, the current data fidelity may be insufficient and lead to an underdetermined optimization problem (Sec. 7.4.2).

The first parameterization of secondary He involved approximation by a second drifting Maxwellian distribution that originates at the HP (at 150 AU from the Sun), not invoking any specific model [78]. Slower and hotter than the ISN flow, the new component was dubbed "Warm Breeze", leaving the door open to alternate interpretations. The same modeling tools as described in Sec. 7.2 apply to the secondary neutrals. However, they originate in the OHS, unlike the primary ISN gas that enters as a homogeneous flow from outside.

Subsequently, a more comprehensive analysis of several years of secondary neutral He followed this iterative approach based on a refined set of ISN parameters. The resulting best-fit drifting Maxwellian ($v_{Sec} = 11.28 \pm 0.48 \pm 0.7\,\text{km/s}$, $\lambda_{Sec} = 71.57 \pm 0.50 \pm 0.9°$, $\beta_{Sec} = -11.95 \pm 0.30 \pm 0.6°$, $T_{Sec} = 9480 \pm 920 \pm 1600\,\text{K}$) is consistent with predictions for a secondary He component produced by CX in the OHS in global heliospheric models [79]. The second set of uncertainties in this result represents systematic model-related uncertainties, in addition to the statistical fit uncertainties in the first set. A flow velocity as used for the ISN flow is adopted here, given in terms of the arrival direction at the HP.

Parallel to the extraction of the He component, the secondary O component attained a quantitative grounding. The much lower counting statistics of the O atoms (by more than two orders of magnitude) require a detailed analysis of the O maps to assess which signatures, in addition to the well visible ISN O flow, are statistically significant. Three different statistical methods were applied, known from image processing with low photon-count statistics: signal-to-noise filter, confidence limit, and cluster-analysis methods [153]. Figure 7.13 shows O maps accumulated from 2009 through 2011 (to improve statistics) analyzed with each of the three methods (from left to right) for E-Step or ESA 5 (0.28 keV), 6 (0.60 keV), 7 (1.2 keV), and 8 (2.4 keV) of *IBEX*-Lo (from top to bottom). The ISN O flow is the most significant feature in the E-Step 5 through 7 maps, whereas E-Step 8 (2.4 keV) solely contains heliospheric ENAs.

- *Signal-to-Noise Filter* (S/N): The S/N for each pixel (i, j) in a map is $S/N(i, j) = C(i, j)/\sigma_C(i, j)$, where C represents the number of counts in each pixel and σ the Poisson standard deviation. The average S/N ratio of each complete map determines a threshold above which the observed counts are significant. The left column of Fig. 7.13 shows the S/N ratios, with averages 1.34, 1.24, 1.10, and 1.05, respectively, and 1.5, 1.4, 1.2, and 1.2 are the thresholds, requiring a signal at least 20% above the noise [153].

- *Confidence-Limit Method* (CLM): This method considers the counts in each pixel and its immediate neighbors and thus is more immune to pixel-to-pixel fluctuations. It was initially developed for low photon count images in X-ray and γ-ray astronomy [154]. Because ENA imaging typically yields very few counts per pixel, we employ Poisson statistics. For the expected number λ (average after infinite repetition) of counts per pixel over a time interval Δt, N counts are samples of the probability distribution

$$P_\lambda(N) = \frac{\lambda^N e^{-\lambda}}{N!} \tag{7.16}$$

The upper λ_u and lower λ_l confidence limits for the accrued counts N in a pixel are defined as [154]

$$1 - CL = \sum_{x=0}^{N} \frac{\lambda_u^x e^{-\lambda_u}}{x!}$$

$$CL(N \neq 0) = \sum_{x=0}^{N-1} \frac{\lambda_l^x e^{\lambda_l}}{x!}$$

(7.17)

where CL is the specified confidence limit (set to 84.13%, or the 1σ Gaussian error) in Fig. 7.13 [153]. Multiple sampling is out of the question for the ENA application. Therefore, a test region around each pixel with radius o pixels (here $o = 2$) is defined. Using multiple pixels in a test region is equivalent to replacing sampling over time with spatial sampling. Normalization for a fixed sampling

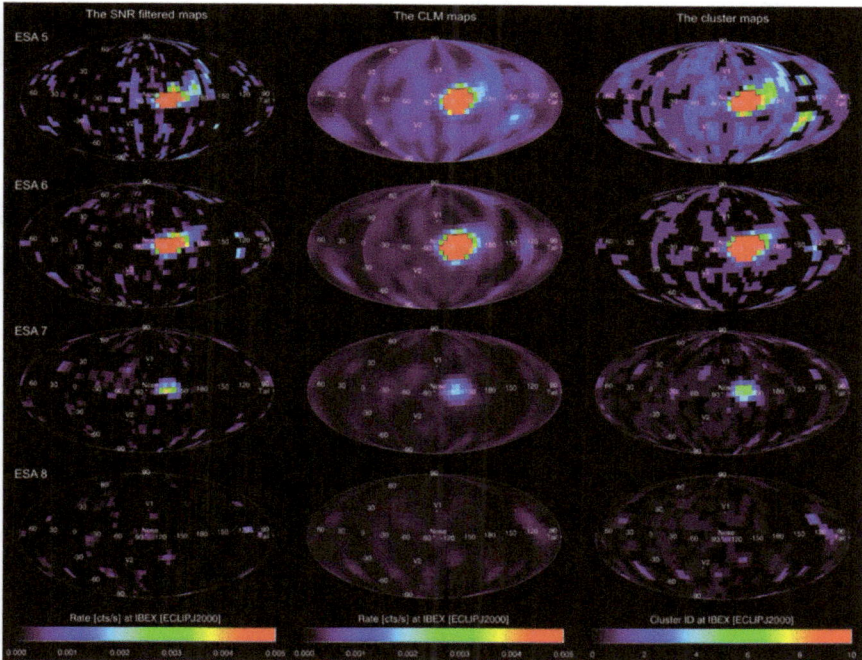

Fig. 7.13. *IBEX*-Lo O-Maps for E-Steps 5, 6, 7, and 8, accumulated for 2009–2011, assessed for their significance with three statistical methods: Signal-to-Noise Filter (left), Confidence-Limit Method (center), Cluster-Analysis Method (right) [153]. (© AAS. Reproduced with permission).

time yields the average number of counts N in the pixels of the test region, thereby adjusting for accumulation time differences between individual pixels due to cleaning the data for any foreground, *e.g.*, parts of the Earth's magnetosphere. The number of counts x in the sampling of any pixel falls below the upper or above the lower limit with the probability CL. The upper limit λ_u is chosen to represent the significant count number in each pixel, shown in the center column of Fig. 7.13 before adjustment for a homogeneous background B. The average number of counts in all pixels with $S/N < 1$ (no statistical significance) represents B.

$$B = \sum_{i_B, j_B} C(i, j)/N_B \tag{7.18}$$

Subtraction of B yields the background adjusted value λ_u in each pixel.

- *Cluster-Analysis Method* (CAM): The CAM evaluates localized regions of a group or cluster of pixels for coordinated variations in counts. It relates to the density-based clustering method in automated data mining, with regions (clusters) of high density separated from regions of low density [155]. Like in the CLM, the normalized counts $C(m, n)$ are summed up for all pixels within a given radius $k(|m - i| \leq k$ and $|n - j| \leq k)$ around the center pixel (i, j) of a cluster resulting in $C_{NCluster}(i, j)$. All potential clusters are compared with a specified count level C_{NSpec}, starting from the maximum $C_{NCluster}(i, j)$ in the map and decreasing C_{NSpec} by 1 in each step until reaching the background level. $C_{NCluster}(i, j) \geq C_{NSpec}$, (i, j) defines a core pixel of a meaningful cluster with C_{NSpec}. Otherwise, it is considered a noise or background pixel. The right column of Fig. 7.13 shows a CAM map for $k = 1$, where each pixel is assigned the highest value of C_{NSpec}, for which it qualifies as a center pixel.

Because of the summing over neighboring pixels, the CLM and CAM achieve increased counting statistics at the expense of smoothing and reduced angular resolution of the map. The CAM does not include a formal statistical significance test like the other two methods. In Fig. 7.13, all three methods consistently expose an extended tail that emerges from the primary ISN O distribution toward larger longitude and higher latitude, identified as the secondary O component. Based on the maps for O and He, the maximum direction of the two secondary components is determined and traced to the HP, using Eqs. 7.1–7.8 for the ISN flow analysis. The arrival directions of the two components lie on a great arc in the sky that connects the ISN flow and ISMF direction based on the center of the ENA Ribbon [79, 156]. Forward modeling within a global heliospheric simulation connects the observed secondary populations to their origin from CX in the OHS. The ISN trajectory relations (Eqs. 7.1–7.8) connect the secondary neutral production in the OHS to the observed distributions.

7.4.2. *Challenges and opportunities with secondary interstellar neutrals*

Unlike the primary ISN flow, the secondary distributions generally feature low counting statistics, which challenges the extraction of characteristic features with reasonable angular resolution from the ENA maps. As stated at the beginning of Sec. 7.4, secondary neutrals emerge from the IS plasma in the OHS, which flows around the HP. The HP shape may be substantially asymmetric due to the orientation of the ISMF, the latitudinal structure of the SW flow, and the temporal variations over the solar cycle. Also, the plasma distribution in the OHS is very likely non-thermal due to the flow diversion and continuous injection of PUIs *via* CX. The distribution probably changes as the flow progresses toward the heliotail. Two approaches produce comparable simulated ENA maps, the simplistic assumption of a second hotter Maxwellian flow distribution arriving from a different direction than the ISN flow outside the HP [79] and the more realistic secondary He generation *via* CX in a global heliospheric model. There are no significant differences in the goodness of the χ^2 fit between either model ENA map and the observations [157]. These results suggest that the current ENA observations with *IBEX* are insufficient to extract information on the flow structure and any non-thermal signatures in the plasma distribution. Minor differences in the modeled ENA maps surface in the outskirts of the secondary neutral distribution, where the S/N is low, suggesting that improved collection power and better background suppression in the future will improve the situation.

Another limitation is that *IBEX* observes ENAs close to 90° from the Earth-Sun line, only allowing a 1D cut through the OHS with limited viewing directions. This limitation may cause a degeneracy, similar to the ISN parameter tube discussed in Sec. 7.2. The upcoming *IMAP* mission will relax this limitation because the *IMAP*-Lo camera can be pointed between 60° and 165° from the Sun, thus providing a tomographic ENA scan of the OHS. This capability should provide a new perception of the 3D flow geometry in the OHS.

In addition, there is no discernible separation between the ISN and secondary H flows. Because of the one-to-four mass ratio relative to He, the thermal speed of H is twice as large, widening its angular distribution in the sky. Thus, the primary and secondary H flows form a combined distribution, due to varying proportions depending on look direction. Also, the intense radiation pressure and much heavier ionization loss substantially modify the H velocity distribution from the HP to the observer at 1 AU. Maximizing the secondary to primary ISN H ratio differences in the observations through a variation in look direction on *IMAP* will provide deeper insight into these distributions [158].

To appreciate the critical secondary neutral diagnostic opportunities, Fig. 7.14 shows the arrival direction of the primary ISN flow (Sec. 7.2), the center of the ENA Ribbon (Sec. 7.3), and the apparent secondary arrival directions at the HP. Within uncertainties, the flow directions of all components appear to fall into a common $\vec{B}_{ISM} \times \vec{v}_{ISN}$ plane. A deflection of the IS plasma flow in the OHS within this plane toward $\vec{B}_{ISM}$ relative to the pristine ISN flow may explain this observation [82, 84, 159]. The neutrals that emerge from the plasma *via* CX carry this information to the observer. The secondary He and O populations are well-resolved [79, 156]. The ISN H contains both components and thus is less deflected [87, 160]. The primary ISN O flow is slightly deflected in the opposite direction [161], indicating preferential CX extinction of slower ISN O.

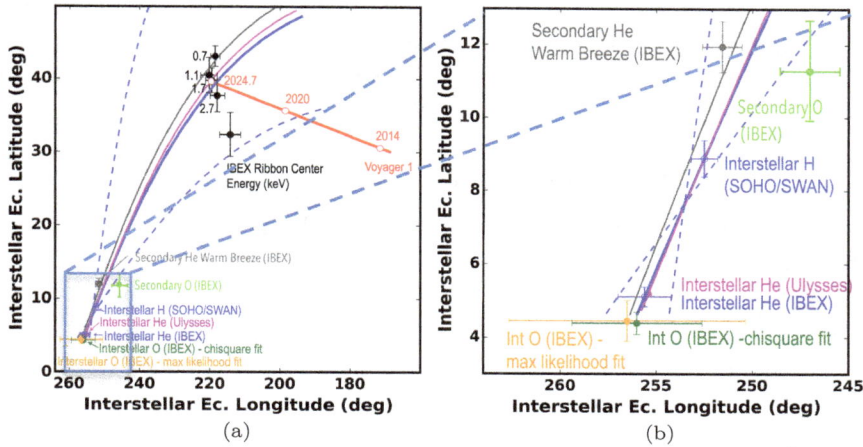

Fig. 7.14. Arrival directions of ISN H based on *SOHO* SWAN [87, 160], ISN He found on *Ulysses* GAS [71] and *IBEX* [72], ISN O [161], and secondary He from *IBEX* [79] and the Ribbon centers at 5 *IBEX* energies [90] with error bars. The red line shows an extrapolated shift of the ISMF direction at *Voyager 1*, based on a trend during quiet ISMF times [141]. Three attempts to derive the projection of the $\vec{B}_{ISM} \times \vec{v}_{ISN}$ plane in the sky use ISN H and He from *Ulysses* (purple), ISN H and He from *IBEX* (blue), and a χ^2 fit to ISN H, He, and secondary He (grey). The blue dashed lines indicate the uncertainty range of the plane. A blow-up of the part of the sky that contains the ISN flow directions is on the right [161]. (© AAS. Reproduced with permission, adapted).

Analyzing the secondary components for different species will test the IS plasma flow deflection models and constrain their valid parameter space. H, He, and O differ in m by factors of four, but their temperatures in the ISM appear equal within errors [47, 63, 67, 161], which implies that their thermal speeds and flow Mach numbers differ by factors of two. These differences should stand out in their deflection strength, likely most substantial for O with the highest Mach number. The plasma flow deflection may not be precisely symmetric about the $\vec{B}_{ISM} \times \vec{v}_{ISN}$ plane, which may become evident in asymmetries of the secondary components with much-improved counting statistics and reduced background on *IMAP*. Tomographic ENA images may reveal the fully 3D flow pattern along with the HP shape and the ISMF draping. In addition, a non-thermal behavior may become visible, likely caused by continuous PUI injection from the CX that produces the secondary neutrals.

Finally, the density of He^+ in the OHS and the ionization fraction of He in the LIC can be deduced from the secondary He flux [157], using the IHS thickness from simulations, the ISN He density outside the heliosphere from PUI [21, 42] and ISN observations [44]. With improved counting statistics of secondary O, the ISN H density at the HP from PUI [41] and SW slowdown observations [25], and the neutral O density from PUI observations [42], also the LIC O ionization fraction could be deduced. Better separation of ISN and secondary H with *IMAP* should enable a similar analysis for LIC H. These results will provide much-needed constraints on the ionization and radiation environment of the solar neighborhood [5, 6].

7.5. Globally Distributed ENAs

Finally, we turn to the GDF of ENAs generated in the IHS and distributed across the sky. Relative to the more localized heliospheric sources, *i.e.*, the ISN flow, the ENA Ribbon, and the secondaries, they can be compared with the scattered light in the fog when trying to surmise mistakenly how far a street stretches, based on the visible number of streetlights (opening of Chap. 1). However, the GDF carries valuable information about the "scatterers" in the IHS. Most importantly, the simulated ENA maps of the GDF sold the *IBEX* mission as the tool to obtain the global structure of the TS and IHS [61]. Now we know that *IBEX*'s top goal was only achievable after removing the unexpected bright Ribbon from the ENA Maps [93]. Sec. 7.5.1 will describe the separation technique and the resulting maps, and Sec. 7.5.2 the challenges and opportunities.

7.5.1. *Extraction of the globally distributed ENAs*

The extraction of the GDF starts with a Ribbon that is almost circular and aligned along a cone around a central location in the sky in ecliptic longitude and latitude $(\lambda_{RC}, \beta_{RC}) = (219.2 \pm 1.3°, 39.9 \pm 2.3°)$ with a half cone angle $\varphi_C = 74.5 \pm 2.0°$ [90]. To make use of this symmetry, the ENA maps are rotated into a coordinate

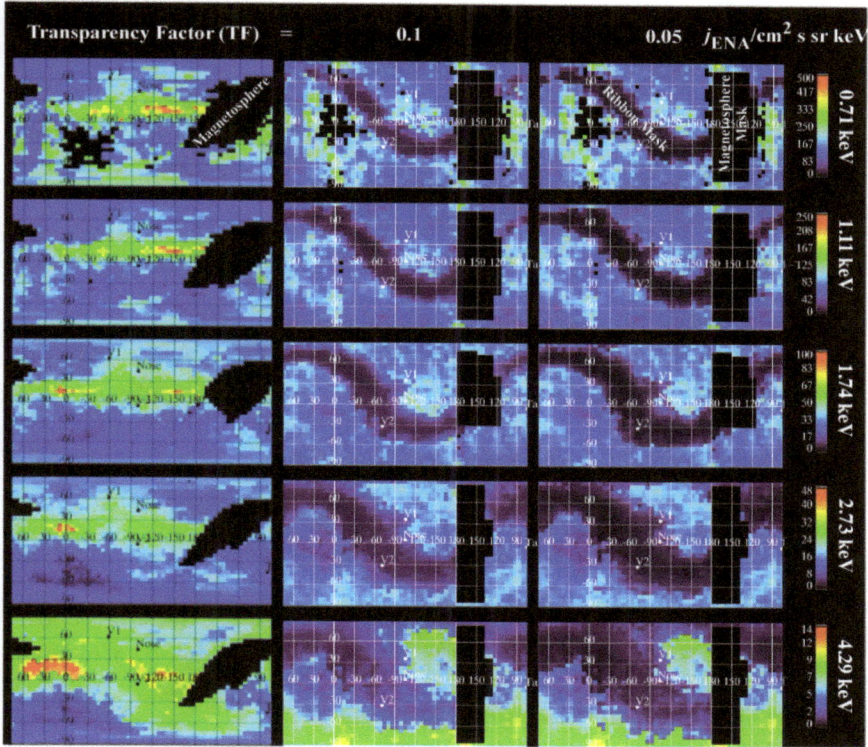

Fig. 7.15. *IBEX*-Hi ENA maps for five energies in the Ribbon coordinate system (left). Magnetospheric $M_{Mag}(\phi, \theta)$, S/N $M_{S/N}(\phi, \theta)$, and Ribbon $M_{Rib}(\phi, \theta)$ masks applied in the heliospheric coordinate system, result in black pixels, for $TF = 0.1$ (center) and 0.05 (right) [93]. (© AAS. Reproduced with permission, adapted).

system with the Ribbon center as one of its poles, as shown in Fig. 7.7 for *IBEX*-Hi maps at 0.71–4.29 keV. In the left column of Fig. 7.15, the Ribbon lines up at a fixed latitude, $\theta \approx +15°$ for 0.71–2.73 keV, but is much broader and less contiguous for 4.29 keV at $\theta \approx +12°$. Erased from this analysis and shown in black are regions with a substantial ENA signal from Earth's magnetosphere and where the statistical uncertainty calculated from a covariance analysis exceeds 36%. A magnetospheric mask $M_{Mag}(\phi, \theta)$ and a low S/N mask $M_{S/N}(\phi, \theta)$ set these pixels to 0. Next, the total ENA flux $j_{Tot}(\phi, \theta)$ in longitude ϕ and latitude θ is fitted to a Ribbon flux $j_{Rib[1]}(\phi, \theta)$

that represents a Gaussian in θ, centered around $\theta \approx 15°$ or $12°$, and a GDF component $j_{GDF[1]}(\phi, \theta)$ approximated by a sloped linear function in θ.

A Ribbon mask $M_{Rib}(\phi, \theta)$ is constructed from the GDF and Ribbon flux ratio with the initial approximations $j_{GDF[1]}(\phi, \theta)$ and $j_{Rib[1]}(\phi, \theta)$.

$$M_{Rib[1]}(\phi, \theta) = Min(1, \{TF \cdot j_{GDF[1]}(\phi, \theta)/j_{Rib[1]}(\phi, \theta)\}) \quad (7.19)$$

TF is a transparency factor that controls and smooths the suppression of the Ribbon fluxes at the edge of the region. After variation between 0.0025 and 1, $TF = 0.05$ provides the most consistent results in the analysis. $M_{Rib}(\phi, \theta)$ is set to 1 far away from the Ribbon and calculated iteratively to improve the separation. All three combine into a single mask [93]:

$$M_{1-GDF}(\phi, \theta) = M_{Rib}(\phi, \theta) \cdot M_{Mag}(\phi, \theta) \cdot M_{S/N}(\phi, \theta) \quad (7.20)$$

To obtain a realistic and smooth 2D interpolation for the GDF across the Ribbon, the differential energy flux for location (ϕ, θ) yields:

$$\langle j(\phi, \theta) \rangle = M_{Tot[1]k} j_k + (1 - M_{Tot[1]k})$$
$$\times \frac{\sum_{i=1}^{N} M_{Tot[1]i} j_i \exp\{-\alpha_i(\phi, \theta)/\Delta\alpha\}}{\sum_{i=1}^{N} M_{Tot[1]i} \exp\{-\alpha_i(\phi, \theta)/\Delta\alpha\}} \quad (7.21)$$

The index k refers to the map pixel that contains the look direction (ϕ, θ), and N is the number of pixels considered around pixel k. j_i is the differential flux observed in pixel i. Eq. (7.21) contains an additional factor $\alpha_i(\phi, \theta)$ that describes how the fluxes observed in each pixel fall off with the angular distance from the center of pixel k, using the halfwidth of the *IBEX* sensor FOV, $\Delta\alpha = 3.5°$. A covariance matrix similar to Eq. (7.21) determines the standard deviations within each pixel:

$$\langle \text{var}(\phi, \theta) \rangle = M_{Tot[1]k} \sigma_k^2 + (1 - M_{Tot[1]k})$$
$$\times \frac{\sum_{i=1}^{N} M_{Tot[1]i} \sigma_i^2 \exp\{-\alpha_i(\phi, \theta)/\Delta\alpha\}}{\sum_{i=1}^{N} M_{Tot[1]i} \exp\{-\alpha_i(\phi, \theta)/\Delta\alpha\}} \quad (7.22)$$

where σ_i is the statistical uncertainty in pixel i. An iteration to obtain $M_{Rib[2]}(\phi, \theta)$ in Eq. (7.19) starts from the results $j_{GDF[2]}(\phi, \theta)$ and $j_{Rib[2]}(\phi, \theta) = j_{kTot} - j_{GDF[2]}(\phi, \theta)$ obtained from Eq. (7.21). This process converges within a few steps to similar stable solutions for a wide range of TF values. $TF = 0.05$ generates a reasonable mask for the Ribbon that is neither too wide (0.025) nor too narrow (0.1), not covering the Ribbon at higher energies.

The final combined mask $M_{Tot[Fin]}$ is the basis for deriving separate GDF $(j_{GDF[Fin]}(k))$ and Ribbon $(j_{Rib[Fin]}(k))$ fluxes for each map pixel k. Figure 7.16 shows two cuts in latitude, summed over $30°$ in longitude (0–30° left, −150–180° right), together with

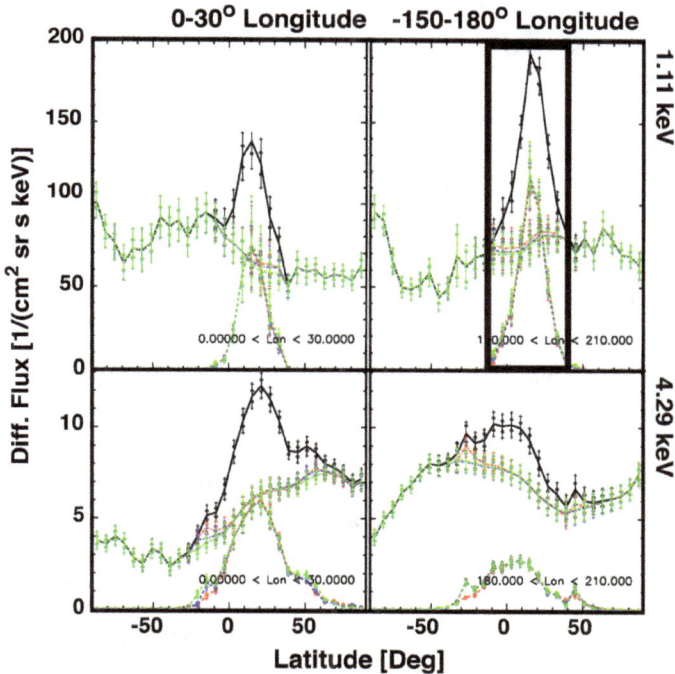

Fig. 7.16.   GDF, Ribbon (both in green), and total (black) fluxes for two energy bands of *IBEX*-Hi as a function of latitude for two longitude ranges in the rotated Ribbon frame with total statistical uncertainties. The final fit results using $TF = 0.1$ (red), 0.05 (blue), and 0.025 (green) are very similar within the uncertainties, demonstrating the robustness of the method [93]. (© AAS. Reproduced with permission, adapted).

the statistical 1σ uncertainty for two energy bands. The iterative procedure produces a reasonable angular distribution for both ENA components beyond the simple initial approximation.

This method for separating the GDF and Ribbon fluxes has facilitated follow-up investigations on each of the two populations, making it an essential technique for analyzing overlaid diffuse and localized object regions in all observational maps. Indistinguishable ENA components in the skymaps would otherwise hamper such studies.

7.5.2. *Challenges and opportunities with the GDF of ENAs*

The GDF is observed simultaneously with other ENA sources in the same FOV, thus making them *a priori* indistinguishable. ENAs do not carry any marker of their origin, such as narrow spectral lines in the case of electromagnetic radiation. Also, none of the ENA sources are point sources, thus requiring methods as described in Sec. 7.5.1 to disentangle them. This inherent challenge to ENA imaging is often compounded by additional foreground sources, such as planetary magnetospheres or atmospheres, and low counting statistics near the sensor background.

Let us illustrate the last challenge based on Fig. 7.16. The GDF attains flux levels as low as 3 $(cm^2 \ sr \ s \ keV)^{-1}$ at 4.29 keV, where the *IBEX*-Hi collecting power is 0.0045 cm^2 sr keV/keV and $\Delta E = 2.8$ keV [92], leading to an instantaneous count rate of $0.04 \, s^{-1}$. With 60 6°-map pixels per spin and six energy steps, 10 ENA counts/day are accrued on average, hitting the statistical limit per pixel. Even though *IBEX* covers the same circle in the sky for 6–7 days over one year, many pixels only feature one or two days with a low enough or well-characterized background. Therefore, the Ribbon separation became feasible only after two years into the mission.

Even with triple coincidence, *IBEX*-Hi is subject to a general background due to penetrating cosmic rays of 0.04–$0.06 \, s^{-1}$ [94], comparable to the lowest GDF rates. Therefore, based on their correlation with neutron monitor rates and high energy cosmic ray fluxes from *LRO* CRaTER, the *IBEX*-Hi ENA fluxes are corrected for these background rates [162].

Interference of more prominent ENA sources (the moon or planetary magnetospheres) is considered, typically as part of the mask, to obtain a clean GDF. Such backgrounds and foregrounds become even more prevalent at the lower energies that *IBEX*-Lo covers. Here, the ISN flow discussed in Sec. 7.2 and the secondary neutrals covered in Sec. 7.4 are so overwhelmingly strong (ISN) or extended across the sky map (secondaries) that there has not been a separation attempt yet. Only a small fraction of the sky is unaffected by the Earth's magnetosphere and these two competing heliospheric sources to obtain the Ribbon and GDF spectra [64, 163, 164]. Despite these challenges, it is worth pushing the limits of the GDF observations with a mission, such as *IMAP*, that avoids the most formidable foreground of the Earth's magnetosphere altogether [134]. Several diagnostic opportunities are more than worth the effort.

- *IHS Thickness from ENA and in-situ Observations*: The GDF originates in the IHS between the TS and the HP and is distributed over the entire sky, making it the harbinger of the 3D shape of the heliospheric boundary region. This shape was one of the two original *IBEX* quests [61]. Because the two *Voyagers* have traversed the IHS simultaneously with the prime and early extended *IBEX* mission, a direct comparison of the GDF in the *Voyager* directions provides ground truth depth and distance information and adds an absolute scale to the *IBEX* maps.

The IHS thickness had been suggested as one of the first ENA map targets, tapping Eq. (3.10). The integration length ΔL or the thickness of the source region remains the unknown quantity after ENA maps ($J_{ENA}(E, \theta, \phi)$) from *SOHO* HSTOF, *Cassini* INCA, and *IBEX*, and *in-situ* measurements of ions ($f_i(v)$) on the *Voyagers*, have become available. With *Voyager* crossings of the TS and the HP, there is even an independent validation. However, not so fast! There has been an ongoing debate about the results. This topic involves the GDF of ENAs, and the removal of any other sources is necessary before using their fluxes to extract ΔL

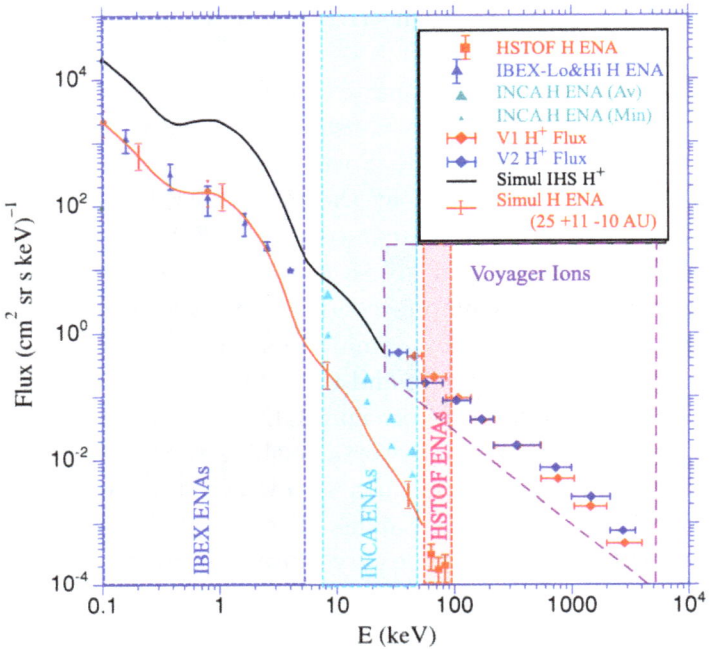

Fig. 7.17. ENA fluxes from *IBEX* in the *Voyager 1* and *2* directions (blue triangles), INCA (cyan triangles, average: large, minimum: small), HSTOF (red squares), *Voyager 1* and *2* H^+ fluxes (red and blue diamonds). Shaded boxes highlight the energy ranges of these observations in the respective colors (purple for both *Voyagers*). The solid red line with four sample error bars shows the estimated ENA spectrum using the simulated (black line) and measured H^+ spectrum for $\Delta L = 25(+11/-10)$ AU and $n_{IHSH} = 0.1\,\mathrm{cm}^{-3}$ [165]. (© AAS. Reproduced with permission, adapted).

unambiguously. The *Voyager* crossings only return valid results for stationary heliospheric boundaries, which is typically not the case. Therefore, the deduced ΔL can only be an estimated value. This technique is illustrated in Fig. 7.17, combining early remote ENA and Voyager *in-situ* H^+ measurements [165]. Shown are the ENA fluxes found by *IBEX* in the *Voyager 1* and *2* directions (blue triangles), by INCA (cyan triangles, average: large, minimum: small), and by HSTOF ±45° about the apex in the ecliptic (red squares), along with *Voyager 1* and *2* H^+ fluxes after crossing the TS and well into IHS (red and blue diamonds). The H^+ and ENA measurements only overlap in a small portion of the spectrum. Simulations of SW and PUIs (black line) at the TS extend the H^+ measurements to lower energies [166]. With the simulated and measured H^+ differential flux (0.1–100 keV) as the sole source of H ENAs *via* CX with ISN H ($n_{IHSH} = 0.1\,\text{cm}^{-3}$) in the IHS, the red line with sample error bars represents the resulting H ENA spectrum for $\Delta L = 25^{+11}_{-10}$ AU.

As an illustration of the challenges, let us estimate ΔL from two overlapping ENA/H^+ measurements, using the flux average from HSTOF at 70 keV and INCA at 45 keV [167]. Combining ENAs and *Voyager* H^+ based on Eq. (3.10) and replacing the integral yields:

$$\frac{j_{H^+}(E)}{j_{ENA}(E)} = \frac{1}{n_H \cdot \sigma_{CX}(E) \cdot \Delta L} = C \qquad (7.22)$$

Using a weighted average for $n_H = 0.083 \pm 0.023\,\text{cm}^{-3}$ [168] and $\sigma_{CX}(70\text{keV}) = 3.77 \cdot 10^{-17}\,\text{cm}^2$ [117], we find $C(70\,\text{keV}) = 600$ for the HSTOF observations at 70 keV and $\Delta L = 35.5 \pm 7.3$ AU. For the *Voyager* and INCA channels, we find $C(45\,\text{keV}) = 60$, using the minimum ENA flux [167]. With $\sigma_{CX}(46\,\text{keV}) = 1.5 \cdot 10^{-16}\,\text{cm}^2$ [117] instead of $\sigma_{CX}(46\,\text{keV}) = 2.32 \cdot 10^{-16}\,\text{cm}^2$ as used earlier [167], we find $\Delta L = 66.9 \pm 14.5$ AU. These values are close to the ones reported by Hsieh *et al.* [2010] [165], slight differences likely due to different ISN H densities at the TS and choosing specific energies in Eq. (7.22).

We note the thickness derived from INCA observations [167] is much larger than that from HSTOF and *IBEX* [165]. The consistently higher ENA fluxes from INCA, seen in Fig. 7.17 even for the minimum fluxes, could be a reason. This higher flux may have to do with the dominance of the INCA Belt [167] in the ENA maps. The *IBEX* maps have shown how important it is to separate the Ribbon and the GDF. Only the latter appears to come from the IHS. If the Belt ENAs come mainly from the OHS — not clear at this writing — their contribution would drive ΔL to the larger values. The Ribbon or Belt ENA flux decreases drastically with E due to the E-dependence of the neutral SW spectrum and the CX cross-sections. At HSTOF energies, the Belt or Ribbon may not contribute at all. Additional reasons for the differences may be that the *IBEX*, INCA, and HSTOF observations stem from different times, and HSTOF with a $\pm 17°$ FOV above and below the ecliptic does not overlap with the *Voyager 1* or *2* directions. At this writing, work to resolve the apparent discrepancy shown in Fig. 7.17 is underway.

- *IHS Thickness Estimate with the Ribbon*: Using the Secondary-ENA model, Ribbon ENAs can provide an estimate for the IHS thickness ΔL that does not depend on an absolute ENA flux calibration. It does not even need a simultaneous *in-situ* plasma observation [169]. This method relies on the Ribbon flux ratio at 0.2 and 1.1 keV. The former stems from the IHS and scales directly with ΔL, while the latter reflects the well-constrained supersonic neutral SW independent of ΔL. In the *Voyager 1* direction, ΔL is consistent with this estimate based on the TS and HP crossings, albeit with significant uncertainties, mainly due to the poor ENA counting statistics at 0.2 keV. *IMAP* will bring substantial improvements with increased collecting power and less foreground at L1 [134].
- *Total Pressure in the IHS*: A comparison of the total dynamic pressure inferred from the GDF with the pressure at the IHS boundaries can serve as another constraint on the IHS thickness [93]. The total plasma pressure in the IHS is:

$$P_{Plasma,r} = 4\pi m \int_{v_p min}^{v_{pmax}} \left(\frac{1}{3}v_p^4 + u_r^2 v_p^2 \right) f_p(\vec{v}_p) dv_p \qquad (7.23)$$

$f_p(\vec{v}_p)$ is the ion distribution, with particle velocities $\vec{v}_p$ in the plasma frame, and u_r is the radial component of the plasma flow velocity. The ion distribution directly relates to the observed ENA distribution $f_{p,ENA}(\vec{v}_p)$ via $f_{p,ENA}(\vec{v}_p) = f_p(\vec{v}_p)/(n_H \cdot \sigma_{CX}(E_p) \cdot \Delta L) \cdot \vec{v}_p$ in the plasma frame relates to the velocity $\vec{v}_0$ in the observer frame via $\vec{v}_p = \vec{v}_0 - \vec{u}$, with the bulk flow velocity $\vec{u}$. Only the radial components are relevant in the transformation between the IHS and the observer at 1 AU, simplifying the relation to speeds. The particle speed in the plasma frame is larger than in the rest frame because the average IHS flow is outward, $v_p = v_0 - u_r$.

Together with Eq. (3.9), we find [93]:

$$P_{Plasma,r} \cdot \Delta L = \frac{2\pi m^2}{n_H} \int_{E_{Min}}^{E_{Max}} \frac{j_{0,ENA}(E_0)}{\sigma_{CX}(E_p)}$$

$$\times \left(\frac{(|\vec{v}_0| + u_r)^4}{3|\vec{v}_0|} + \frac{2u_r^2 \left(|\vec{v}_0| + u_r\right)^2}{|\vec{v}_0|} \right) \frac{dE_0}{E_0}$$

$$(7.24)$$

The first term in the parentheses on the RHS represents the isotropic pressure of the particles in the plasma frame, whereas the second term contains the ram pressure in the radial direction from a near-Earth vantage point. However, parts of the distribution outside the ENA energy range of the sensors may also contribute to the total pressure. The ENAs suffer losses from the outer heliosphere to the observer, similar to the ionization losses of the ISN gas. Solar gravitation and radiation pressure deflect low-energy H ENAs, but only insignificantly at energies above 100 eV.

Taking an average radial bulk speed of about 70 km/s in the IHS based on *Voyager* observations, pressure balance with the SW at the TS yields an IHS thickness of $\Delta L \approx 38$ AU [93], which is, within uncertainties, comparable with the value derived above based on Eq. (7.23). While each method is subject to different systematic uncertainties, it is worth noting that the ones based on Eqs. (7.23) and (7.24) hinge upon the absolute flux calibration of the ENA sensors, which is always a formidable challenge.

After using pressure balance to deduce ΔL, it appears logical to produce pressure maps from the ENA fluxes to reveal relative pressure variations in the IHS across the sky. They revealed a maximum pressure region about 20° south of the nose as a significant result [170]. This asymmetry is consistent with the effects of the ISMF on the heliosphere. The location explains the plasma flow directions observed by the two *Voyagers* as away from the maximum pressure and not from the heliospheric nose. In other words, ENA-generated pressure maps provide essential information about the large-scale dynamic structure of the heliospheric boundary regions, and the pressure maps provide the IHS thickness in most directions.

When analyzing the heliotail, it turns out that CX extinction limits the reach of the ENA signal, and cooling or heating due to SW pressure changes affects the total pressure [171].

- *Kinetic Properties of the IHS Plasma*: Beyond the second moment of the ion distribution in the IHS, the shape of the ENA spectra and their variation across the sky provide insight into the kinetic structure. It has become clear that the IHS ion distribution is far from thermodynamic equilibrium. Other populations, including PUIs, or a description as κ-distribution is needed to interpret the spectra physically [172–174].

Here, we conclude our narrative, pushing the limits of neutral-atom diagnostics by studying the outer regions of the heliosphere and the ISM. This push includes their interaction and related consequences, along with a view into the Sun's immediate neighborhood in a manner that was not possible before. A summary of neutral-atom diagnostics and a brief forecast of what is to come in neutral-atom astronomy will conclude the book in Chap. 8.

References

1. Cox, D.P., & Reynolds, R. J. (1987). The local interstellar medium, *Ann. Rev. Astr. Ap.*, **25**, 303–344.
2. Frisch, P. C., Bzowski, M., Grün, E. *et al.* (2009). The galactic environment of the sun: interstellar material inside and outside of the heliosphere, *Space Sci. Rev.*, **146**, 235–273.
3. Cheng, K.-P., & Bruhweiler, F. C. (1990). Ionization processes in the local interstellar medium: effects of the hot coronal substrate, *Astrophys. J.*, **364**, 573–581.
4. Wolff, B., Koester, D., & Lallement, R. (1999). Evidence for an ionization gradient in the local interstellar medium: EUVE observations of white dwarfs, *Astron. Astrophys.*, **346**, 969–978.
5. Slavin, J., & Frisch P.C. (2002). The ionization of nearby interstellar gas, *Astrophys. J.*, **565**, 364–379.
6. Slavin, J. D., & Frisch, P. C. (2008). The boundary conditions of the heliosphere: photoionization models constrained by interstellar and *in situ* data, *Astron. Astrophys.*, **491**, 53–68.

7. McClintock, W., Henry, R. C., Linsky, J. L., & Moos, W. H. (1978). Ultraviolet observations of cool stars. VII. local interstellar hydrogen and deuterium Lyman-alpha, *Astrophys. J.*, **225**, 465–481.

8. Frisch, P. C. (1981). The nearby interstellar medium, *Nature*, **293**, 377–379.

9. Crutcher, R. M. (1982). The local interstellar medium, *Astrophys. J.*, **254**, 82–87.

10. Lallement, R., & Bertin, P. (1992). Northern-hemisphere observations of nearby interstellar gas: possible detection of the local cloud, *Astron. Astrophys.*, **266**, 479–485.

11. Linsky, J. L., Brown, A., Gayley, K. *et al.* (1993). Goddard high-resolution spectrograph observations of the local interstellar medium and the deuterium/hydrogen ratio along the line of sight toward Capella, *Astrophys. J.*, **402**, 694–709.

12. Redfield, S., & Linsky, J. L. (2008). The structure of the local interstellar medium. IV. Dynamics, morphology, physical properties, and implications of cloud-cloud interactions, *Astrophys. J.*, **673**, 283–314.

13. Breitschwerdt, D., de Avillez, M. A., Fuchs, B., & Dettbarn, C. (2009). What physical processes drive the interstellar medium in the Local Bubble? *Space Sci. Rev.*, **143**, 263–276.

14. Fuchs, B., Breitschwerdt, D., de Avillez, M. A., & Dettbarn, C. (2009). Origin of the Local Bubble, *Space Sci. Rev.*, **143**, 437–448.

15. *Solar Journey: The Significance of our Galactic Environment for the Heliosphere and Earth*, Ed. Frisch, P. C., *Astrophysics and Space Science Library*, **Vol. 33** (Springer, 2006).

16. Redfield, S. (2009). Physical properties of the local interstellar medium, *Space Sci. Rev.*, **143**, 323–331.

17. Strömgren, B. (1939). The physical state of interstellar hydrogen, *Astrophys. J.*, **89**, 526–547.

18. Fahr, H. J. (1968). Charge-transfer interactions between solar wind protons and neutral particles in the vicinity of the Sun, *Nature*, **219**, 473–474.

19. Siscoe, G. L., & Mukherjee, N. R. (1972). Upper limits on the lunar atmosphere determined from solar wind measurements, *J. Geophys. Res.*, **77**, 6042–6051.

20. Möbius, E., Hovestadt, D., Klecker, B., Scholer, M., Gloeckler, G., & Ipavich, F. M. (1985). Direct observation of He^+ pick-up ions of interstellar origin in the solar wind, *Nature*, **318**, 426–429.

21. Gloeckler, G., & Geiss, J. (1998). Interstellar and inner source pickup ions observed with SWICS on Ulysses, *Space Sci. Rev.*, **86**, 127–159.

22. Klecker, B. (1995). The anomalous component of cosmic rays in the 3-D heliosphere, *Space Sci. Rev.*, **72**, 419–430.

23. Jokipii, J. R. (1998). Insights into cosmic-ray acceleration from the study of anomalous cosmic rays, *Space Sci. Rev.*, **86**, 161–178.

24. Richardson, J. D., Paularena, K. I., Lazarus, A. J., & Belcher, J. W. (1995). Evidence for a solar wind slowdown in the outer heliosphere? *Geophys Res. Lett.*, **22**, 1469–1472.

25. Richardson, J., Liu, Y., Wang, C., & McComas, D. J. (2008). Determining the LIC H density from the solar wind slowdown, *Astron. Astrophys.*, **491**, 1–5.

26. Möbius, E., Bzowski, M., Müller, H.-R., & P. Wurz (2006). Effects in the inner heliosphere caused by changing conditions in the galactic environment, in *Solar Journey: The Significance of our Galactic Environment for the Heliosphere and Earth*, Ed. P.C. Frisch (Springer) 209–258.

27. Müller, H.-R., & Zank, G. P. (2004). Heliospheric filtration of interstellar heavy atoms: Sensitivity to hydrogen background, *J. Geophys. Res.*, **109**, A07104.

28. Zank, G.P., Müller, H.-R., Florinski, V., & Frisch, P.C. (2006). Heliospheric variation in response to changing interstellar environmen, in *Solar Journey: The Significance of our Galactic Environment for the Heliosphere and Earth*, Ed. Frisch, P.C. (Springer) 23–51.

29. Danby, J. M. A., & Camm, G. L. (1957). Statistical dynamics and accretion, *MNRAS*, **117**, 50–71.

30. Fahr, H. J. (1974). The extraterrestrial UV-background and the nearby interstellar medium, *Space Sci. Rev.*, **15**, 483–540.

31. Fahr, H. J., Lay, G., & Wulf-Mathies, C. (1978). Derivation of interstellar helium gas parameters from an EUV-rocket observation, in *Space Research XVIII* (A79-13382 03-8, Oxford, Pergamon Press, Ltd.), 393–396.

32. Wu, F. M., & Judge, D. L. (1979). Temperature and flow velocity of the interplanetary gases along solar radii, *Astrophys. J.*, **231**, 594–605.

33. Lee, M. A., Kucharek, H., Möbius, E. *et al.* (2012). An analytical model of interstellar gas in the heliosphere tailored to IBEX observations, *Astrophys. J. Suppl.*, **198**:10.

34. Bertaux, J. L., & Blamont, J. E. (1971). Evidence for a source of an extraterrestrial hydrogen Lyman-alpha emission, *Astron. Astrophys.*, **11**, 200–217.

35. Thomas, G. E., & Krassa, R. F. (1971). OGO 5 measurements of the Lyman Alpha sky background, *Astron. Astrophys.*, **11**, 218–233.

36. Adams, T. F., & Frisch, P. C. (1971). High-resolution observations of the Lyman alpha sky background, *Astrophys. J.*, **212**, 300–308.
37. Bertaux, J. L., Lallement, R., Kurt, V. G., & Mironova, E. N. (1985). Characteristics of the local interstellar hydrogen determined from PROGNOZ 5 and 6 interplanetary Lyman-alpha line profile measurements with a hydrogen absorption cell, *Astron. Astrophys.*, **150**, 1–20.
38. Weller, C. S., & Meier, R. R. (1974). Observations of helium in the interplanetary/interstellar wind: the solar-wake effect, *Astrophys. J.*, **193**, 471–476.
39. Möbius, E., Ruciński, D., Hovestadt, D., & Klecker, B. (1995). The Helium Parameters of the Very Local Interstellar Medium as Derived from the Distribution of He$^+$ Pickup Ions in the Solar Wind, *Astron. Astrophys.*, **304**, 505–519.
40. Gloeckler, G., Fisk, L. A., & Geiss, J. (1997). Anomalously small magnetic field in the local interstellar cloud, *Nature*, **386**, 374–377.
41. Bzowski, M., Möbius, E., Tarnopolski, S., Izmodenov, V., & Gloeckler, G. (2008). Density of neutral interstellar hydrogen at the termination shock from Ulysses pickup ion observation, *Astron. Astrophys.*, **491**, 7–19.
42. Gloeckler, G., & Geiss, J. (2001). Composition of the local interstellar cloud from observations of interstellar pickup ions, *AIP Conf. Proc.*, **598**, 281–289.
43. Witte, M., Banaszkiewicz, M., & Rosenbauer, H. (1996). Recent results on the parameters of the interstellar helium from the ULYSSES/GAS experiment, *Space Sci. Rev.*, **78**, 289–296.
44. Witte, M. (2004). Kinetic parameters of interstellar neutral helium: review of results obtained during one solar cycle with the Ulysses/GAS-instrument, *Astron. Astrophys.*, **426**, 835–844.
45. Möbius, E., Kucharek, H., Clark, G. *et al.* (2009). Diagnosing the neutral interstellar gas flow at 1 AU with IBEX-Lo, *Space Sci. Rev.*, **146**, 149–172.
46. Möbius, E., Bochsler, P., Bzowski, M. *et al.* (2009). Direct observations of interstellar H, He, and O by the Interstellar Boundary Explorer, *Science*, **326**, 969–971.
47. Möbius, E., Bochsler, P., Bzowski, M. *et al.* (2012). Interstellar gas flow parameters derived from IBEX-Lo observations in 2009 and 2010 — analytical analysis, *Astrophys. J. Suppl.*, **198**(2):11.
48. Bzowski, M., Kubiak, M. A., Möbius, E. *et al.* (2012). Neutral interstellar helium parameters based on IBEX-Lo observations and test particle calculations, *Astrophys. J. Suppl.*, **198**(2):12.

49. Saul, L., Wurz, P., Rodriguez, D. *et al.* (2012). Local interstellar neutral hydrogen sampled *in-situ* by IBEX, *Astrophys. J. Suppl.*, **198**(2):14.

50. Schwadron, N.A., Moebius, E., Kucharek, H. *et al.* (2013). Solar radiation pressure and local interstellar medium flow parameters from Interstellar Boundary Explorer low energy hydrogen measurements, *Astrophys. J.*, **775**(2):86.

51. Bochsler, P., Petersen, L., Möbius, E. *et al.* (2012). Estimation of the neon/oxygen abundance ratio at the heliospheric termination shock and in the local interstellar medium from IBEX observations, *Astrophys. J. Suppl.*, **198**(2):13.

52. Park, J., Kucharek, H., Möbius, E. *et al.* (2014). The Ne to O abundance ratio of the interstellar medium from the IBEX-Lo observations, *Astrophys. J.*, **795**(1):97.

53. Chalov, S. V., & Fahr, H. J. (2006). Pickup interstellar helium ions in the region of the solar gravitational Cone, *Astron. Lett.*, **32**(7), 487–494.

54. Quinn, P. R., Schwadron, N. A., & Möbius, E. (2016). Transport of helium pickup ions within the focusing cone: reconciling STEREO observations with IBEX, *Astrophys. J.*, **824**(2), 142.

55. Gloeckler, G., Möbius, E., Geiss, J. *et al.* (2004). Observations of the helium focusing cone with pickup ions, *Astron. Astrophys.*, **426**, 845–854.

56. Möbius, E., Klecker, B., Bochsler, P. *et al.* (2010). He pickup ions in the inner heliosphere — diagnostics of the local interstellar gas and of interplanetary conditions, *AIP Conf. Proc.*, **1302**, 37–43.

57. Möbius, E., Lee, M. A., & Drews, C. (2015). Interstellar flow longitude from the symmetry of the pickup ion cutoff at 1 AU, *Astrophys. J.*, **815**(1):20.

58. Taut, A., Berger, L., Möbius, E. *et al.* (2018). Challenges in the determination of the interstellar flow longitude from the pickup ion cutoff, *Astron. Astrophys.*, **611**:61.

59. Bower, J., Moebius, E., Berger, L. *et al.* (2019). Effect of rapid changes of solar wind conditions on the pickup ion velocity distribution, *J. Geophys. Res.*, **124**, 6418–6437.

60. Kóta, J., Hsieh, K. C., Jokipii, J. R., Czechowski, A., & Hilchenbach, M. (2001). Viewing co-rotating interaction regions globally using energetic neutral atoms, *J. Geophys. Res.*, **106**, 24,907–24,914.

61. McComas, D. J., Allegrini, F., Bochsler, P. *et al.* (2009). IBEX — Interstellar Boundary Explorer, *Space Sci. Rev.*, **146**, 11–33.

62. Lee, M. A., Möbius, E., & Leonard, T. (2015). The analytical structure of the primary interstellar helium distribution function in the heliosphere, *Astrophys. J. Suppl.*, **220**(2):23.p
63. Möbius, E., Bzowski, M. P.C. *et al.* (2015). Interstellar flow and temperature determination with IBEX: Robustness and sensitivity to systematic effects, *Astrophys. J. Suppl.*, **220**(2):24.
64. Galli, A., Wurz, P., Park, J. *et al.* (2015). Can IBEX detect interstellar neutral helium or oxygen from anti-ram directions? *Astrophys. J. Suppl.*, **220**(2):30.
65. Schwadron, N. A., Möbius, E., McComas, D. J. *et al.* (2021) Interstellar Neutral He Parameters from Crossing Parameter Tubes with the Interstellar Mapping and Acceleration Probe (IMAP) informed by 10 Years of Interstellar Boundary Explorer (IBEX) Observations, *Astrophys. J. Suppl.*, in press.
66. McComas, D. J., Alexashov, D., Bzowski, M. *et al.* (2012). The heliosphere's subsonic interstellar interaction: no bow shock, *Science*, **336**, 1291–1293.
67. Schwadron, N. A., Möbius, E., Leonard, T. *et al.* (2015). Determination of interstellar He parameters using 5 years of data from the Interstellar Boundary Explorer — beyond closed form approximations, *Astrophys. J. Suppl.*, **220**(2):25.
68. Bzowski, M., Swaczyna, P., Kubiak, M. A. *et al.* (2015). Interstellar neutral helium in the heliosphere from Interstellar Boundary Explorer observations III. Mach number of the flow, velocity vector, and temperature from the first six years of measurements, *Astrophys. J. Suppl.*, **220**(2):28.
69. Swaczyna, P., Bzowski, M., Kubiak, M. A. *et al.* (2015). Interstellar neutral helium in the heliosphere from Interstellar Boundary Explorer observations I. Uncertainties and backgrounds in the data and parameter determination method, *Astrophys. J. Suppl.*, **220**(2):26.
70. Swaczyna, P., Bzowski, M., Kubiak, M. A. *et al.* (2018). Interstellar neutral helium in the heliosphere from IBEX Observations. V. Observations in IBEX-Lo ESA steps 1, 2, and 3, *Astrophys. J.*, **854**(2):119.
71. Wood, B. E., Müller, H.-R., & Witte, M. (2015) Revisiting Ulysses observations of interstellar helium, *Astrophys. J.*, **801**(1):62.
72. McComas, D. J., Bzowski, M., Frisch, P. *et al.* (2015). Warmer local interstellar medium: a possible resolution of the Ulysses-IBEX enigma, *Astrophys. J.*, **801**(1):28.
73. Möbius, E., Bzowski, M., Fuselier, S. A. *et al.* (2015). Interstellar gas flow vector and temperature determination over 5 years of IBEX observations, *J. Phys. Conf. Series*, **577**(1):012019.

74. Bzowski, M., Kubiak, M. A., Hłond, M. *et al.* (2014). Neutral interstellar He parameters in front of the heliosphere 1994–2007, *Astron. Astrophys.*, **569**:A8.

75. Fuselier, S. A., Ghielmetti, A. G., Hertzberg, E. *et al.* (2009). IBEX-Lo Sensor, *Space Sci. Rev.*, **146**, 117–147.

76. Hłond, M., Bzowski, M., Möbius, E. *et al.* (2012). Precision pointing of IBEX-Lo observations, *Astrophys. J. Suppl.*, **198**(2):9.

77. Wood, B., Müller, H.-R., Bzowski, M. *et al.* (2015). Exploring the possibility of O and Ne contamination in Ulysses observations of interstellar helium, *Astrophys. J. Suppl.*, **220**:31.

78. Kubiak, M. A., Bzowski, M., Sokòl, J.M. *et al.* (2014). Warm Breeze from the starboard bow: a new population of neutral helium in the heliosphere, *Astrophys. J. Suppl.*, **213**:29.

79. Kubiak, M. A., Swaczyna, P., Bzowski, M. *et al.* (2016). Interstellar neutral helium in the heliosphere from IBEX observations. IV. Flow vector, Mach number, and abundance of the warm breeze, *Astrophys. J. suppl.*, **223**:25.

80. Livadiotis, G., & McComas, D. J. (2013). Understanding kappa distributions: a toolbox for space science and astrophysics, *Space Sci. Rev.*, **175**, 183–214.

81. Stone, E. C., Cummings, A. C., McDonald, F. B., Heikkila, B. C., Lal, N., & Webber, W. R. (2008). An asymmetric solar wind termination shock, *Nature*, **454**, 71–74.

82. Opher, M., Stone, E.C., & Liewer, P. C. (2006). The effects of a local interstellar mgnetic field on Voyager 1 and 2 observations, *Astrophys. J.*, **640**, L71–L74.

83. Izmodenov, V.V. (2009). Local interstellar parameters as they are inferred from analysis of observations inside the heliosphere, *Space Sci. Rev.*, **143**, 139–150.

84. Pogorelov, N. V., Borovikov, S. N., Zank, G. P., & Ogino, T. (2009). Three-dimensional features of the outer heliosphere due to coupling between the interstellar and interplanetary magnetic fields. III. The effects of solar rotation and activity cycle, *Astrophys. J.*, **696**, 1478–1490.

85. McComas, D. J., Allegrini, F., Bochsler, P. *et al.* (2009). Global observations of the interstellar interaction from the Interstellar Boundary Explorer (IBEX), *Science*, **326**, 959–962.

86. Schwadron, N. A., Bzowski, M., Crew, G. B. *et al.* (2009). Comparison of Interstellar Boundary Explorer observations with 3D global heliospheric models, *Science*, **326**, 966–968.

87. Lallement, R., Quemerais, E., Bertaux, J. L. *et al.* (2005). Deflection of the interstellar neutral hydrogen flow across the heliospheric interface, *Science*, **307**, 1447–1449.

88. Opher, M., Richardson, J. D., Toth, G., & Gombosi, T. I. (2009). Confronting observations and modeling: the role of the interstellar magnetic field in Voyager 1 and 2 asymmetries, *Space Sci. Rev.*, **143**, 43–55.

89. Parker, E. N. (1961). The stellar-wind regions, *Astrophys. J.*, **134**, 20–27.

90. Funsten, H. O., DeMajistre, R., Frisch, P. C. *et al.* (2013). Circularity of the IBEX ribbon of enhanced energetic neutral atom (ENA) flux, *Astrophys. J.*, **776**:30.

91. Swaczyna, P., Bzowski, M., & Sokol, J. (2016). The energy-dependent position of the IBEX ribbon due to the solar wind structure, *Astrophys. J.*, **827**:71.

92. Funsten, H. O., Allegrini, F., Crew, G. B. *et al.* (2009). Structures and spectral variations of the outer heliosphere in IBEX energetic neutral atom maps, *Science*, **326**, 964–966.

93. Schwadron, N. A., Allegrini, F., Bzowski, M. *et al.* (2011). Separation of the IBEX ribbon from globally distributed energetic neutral atom flux, *Astrophys J*, **731**:56.

94. McComas, D. J., Dayeh, M.A., Allegrini, F. *et al.* (2012). The first three years of IBEX observations and our evolving heliosphere, *Astrophys. J. Suppl.*, **203**:1.

95. McComas, D. J., Zirnstein, E. J., Bzowski, M. *et al.* (2017). Seven years of imaging the global heliosphere with IBEX, *Astrophys. J. Suppl.*, **229**:41.

96. McComas, D. J., Ebert, R. W., Elliott, H. A. *et al.* (2008). Weaker solar wind from the polar coronal holes and the whole Sun, *Geophys. Res. Lett.*, **35**:L18103.

97. Prusti, T., de Bruijne, J. H. J., Brown, A. G. A. *et al.* (2016). The GAIA Mission, *Astron. Astrophys.*, **595**:1.

98. McComas, D. J., Lewis, W.S., & Schwadron, N.A. (2014). IBEX's enigmatic ribbon in the sky and its many possible sources, *Rev. Geophys.*, **52**, 118–155.

99. Swaczyna, P., Bzowski, M., Christian, E. R., Funsten, H. O., McComas, D. J., & Schwadron, N. A. (2016). Distance to the IBEX ribbon source inferred from parallax, *Astrophys. J.*, **823**:119.

100. Kucharek, H., Fuselier, S. A., Wurz, P. *et al.* (2013). Solar wind as a possible source for fast temporal variations of the heliospheric Ribbon, *Astrophys. J.*, **776**:109.

101. Fahr, H.-J., SieWert, M., McComas, D.J., & Swadron, N. A. (2011). The inner heliospheric source for keV-energetic IBEX ENAs: The anomalous cosmic ray-induced component, *Astron. Astrophys.*, **531**:A77.

102. Siewert, M., Fahr, H.-J., McComas, D. J., & Schwadron, N. A. (2012). The inner heliosheath source for keV-ENAs observed with IBE: Shock-processed downstream pick-up ions, *Astron. Astrophys.*, **539**:A75.

103. Siewert, M., Fahr, H.-J., McComas, D. J., & Schwadron, N. A. (2013). Spectral properties of keV-energetic ion populations inside the heliopause reflected by IBEX-relevant energetic neutral atoms, *Astron. Astrophys.*, **551**:A58.

104. Fitchtner, H., Scherer, K., Effenberger, F., Zönnchen, J. N., & McComas, J. (2014). The IBEX ribbon as a signature of the inhomogeneity of the local interstellar medium, *Astron. Astrophys.*, **561**:A74.

105. Kivelson, M. G., & Jia, X. (2013). An MHD model of Ganymede's mini-magnetosphere suggests that the heliosphere forms in a sub-Alfvénic flow, *J. Geophys. Res.*, **118**, 6, 839–6, 846.

106. Heerikhuisen, J., Pogorelov, N. V., Zan, G. P. *et al.* (2010). Pick-up ions in the outer heliosheath: a possible mechanism for the Interstellar Boundary EXplorer ribbon, *Astrophys. J. Lett.*, **708**, L126–L130.

107. Chalov, S. V., Alexashov, D. B., McComas, D., Izmodenov, V. V., Malama, Y. G., & Schwadron, N. (2010). Scatter-Free Pickup Ions Beyond the Heliopause as a Model for the Interstellar Boundary Explorer Ribbon, *Astrophys. J. Lett.*, **716**, L99–L102.

108. Möbius, E., Liu, K., Funsten, H, Gary, S. P., & Winske, D. (2013). Analytic model of the IBEX ribbon with neutral solar wind based ion pickup beyond the heliopause, *Astrophys. J.*, **766**:129.

109. Zirnstein, E. J., Heerikhuisen, J., Funsten, H. O., Livadiotis, G., McComas, D. J., & Pogorelov, N. V. (2016). Local interstellar magnetic field determined from the Interstellar Boundary Explorer ribbon, *Astrophys. J. Lett.*, **818**:L18.

110. Zirnstein, E. J., Funsten, H. O., Heerikhuisen, J., & McComas, D. J. (2016). Effects of solar wind speed on the secondary energetic neutral source of the Interstellar Boundary Explorer ribbon, *Astron. Astrophys.*, **586**:A31.

111. Gamayunov, K., Zhang, M., & Rassoul, H. (2010). Pitch angle scattering in the outer heliosheath and formation of the Interstellar Boundary Explorer ribbon, *Astrophys. J.*, **725**, 2,251–2,261.

112. Liu, K., Möbius, E., Gary S. P., & Winske, D. (2012). Pickup proton instabilities and scattering in the distant solar wind and the outer heliosheath: Hybrid simulations, *J. Geophys. Res.*, **117**:A10307.

113. Schwadron, N. A., & McComas, D. J. (2013). Spatial retention of ions producing the IBEX ribbon, *Astrophys. J.*, **264**:92.

114. Isenberg, P. A. (2014). Spatial confinement of the IBEX ribbon: A dominant turbulence mechanism, *Astrophys. J.*, **787**:76.

115. Giacalone, J., & Jokipii, J. R. (2015). A new model for the heliosphere's "IBEX ribbon", *Astrophys. J. Lett.*, **812**:L9.

116. Grzedzielski, S., Bzowski, M., Czechowski, A., Funsten, H. O., McComas, D. J., & Schwadron, N. A. (2010). A possible generation mechanism for the IBEX ribbon from outside the heliosphere, *Astrophys. J. Lett.*, **715**, L84–L87.

117. Lindsay, B. G., & Stebbings, R. F. (2005). Charge transfer cross sections for energetic neutral atom data analysis, *J. Geophys. Res.*, **110**:A12213.

118. Sokòl, J. M., Bzowski, M., Tokumaru, M., Fujiki, K., & McComas, D. J. (2013). Heliolatitude and time variations of solar wind structure from *in situ* measurements and interplanetary scintillation observations, *Solar Physics*, **285**, 167–200.

119. Sokòl, J. M., Swaczyna, P., Bzowski, M., & Tokumaru, M. (2015). Reconstruction of helio-latitudinal structure of the solar wind proton speed and density, *Solar Physics*, **290**, 2,589–2,615.

120. McComas, D. J., Barraclough, B. L., Funsten, H. O. *et al.* (2000). Solar wind observations over Ulysses' first full polar orbit, *J. Geophys. Res.*, **105**, 10,419–10,433.

121. McComas, D. J., Elliott, H. A., Schwadron, N. A., Gosling, J. T., Skoug, R. M., & Goldstein, B. E. (2003). The three-dimensional solar wind around solar maximum, *Geophys. Res. Lett.*, **30**(10):1,517.

122. Stone, E. C., Cummings, A. C., McDonald, F. B., Heikkila, B. C., Lal, N., & Webber, W. R. (2013). Voyager 1 observes low-energy galactic cosmic rays in a region depleted of heliospheric ions, *Science*, **341**, 150–153.

123. Gurnett, D. A., Kurth, W. S., Burlaga, L. F., & Ness, N. F. (2013). *In situ* observations of interstellar plasma with Voyager 1, *Science*, **341**, 1,489–1,492.

124. Wu, C. S., Winske, D., & Gaffey Jr., J. D. (1986). Rapid pickup of cometary ions due to strong magnetic turbulence, *Geophys. Res. Lett.*, **13**, 865–868.

125. Gary, S. P., Hinata, S., Madland, C. D., & Winske, D. (1986). The development of shell-like distributions from newborn cometary ions, *Geophys. Res. Lett.*, **13**, 1364–1367.

126. Gary, S. P., & Madland, C. D. (1988). Electromagnetic ion instabilities in a cometary environment, *J. Geophys. Res.*, **93**, 235–241.

127. Florinski, V., Zank, G. P., Heerikhuisen, J., Hu, Q., & Khazanov, I. (2010). Stability of a pickup ion ring-beam population in the outer heliosheath: implications for the IBEX Ribbon, *Astrophys. J.*, **719**, 1097–1103.

128. Summerlin, E. J., Viñas, A. F., Moore, T. E., Christian, E. R., & Cooper, J. F. (2014). On the stability of pick-up ion ring distributions in the outer heliosheath, *Astrophys. J.*, **793**:93.

129. Florinski, V., Heerikhuisen, J., Niemiec, J., & Ernst, A. (2016). The IBEX ribbon and the pickup ion ring stability in the outer heliosheath. I. Theory and hybrid simulations, *Astrophys. J.*, **826**:197.

130. Fuselier, S. A., Allegrini, F., Funsten, H. O. *et al.* (2009b). Width and variation of the ENA flux ribbon observed by the Interstellar Boundary Explorer, *Science*, **326**, 962–964. 10.1126/science.1180981.

131. Burlaga, L. F., & Ness, N. F. (2014). Interstellar magnetic fields observed by Voyager 1 beyond the heliopause, *Astrophys. J Lett.*, **795**:L19.

132. Matloff, G. L. (2005). *Deep Space Probes: To the Outer Solar System and Beyond*. (Springer Praxis Books. ISBN 978-3540247722.)

133. Burlaga, L. F., Florinski, V., & Ness, N. F. (2015). In situ observations of magnetic turbulence in the local interstellar medium, *Astrophys. J Lett.*, **804**:L31.

134. McComas, D. J., Christian, E. R., Schwadron, N. A. *et al.* (2018). Interstellar Mapping and Acceleration Probe (IMAP): a new NASA mission, *Space Sci. Rev.*, **214**:116.

135. Isenberg, P. A., Forbes, T. G., & Möbius, E. (2015). Draping of the interstellar magnetic field over the heliopause: a passive field model, *Astrophys. J.*, **805**:153.

136. Schwadron, N. A., Adams, F. C., Christian, E. R. *et al.* (2014). Global anisotropies in TeV cosmic rays related to the Sun's local galactic environment from IBEX, *Science*, **343**, 988–990.

137. Frisch, P.C., Berdyugin, A., Piirola, V. *et al.* (2015). Charting the interstellar magnetic field causing the Interstellar Boundary Explorer (IBEX) ribbon of energetic neutral atoms, *Astrophys. J.*, **814**:112.

138. Burlaga, L. F., Ness, N. F., & Stone, E. C. (2013). Magnetic field observations of Voyager 1 entered the heliosheath depletion region, *Science*, **341**, 147–150.

139. Opher, M., & Drake, J. F. (2013). On the rotation of the magnetic field across the heliopause, *Astrophys. J. Lett.*, **778**:L26.

140. Gloeckler, G., & Fisk, L. A. (2014). A test for whether or not Voyager 1 has crossed the heliopause, *Geophys. Res. Lett.*, **41**, 5325–5330.

141. Schwadron, N. A., Richardson, J. D., Burlaga, L. F., McComas, D. J., & Moebius, E. (2015). Triangulation of the interstellar magnetic field, *Astrophys. J. Lett.*, **813**:L20.

142. Rankin, J. S., Stone, E. C., Cummings, A. C., McComas, D. J., Lal, N., & Heikkila, B. C. (2019). Galactic cosmic-ray anisotropies: Voyager 1 in the local interstellar medium, *Astrophys. J.*, **873**:46.

143. Schwadron, N. A., & McComas, D. J. (2017). Effects of solar activity on the local interstellar magnetic field observed by Voyager 1 and IBEX, *Astrophys. J.*, **849**:135.

144. Wallis, M. K. (1973). Interaction between the interstellar medium and the solar wind plasma, *Astrophys. Space Sci.*, **20**, 3–18.

145. Baranov, V. B., Lebedev, M. G., & Ruderman, M. S. (1979). Structure of the region of solar wind-interstellar medium interaction and its influence on H atoms penetrating the solar wind, *Astrophys. Space Sci.*, **66**, 441–451.

146. Ripken, H. W., & Fahr, H. J. (1983). Modification of the local interstellar gas properties in the heliospheric interface, *Astron. Astrophys.*, **122**, 182–191.

147. Ripken, H. W., & Fahr, H. J. (1984). The physics of the heliospheric interface and its implications for LISM diagnostics, *Astron. Astrophys.*, **139**, 551–554.

148. Alexashov, D., & Izmodenov, V. (2005). Kinetic *vs.* multi-fluid models of H atoms in the heliospheric interface: a comparison, *Astron. Astrophys.*, **439**, 1171–1181.

149. Izmodenov, V. V. (2009). Local interstellar parameters as they are inferred from analysis of observations inside the heliosphere, *Space Sci. Rev.*, **143**, 139–150.

150. Malama, Yu. G., Izmodenov, V. V., & Chalov, S. V. (2006). Modeling of the heliospheric interface: multi-component nature of the heliospheric plasma, *Astron. Astrophys.*, **445**, 693–701.

151. Müller, H.-R., Florinski, V., Heerikhuisen, J. *et al.* (2008). Comparing various multi-component global heliosphere models, *Astron. Astrophys.*, **491**, 43–51.

152. Zank, G. P., Pogorelov, N. V., Heerikhuisen, J. *et al.* (2009). Physics of the solar wind–local interstellar medium interaction: role of magnetic fields, *Space Sci. Rev.*, **146**, 295–327.

153. Park, J., Kucharek, H., Möbius, E. *et al.* (2015). Statistical analysis of the heavy neutral atoms measured by IBEX, *Astrophys. J. Suppl.*, **220**:34.

154. Gehrels, N. (1986). Confidence limits for small numbers of events in astrophysical data, *Astrophys. J.*, **303**, 336–346.

155. Tan, P.-N., Steinbach, M., Karpane, A., & Kumar, V., *Introduction to Data Mining* (2nd ed.), Ch. 7 (Pearson Education, New-York, NY, 2019).

156. Park, J., Kucharek, H., Möbius, E. *et al.* (2016). IBEX observations of secondary interstellar helium and oxygen distributions, *Astrophys. J.*, **833**:130.

157. Bzowski, M., Kubiak, M. A., Czechowski, A., Grygorczuk, J. (2017). The helium warm breeze in IBEX observations as a result of charge-exchange collisions in the outer heliosheath, *Astrophys. J.*, **845**:15.

158. Sokòl, J. M., Kubiak, M. A., Bzowski, M., Möbius, E., & Schwadron, N. A. (2019). Science opportunities from observations of the interstellar neutral gas with adjustable boresight direction, *Astrophys. J. Suppl.*, **245**:28.

159. Izmodenov, V. V., Alexashov, D. B., & Myasnikov, A. V. (2005). Direction of the interstellar H atom inflow in the heliosphere: role of the interstellar magnetic field, *Astron. Astrophys.*, **437**, L35–L38.

160. Lallement, R., Quemerais, E., Koutroumpa, D. *et al.* (2010). The interstellar H flow: updated analysis of SOHO/SWAN data, in *12th International Solar Wind Conf.*, Eds. Maksimeric, M., Meyer-Vernet, N., Moncuquet, M., & Panetellini, F., *AIP Conf. Proc.*, **1216**, 555–558.

161. Schwadron, N. A., Möbius, E., McComas, D. J. *et al.* (2016). Determination of interstellar O parameters using the first two years of data from the Interstellar Boundary Explorer, *Astrophys. J.*, **828**(81).

162. Spence, H. E., Case, A. W., Golightly, M. J. *et al.* (2010). CRaTER: The cosmic ray telescope for the effects of radiation experiment on the Lunar Reconnaissance Orbiter mission, *Space Sci. Rev.*, **150**, 243–284.

163. Fuselier, S. A., Allegrini, F., Bzowski, M. *et al.* (2012). Heliospheric neutral atom spectra between 0.01 and 6 keV from IBEX, *Astrophys. J.*, **754**:14.

164. Galli, A., Wurz, P., Schwadron, N. A. *et al.* (2016). The roll-over of heliospheric neutral hydrogen below 100 eV: observations and implications, *Astrophys. J.*, **821**:107.

165. Hsieh, K. C., Giacalone, J., Czechowski, A. *et al.* (2010). Thickness of the heliosheath, return of the pick-up ions, and Voyager 1's crossing the heliopause, *Astrophys. J. Lett.*, **718**, L185–L188.

166. Giacalone, J., & Decker, R. (2010). The origin of low-energy anomalous cosmic rays at the solar-wind termination shock, *Astrophys. J.*, **710**, 91–96.

167. Krimigis, S. M., Mitchell, D. G., Roelof, E. C., Hsieh, K. C., & McComas, D. J. (2009). Imaging the interaction of the heliosphere with the interstellar medium from Saturn with Cassini, *Science*, **326**, 971–973.

168. Möbius, E. (2009). From the heliosphere to the local bubble — what have we learned? *Space Sci. Rev.*, **143**, 465–473.
169. Fuselier, S. A., Dayeh, M. A., & Möbius, E. (2018). The IBEX ribbon and the thickness of the inner heliosheath, *Astrophys. J.*, **861**:109.
170. McComas, D. J., & Schwadron, N. A. (2014). Plasma flows at Voyager 2 away from the measured suprathermal pressures, *Astrophys. J. Lett.*, **795**:L17.
171. Schwadron, N. A., & Bzowski, M. (2018). The heliosphere is not round, *Astrophys. J.*, **862**:11.
172. Desai, M. I., Allegrini, F., Dayeh, M. A. *et al.* (2015). Latitudinal and energy dependence of energetic neutral atom spectral indices measured by the Interstellar Boundary Explorer, *Astrophys. J.*, **802**:100.
173. Desai, M. I., Dayeh, M. A., Allegrini, F. *et al.* (2016). Latitude, energy, and time variations in the energetic neutral atom spectral indices measured by the Interstellar Boundary Explorer (IBEX), *Astrophys. J.*, **832**, 116.
174. Livadiotis, G., McComas, D. J., Randol, B. M. *et al.* (2012). Pick-up ion distributions and their influence on ENA spectral curvature, *Astrophys. J.*, **751**:64.

Chapter 8

What Have ENAs Revealed
and What is Next?

"Unanticipated novelty, the new discovery, can emerge only to the extent that his anticipations about nature and his instruments prove wrong."

From *The Structure of Scientific Revolutions*
Thomas S. Kuhn, 1962

Although one of the youngest tools in the portfolio of space plasma physics, *neutral-atom astronomy* has brought about an impressive array of successes. For the Earth's magnetosphere, ENAs have given us a global view of the ring current, magnetosheath, and tail, as well as a top view of aurorae, along with their spatial and temporal evolution. They have also shown us the lunar interaction with the SW. Carrying the method to its limit, HSTOF on *SOHO* detected heliospheric ENAs, followed by neutral-atom cameras on *Cassini* and, most notably, on *IBEX* that returned the first maps of the heliospheric boundary. Also, *Ulysses* GAS and later *IBEX*-Lo have enabled *in-situ* sampling of the ISN. Out of these, *IBEX* is likely to continue its mission into the new era of *IMAP*.

As a surprise, *IBEX* has delivered the completely unexpected discovery of the ENA Ribbon, which provides a compass for the ISMF outside the solar system and suggests equally strong interstellar ram and magnetic pressure on the heliosphere. Secondary IS neutrals are now used as a probe of the IS plasma flow around the HP, and ENAs from the IHS provide an image of the pressure distribution in the boundary regions. This chapter will outline where ENA imaging may soon lead to further studies of these key objects and domains of heliophysics, in concert with further refined *in-situ* observations and increasingly sophisticated simulations.

8.1. Understanding the Cross-Scale Coupling of the Magnetospheric Systems

IMAGE, TWINS, and to some extent, even the heliospheric imager *IBEX* have demonstrated that ENA imaging can provide the much-needed global view of the dynamically interacting magnetospheric regions. Multi-spacecraft missions, equipped with comprehensive high time and spatial resolution *in-situ* plasma-physics instrumentation on missions, such as *Cluster, Themis*, and *MMS*, have enabled unraveling the small-scale dynamics of essential plasma-physics processes, like reconnection and particle acceleration. Together with increasingly sophisticated MHD, Hybrid, and Particle-in-Cell numerical simulations, observations from the widely distributed S/C collectively referred to as the *Heliophysics Systems Observatory* are starting to provide some holistic insight into the dynamics of the entire magnetospheric system and its response to the varying solar and interplanetary drivers (CMEs and SW interaction regions).

At this juncture, it should come as no surprise that the *2013 Heliophysics Decadal Survey* [1] recommends, as the highest priority magnetospheric Science Target, a *Magnetosphere Energetics, Dynamics, and Ionospheric Coupling Investigation* (*MEDICI*) mission. It intimately combines detailed *in-situ* measurements with stereoscopic ENA imaging over the entire energy range of the magnetospheric ion populations and multi-spectral UV imaging. This mission includes a wide array of low-altitude and ground-based observations to tie the images to the ENA and UV source regions, either as parts of this mission or by partnering with other observation campaigns.

The key improvements of the ENA instrumentation for a *MEDICI*-like mission over previous programs are greatly enhanced angular resolution to $\approx 2°$ and much-enlarged collecting power to facilitate high time resolution. Their implementation will result in high spatial resolution of the ENA source regions from a moving S/C.

8.2. Surveying Exposed Surfaces in the Solar System: Planets to Dust

Concerning atoms ejected from largely airless surfaces in the solar system, thus far, only ENAs from our Moon have been observed. Together with PUI studies, we start to gain insight into the interaction of the SW with the lunar surface (Sec. 6.4.1). More recently, ENA sensors on the lunar orbiter *Chandrayaan-1* and rover *Yutu-2* (Sec. 5.3.3) provide new data. These achievements may serve as a template for similar investigations of surfaces of airless planets like Mercury, other moons, and asteroids.

At the time of this writing, the *ESA* S/C *Bepi Colombo* is *en route* to Mercury. This mission's important science targets are Mercury's exosphere and the SW interaction with the Hermean surface, thereby evaluating the surface composition. An ELENA sensor will probe the altitude distribution of the exosphere for the total density and STROFIO for its composition (Sec. 5.3.3). Since this S/C has neutral atom and ion sensors, it will provide a unique view of the near-Mercury environment and surface itself. Consequently, such an instrument package may be high on the priority list for missions to the icy moons, asteroids, and Kuiper Belt objects.

At the microscopic end of the size spectrum of solar-system objects are IP dust grains, whose importance we should not underestimate. When comets and asteroids suffer close encounters with the Sun, they end up as tiny debris to join the IP dust population [2, 3]. Inner source PUIs were found in the 1990s, whose source is most likely the IP dust (Sec. 6.3.2). However, to explain the total flux of these PUIs, the majority of the dust grains would have to be of nanometer size, which then likely also affect the SW dynamics close to the Sun [4]. There have been potential observations of *nano* dust [5], but significant questions remain [6, 7]. Besides PUIs, such a dust population would also reveal itself *via* a substantial neutral SW component in the inner heliosphere, which would eventually be diagnosed directly. Only a future mission with the combination of a dedicated *nano* dust instrument [8], neutral SW sensor, and PUI sensor, along with SW and IMF monitors, may unravel this mystery. The ENA sensor must reject SW and photons effectively to observe the neutral SW and enable the unambiguous detection and analysis of the ENAs of dust origin.

8.3. Understanding the Sun-ISM Interactions and Particle Acceleration on Large Scales

Six decades after the dawn of the Space Age and five decades after the first human set foot on the Moon, we are sticking our head out of the heliosphere. The two *Voyagers* venture into the ISM, and simultaneously *IBEX* provides all-sky maps of its boundary and samples the IS wind. Together with a broad modeling effort, this exploration has given us a first deep look into the surrounding ISM and its interaction with the heliosphere. Still, it also has thrown fascinating new questions at us. Consequently, a much more comprehensive mission built on *IBEX*, the *IMAP* mission emerged as the top priority in the *2013 Heliophysics Decadal Survey* [1], and is now being implemented by *NASA* while the *Voyagers* still function and the anticipated observations meet up with the timeline of the returning ENAs.

IMAP, with its comprehensive array of ENA imagers and *in-situ* instruments, addresses four key objectives [9]:

1. improve understanding of the composition and properties of the ISM;
2. advance understanding of the temporal and spatial evolution of the boundary region in which the SW and ISM interact;
3. identify and advance the understanding of processes related to the interactions of the Sun's magnetic field and ISM;
4. identify and advance the understanding of particle injection and acceleration processes near the Sun, in the heliosphere, the IHS, and OHS.

Let us briefly highlight how IMAP will achieve these goals through its mission concept and ENA imaging suite. First and foremost, *IMAP* will be at L1 like *SOHO*, where the ENA imaging instruments can operate with a substantially lower background coming from the magnetosphere and thus at a substantially increased duty cycle.

The ENA suite continuously covers the 10 eV–300 keV energy range with substantial overlap, *i.e.*, the entire range of heliospheric ANAs and ENAs. The angular resolution is increased to 4° FWHM at energies above a few 100 eV and to 2° above 3 keV to resolve the critical features in the *IBEX* Ribbon and INCA Belt unambiguously (Secs. 7.3 and 7.5.2). The *IMAP*-Lo boresight relative to the Sun is adjustable from 60° to 165° on a pivot platform to track the ISN flow over most of the orbit, breaking the degeneracy of the ISN parameter tube (Sec. 7.2.2). This capability also improves the ISN composition measurements (Sec. 7.2.3) and provides a tomographic view of the secondary neutrals (Sec. 7.4). Additionally, the geometric factor will increase by three- to tenfold. *IMAP* includes a full suite of SW, PUI, energetic-particle, and magnetic field instruments to probe the particle populations *en route* to the outer heliosphere and their acceleration processes, along with a Ly-α imager and an IS dust instrument. The mission will enable a leap in understanding our heliospheric home and its galactic neighborhood.

Let us conclude with the ultimate dream, *i.e.*, a foray into the ISM proper. An *interstellar probe* (*ISP*) has been on the wish list of the heliophysics community since the 1970s. The two *Voyagers* can be considered the first, but they lack adequate instrumentation to probe the IS plasma and neutrals. Most importantly, they will stop their operation long before reaching the pristine ISM (Sec. 1.3). It was always clear that any *ISP* must include instrumentation that can return comprehensive information on the distribution and composition of the charged and neutral components of the ISM. Based on the *IBEX* findings, we should send the *ISP* through the Ribbon region, which the *Voyagers* will miss, to identify *in-situ* the physical processes that lead to the unexpected stability of the Ribbon particle populations (Sec. 7.3.2). It will also be important that the *ISP* payload includes an ENA imager, which will scan the plane perpendicular to the escape trajectory from the heliosphere. Such an *ISP*, collaborating with simultaneous imaging from 1 AU with *IMAP* or a similar mission, would provide the full 3D distribution of the ENA emission regions and thus the global structure of the critical heliospheric boundary regions.

8.4. Bridging the Temporal and Spatial Divide

We would like to highlight two discoveries separated by five decades in time and 10^2 AU in distance to bring our story to a close. They share similar roles that ENAs have played: the injection into the upper atmosphere of protons stemming from Earth's outer radiation belt (Sec. 2.3, Fig. 2.6) and the most likely production of the ENA Ribbon by neutral SW *via* PUIs in the OHS (Sec. 7.3, Fig. 7.9). Both discoveries involve ions trapped in a magnetic field near a 90° pitch angle. They become ENAs by another CX and reach the remotely monitoring S/C, either indirectly as local protons in the inner magnetosphere or directly as ENAs in the inner heliosphere. Had the Earth-orbiting S/C in the 1960s carried ENA sensors, the two discoveries would have been entirely analogous, differing only by the regions where the ENAs were formed and detected.

The still-unfolding story of investigating space plasmas beyond the reach of S/C with ENAs has come a full circle, and we hope that it will inform and inspire our readers to continue the endeavor, as it has undoubtedly enlivened the two of us.

References

1. Baker, D. N., Charo, A., & Zurbuchen, T. (2013). *Solar and Space Physics: A Science for a Technological Society* (National Academy Press, Washington, D.C.).
2. Grün, E., Zook, H.A., Fechtig, H., & Giese, R.H. (1985). Mass input into and output from the meteoritic complex, *Properties and Interactions of Interplanetary Dust, Prec.* Eds. Giese R. H. and Lamy, P., *IAU Coll.*, **85**, 411–415. Doi: 10.1007/978-94-009-5464-9_78
3. Leinert, Ch., & Grün, E. (1990). Interplanetary dust, Ch. 5 in *Physics of the Inner Heliosphere I*, eds. Schwenn R. & Marsc, E. 207–275 (Springer Verlag Berlin).
4. Schwadron, N. A., Geiss, J., Fisk, L. A. *et al.* (2000). Inner source distributions: theoretical interpretation, implications, and evidence for inner source protons, *J. Geophys. Res.* **105**(A4), 7465–7472. Doi: 10.1029/1999JA000225.
5. Meyer-Vernet, N., Maksimovic, M., Czechowski, A. *et al.* (2009). Dust Detection by the Wave Instrument on STEREO: Nanoparticles Picked up by the Solar Wind? *Solar Physics*, **256**(1–2), 463–474. Doi: 10.1007/s11207-009-9349-2.

6. Zaslavsky, A., Meyer-Vernet, N., Mann, I. *et al.* (2012). Interplanetary dust detection by radio antennas: mass calibration and fluxes measured by STEREO/WAVES, *J. Geophys. Res.*, **117**(A5), A05102. Doi: 10.1029/2011JA017480.
7. Juhász, A., & Horányi, M. (2013). Dynamics and distribution of nano-dust particles in the inner solar system, *Geophys. Res. Lett.*, **40**(11), 2500–2504. Doi: 10.1002/grl.50535.
8. O'Brien, L., Auer, S., Gemer, A. *et al.* (2014). Development of the nano-dust analyzer (NDA) for detection and compositional analysis of nanometer-size dust particles originating in the inner heliosphere, *Rev. Sci. Instr.*, **85**(3), 035113. Doi: https://doi.org/10.1063/1.4868506.
9. McComas, D. J., Christian, E. R., Schwadron, N. A. *et al.* (2018). Interstellar Mapping and Acceleration Probe (IMAP): A New NASA Mission, *Space Sci. Rev.*, **214**, 116. Doi: 10.1007/s11214-018-0550-1.

Appendix A

Geometrical Factor and Angular Response

A.1. Geometrical Factor

Figure A.1 illustrates the geometric factor G of a pair of parallel entrance and exit windows (forming a collimator or an aperture-detector pair), which determines the sensor's capability to collect particles within its FOV, regardless of its detection efficiency $\eta(E)$. Together, with the sensor's energy bandwidth, ΔE, $G \cdot \eta(E) \cdot \Delta E$ defines the *total collecting power*, where E is the central energy of the respective energy band. The central concept for G, especially for sensors with a wide FOV, is *projection*, which defines the portion of the incident flux passing the aperture at an angle ξ according to

$$B = \vec{A} \cdot \hat{r} = A \cos \xi \tag{A.1}$$

$\vec{A}$ is the area vector and $\hat{r}$ the unit vector along $\vec{r}$ (Fig. A.1, left).

Figure A.1 (right) shows a perspective view of two co-axial parallel windows. For convenience, we use circles of radius R_1 and R_2. An area element dA_2 in the exit plane is illuminated through the solid angle element $d\Omega$ subtended by $d\vec{A}_1 \cdot \hat{r}$ at distance r, thus $d\Omega \equiv (d\vec{A}_1 \cdot \hat{r})/r^2$. Likewise, the *projected* area of dA_2 that accepts particles arriving along the vector $\vec{r}$ is $d\vec{A}_2 \cdot \hat{r}$. G is now a double integral

$$G \equiv \iint d\Omega dA = \iint \frac{(d\vec{A}_1 \cdot \hat{r})(d\vec{A}_2 \cdot \hat{r})}{r^2} \tag{A.2}$$

Equation (A.2) holds for two surfaces of any shape. Any symmetry greatly simplifies the computation. In practice, there is no need to

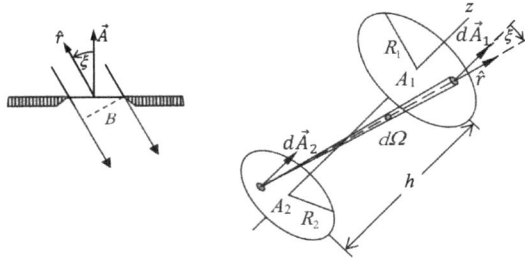

Fig. A.1. Viewing geometry of a sensor. Left: An entrance area A limits the particle flux incident at the angle ξ to a cross-sectional area B, the projection of A in the direction of incidence. Right: Ray geometry through two co-axial circular windows separated by distance h.

deviate from axial symmetry for sensors. Hence, we only consider such cases. Here, we refer to three general cases originally derived for two-detector coincidence cosmic-ray telescopes [1]: 1) two rectangular co-axial windows, 2) a single window of area A exposed to a hemisphere above, which we will consider shortly, and 3) two circular co-axial windows (Fig. A.1, right). For 3), Eq. (A.2) yields

$$G = \frac{1}{2}\pi^2\{R_1^2 + R_2^2 + h^2 - [(R_1^2 + R_2^2 + h^2)^2 - 4R_1^2 R_2^2]^{1/2}\} \quad (A.3)$$

With $A = \pi R^2$, it is evident that Eq. (A.3) falls between two convenient limits:

$$\frac{A_1 A_2}{R_1^2 + R_2^2 + h^2} < G < \frac{A_1 A_2}{h^2} = \Omega A_2 \quad (A.3a)$$

For sensors with high angular resolution ($A \ll h^2$) that are suitable for ENA imaging, the estimate on the right-hand side is adequate.

A.2. Angular Response

For axially symmetric sensors (Fig. A.1, right), the angular response $T(\xi, \phi)$ is defined by the FOV in ξ and ϕ. Thus, we rewrite Eq. (A.2) as

$$G = A \iint T(\xi, \phi) \sin \xi \, d\xi d\phi \quad (A.4)$$

A is the maximum exposable area of A_2 at $\xi = 0°$, ξ the polar angle measured from the z-axis, ϕ the azimuthal angle in the aperture plane, and $\sin\xi d\xi d\phi = d\Omega$ the differential solid angle. The *angular response* $T(\xi, \phi)$ is the portion of A_2 *(exit aperture or detector)* exposed to the incident flux through A_1 *(entrance aperture)* at the *incident angle* (ξ, ϕ), or the *projection* of A_1 onto A_2 at (ξ, ϕ), normalized to that at $(\xi, \phi) = (0°, 0°)$. For a given ϕ or in a rotationally symmetric case of two circular apertures, $T(\xi)$ suffices.

For a circular aperture of area A, exposed to the sky above, Eq. (A.1) yields $T(\xi) = \cos\xi$ with the integration limits $[0, \pi/2]$ for θ and $[0, 2\pi]$ for ϕ, then:

$$G = A \iint T(\xi, \phi) \sin\xi d\xi d\phi = A \iint \cos\xi \sin\xi d\xi d\phi = \pi A$$

(A.4a)

only half of the expected upper limit $\Omega A = 2\pi A$ in Eq. (A.3a).

$T(\xi, \phi)$ does not explicitly appear in Eq. (A.2) but is part of the integral in G and thus is crucial in data interpretation (Sec. A.2.2). Therefore, we show how to obtain the *angular resolution* $\Delta\xi$ and $T(\xi, \phi)$. $\Delta\xi$ is a valid angular resolution only if $T(\xi, \phi)$ is constant in ϕ or when restricted to a single plane of fixed ϕ with 1D symmetry in that plane. Figure A.2 (left) shows such a cut with a pair of parallel windows sharing the central z-axis orthogonal to both. The second window may be a detector. h is the separation of the windows along z. w_1 is the width of the entrance, and w_2 is that of the collimator exit or detector. These parameters determine the *limiting half-angle* ξ_m and the FOV in this plane. As detailed in the caption, Fig. A.2 (left) shows two cases: $w_2 = 2w_1$ in red and $w_2 = w_1$ in blue. Figure A.2 (right) shows the resulting $T(\xi)$ for the two cases, an isosceles triangle for $w_2 = w_1$ (blue circles), which peaks at $\xi = 0°$, where A_2 is exposed completely; thus $\xi_{FWHM} = \xi_m$. For $w_2 = 2w_1$ (red triangles), the entire entrance A_1 illuminates A_2 for the center range in ξ; hence the plateau modified by $\cos\xi$. ξ_{FWHM} is now about twice that for $w_2 = w_1$, but ξ_m is up by only $\approx 50\%$. G is less than doubled because of the projection effect. With appropriate combinations of w_1, w_2,

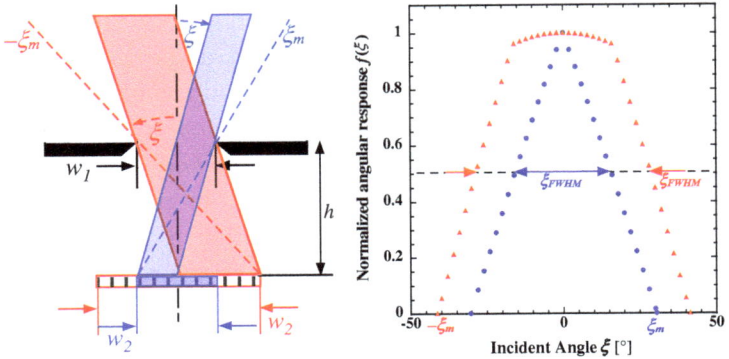

Fig. A.2. Schematic view for two cases of an entrance-detector pair in the x-z plane, separated by height h (left) and the resulting angular response $T(\xi)$ as a function of incident angle ξ (right). In one case, the detector width is $w_2 = w_1$ (blue) and in the other $w_2 = 2w_1$ (red). The throughput at an angle ξ is shown as a transparent parallelogram in blue for $w_2 = w_1$ and in red for $w_2 = 2w_1$. They are limited by one edge of the entrance and the opposite edge of the detector. The former (blue) is at about half that of the latter (red). The extreme trajectories at the limiting half-angle $\pm\xi_m$ end at the respective opposite edge of the detector. Due to symmetry about the z-axis showing one-half of each case is sufficient. The right panel shows the angular response $T(\xi) = (w_2/2 - x(\xi)) \cdot \cos\xi/w_2$ or throughput as a function of ξ (blue circles for $w_2 = w_1$ and red triangles for $w_2 = 2w_1$), along with the FWHM angles ξ_{FWHM} and the liming half angles ξ_m. For $w_2 = w_1$, the FWHM is almost equal to ξ_m, and for $w_2 = 2w_1$, the width of the flattop is $\approx \xi_m$.

and h, two sensors of identical ξ_m or the same FOV can have a different FWHM or angular resolution in that plane.

The FWHM of a sensor's angular response $T(\xi)$ is its angular resolution $\Delta\xi$, *i.e.*, the *minimum angular separation* a sensor can resolve. Figure A.3 illustrates the meaning of a widely used criterion for resolution for the case $w_2 = w_1$, which is typical for a collimator with identical entrance and exit apertures (*e.g.*, *IBEX*, see Sec. 5.3.2 and Fig. 5.15).

For a point source, the image of the flux taken by scanning the sensor FOV across has the shape $T(\xi)$, also referred to as the *point-spread function*. When two point-sources of equal brightness are separated by angles smaller than ξ_{FWHM} (Δ in Fig. A.3), superposition produces a single peak centered between the two

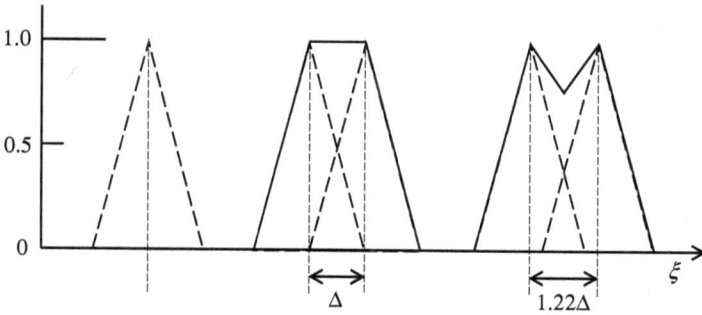

Fig. A.3. Schematics of a criterion for resolving point sources. The angular response $T(\xi)$ to a single point source or the point-spread function is on the left. The center shows an image for two point-sources of equal brightness separated by the angle $\Delta = \xi_{FWHM}$ (the solid line represents the sum of the two). On the right is the result for an angular separation $>\Delta$.

sources. As the angular separation increases, the central peak shrinks until flattened when the separation reaches ξ_{FWHM}, with a dip emerging in the center for even larger separations. We refer to the *critical point* between forming a plateau and indicating a dip as a reasonable criterion for resolving two sources. This criterion is less stringent than the astronomers' criterion for resolving two point-sources with Fresnel diffraction (1.22Δ in Fig. A.3).

For a circular aperture (Fig. A.1, right), although rotationally symmetric, the exposed area in the exit (or detector) plane at an arbitrary incident angle ξ is bounded by two arcs, making the determination of $T(\xi)$ and the important FWHM more involved [2]. For ENA imagers that use a collimator to control the FOV, rectangular or hexagonal grids that provide a 2D pattern without gaps or overlaps are preferred. A collimator with hexagonal holes, as used for the *IBEX* sensors (Sec. 5.3.2 and Sec. A.2.1), provides the optimal compromise between high transparency (or geometric factor) of the collimator and even angular resolution over ϕ. The hexagon is the polygon with the highest number of corners that provides a closed-pattern coverage of a plane. Similarly, a collimator with identical entrance and exit holes ($w_2 = w_1$) provides the optimal compromise between absolute angular resolution and geometric factor, as indicated in the comparison of the two cases in Fig. A.2.

A.2.1. *Example of a calibrated angular response*

During implementation, the sensor's G and $T(\xi)$ of a collimator or aperture and detector system must be verified by a traceable simulation and calibration. Figure A.4 shows results of a simulation of the angular response across the base and the corners of the hexagon apertures (left) and calibration for the particle response and leakage (right) with the *IBEX*-Lo collimator, along with the hexagon geometry and a photo of a collimator portion as insets [3].

The simulated angular response is based on the hexagon width w and collimator height h, as verified for the as-built collimator in an optical comparator. The simulation uses the illuminated area at the collimator exit according to Fig. A.2 (left). Together with the optical verification, the simulation result reflects the collimator response more precisely than any calibration at an accelerator, as shown in Fig. A.4 (right) for a 20-keV Ar^+ beam [3]. Note that the yaw angle features an offset from the beam axis. The angle dependence is close to a Gaussian because the observed distribution is a convolution of the collimator point-spread function and the approximately Gaussian beam profile. More importantly, the value of the beam calibration lies in the ability to verify the required suppression of off-axis leakage

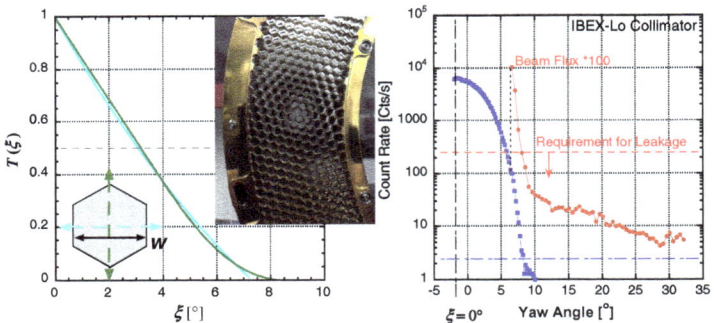

Fig. A.4. Left: Simulated angular response $T(\xi)$ across the base (cyan) and corners (green) of the hexagon with width w, as shown in the inset and the photo. Right: Response of the *IBEX*-Lo collimator to a 20 keV Ar^+ beam as a function of yaw angle of the calibration system around $\xi = 0°$ (blue) and at large angles (red). The beam flux was increased by a factor of 100 to test for leakage and scattering [3]. (Right panel: © Springer. Reproduced with permission).

between neighboring collimator channels and that of scattered particles. These tasks require calibration with an actual particle beam. Appx. B describes the calibration system used.

A.2.2. *Ignoring the angular response, a potential problem*

In conjunction with the TS crossing of *Voyager 2*, the special section of *Nature* (**454**, 2008) also reported the detection of ENAs coming from the frontal lobe of the IHS by the most advanced electron detector STE [4] on the twin spacecraft *STEREO* orbiting the Sun at 1 AU [5].

The reported isosceles-triangular distribution of the flux in the ecliptic plane peaking at about 8° away from the ISN flow direction and the high flux were both perplexing. However, a careful analysis of

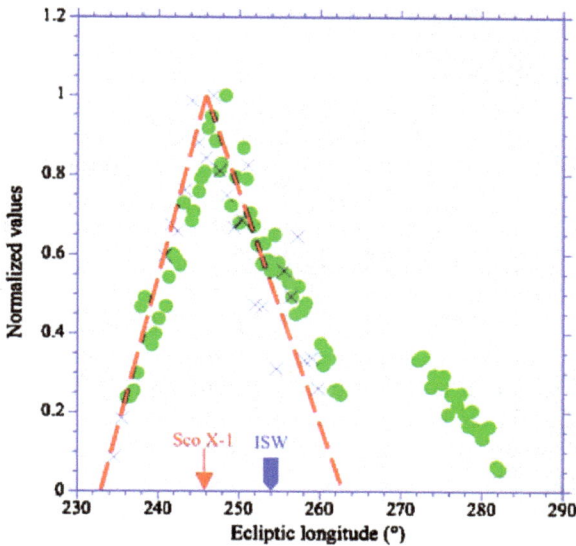

Fig. A.5. Angular distribution of X-rays from Sco X-1 mistaken for ENAs from the ISN flow or the interstellar wind direction [6]. The green dots represent the normalized STE signal. The red dashed line indicates the result of a scan with the angular response of STE across the point source Sco X-1. The blue crosses are normalized X-ray daily averaged intensities based on *Swift*/BAT [7] as Sco X-1 transits the STE FOV over the days of the reported observation. (© AAS. Reproduced with permission).

the STE angular response and the direction of the peak flux suggested that STE possibly detected X-rays from the strong source Sco X-1. Comparing the reported differential energy spectrum with the Sco X-1 X-ray spectrum and the STE point-spread function with the observed angular profile confirmed that STE did not detect ENAs, but X-rays from Sco X-1 (Fig. A.5) [6]. This episode issues a two-point warning.

- Photons and ENAs are indistinguishable unless ENAs are identified *via* coincidence measurements.
- For directional information, the angular response $T(\xi)$ of the sensor must not be ignored.

References

1. Sullivan, J. D. (1971). Geometrical factor and directional response of angle and multi-element particle telescopes, *Nucl. Instr. & Meth.*, **96**, 5–11.
2. Thomas, G. R., & Willis, D. M. (1972). Analytical derivation of the geometrical factor of a particle detector having circular or rectangular geometry, *J. Phys. E: Sci. Instr.*, **5**, 260–263.
3. Fuselier, S. A., Bochsler, P., Chornay, D. *et al.* (2009). IBEX-Lo sensor, *Space Sci. Rev.*, **146**, 117–147.
4. Lin, R. P., Curtis, D. W., Larsen, D. E. *et al.* (2008). The STEREO IMPACT Suprathermal Electron (STE) Instrument, *Space Sci. Rev.*, **136**(1–4), 241–255.
5. Wang, L., Lin, R. P., Larson, D. E., & Luhmann, J. G. (2008). Domination of heliosheath pressure by shock-accelerated pickup ions from observation of neutral atoms, *Nature*, **454**(7), 81–83.
6. Hsieh, K. C., Frisch, P. C., Giacalone, J. *et al.* (2009). A re-interpretation of the STEREO/STE observations and its consequences, *Astrophys. J. Lett.* **694**(1), L79–L82.
7. http://heasarc.gsfc.nasa.gov/docs/swift/results/transients/.

Appendix B

ENA-Beam Calibration Facilities

B.1. Requirements and General Principles of Operation

Like all space instruments, ENA sensors must be calibrated for their angular, energy, and particle-flux responses, and their sensitivity to various ENA species, when equipped with a mass-spectrograph section. This effort requires the use of neutral-atom beam facilities that can deliver calibrated and monitored beams of choice to an ENA sensor during its calibration campaign before integration onto the S/C. Commensurate with the ENA instruments described in Chap. 5, ENA beams over an E range of 10 eV–100 keV, typically for H, D, He, O, and Ne, are required.

While ion-beam facilities have been available for a long time for applications in nuclear physics, solid-state physics, and space particle instruments, calibrated neutral-atom beams are harder to find. To obtain a neutral-atom beam with the needed E and m ranges, one starts with an ion beam, for which E and v filters are readily available. The selected ions are then converted into neutral atoms, with all remaining ions in the beam removed electrostatically. Figure B.1 shows the key subsystems schematically for an ion accelerator covering the needed E range (left panel) and two methods to turn the ion beam into a neutral-atom beam (right panel), as described in the following two sub-sections. Ions originate in a plasma discharge ion source. They are electrostatically extracted and focused, then filtered for their m (selected for their m/q or indirectly A/Q) in a magnetic v filter. Because typical ion extraction voltages

281

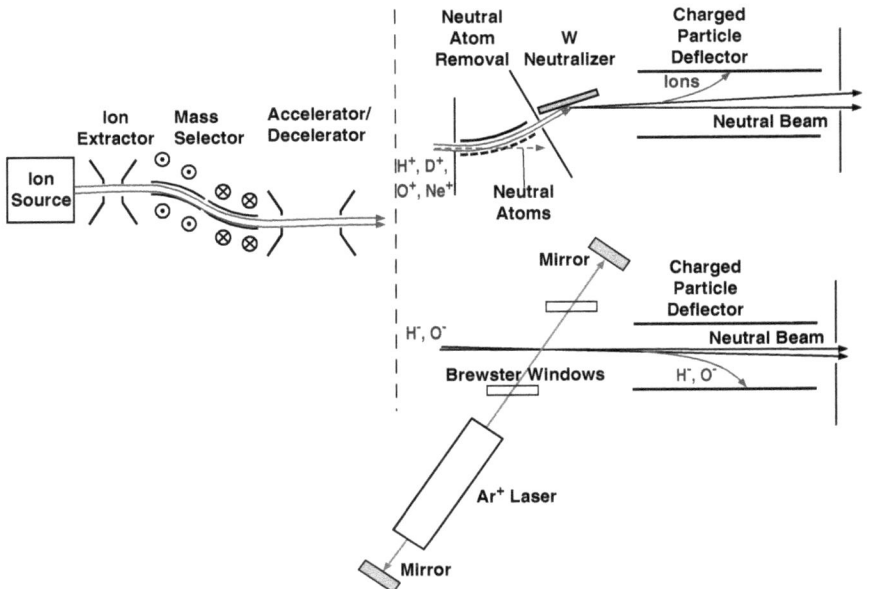

Fig. B.1.   Schematic view of neutral-atom beam facilities. Left: Ion accelerator, which provides an ion beam (positive or negative) of required m and E. Right: Conversion to a neutral-atom beam. Top: After deflection to remove neutral-atom contamination, a positive ion beam strikes a neutralizer plate. Another ion deflector cleans the emerging neutral beam of remaining ions. Bottom: Negative ions are focused into a chamber where an Ar^+ laser neutralizes them. Ion deflection removes the remaining ions.

from the source are 1–2 kV, they are either accelerated to 2–100 keV or decelerated to 10–1000 eV in a final step to meet the beam energy requirement. For the energies just mentioned, a single-stage electrostatic acceleration is usually sufficient. It is more challenging to provide a stable low-E beam, which requires deceleration. Firstly, ions emerging from the ion source typically have an energy spread of 1–3 eV, which the ion beam inherits, thus rendering measurements at the lowest energies the least precise. Second, any decelerating ion-optics diverges the beam, with challenges to provide a focused beam. Finally, low-E (<100 eV) beams are extremely susceptible to electric charges on any insulating surfaces in the beam chamber. For further information on ion-beam facilities that can serve as feeders for the neutral-atom beams, we refer the reader to two articles on widely used calibration facilities in the energy range of interest [1, 2].

B.1.1. *Neutral-atom beams through physical contact*

Turning positive ions into neutral atoms in the E range of interest often relies on ion interaction with a solid. Penetration of a thin foil or grazing incidence on a highly polished surface results in a sizeable fraction of the particles that emerge as neutrals. The latter method is more applicable to $E < 1\,\mathrm{keV}$/nucleon particles. The Q distribution of particles exiting thin C foils has been studied extensively for use in TOF spectrographs. We refer the interested reader to a review paper [3] containing an extensive list of references on the subject. These Q distributions also apply to reflection off a solid surface, typically after penetration of $<100\,\text{Å}$ into the material. Like after passing through a thin foil, the distributions only depend on the projectile's nuclear charge Z and v.

Figure B.1 (top right) shows the key subsystems for the neutral-atom beam generation *via* grazing incidence on a W surface, as used for the Time-Of-flight Calibration Facility MEFISTO at the University of Bern [2]. In this facility, another ESA first filters the $Q = +1$ ion beam for the desired E band. Simultaneously, any neutral atoms in the beam that stem from CX with the rest gas in the ion accelerator are removed. This contamination can be of the order of 1% and contains atoms with a range of energies along the acceleration/deceleration path. So, its removal is essential for a clean and calibrated neutral-atom beam. After emerging from the W surface, the beam passes another electrostatic deflector, which sweeps out any ions ($+$ and $-$) that leave the surface. This atom-beam facility has been described in detail [4]. The use of surface reflection is essential for $E < 1\,\mathrm{keV}$/nucleon because the penetration of even the thinnest C foils becomes impossible at such energies. We note here that any contact with a solid (surface reflection or foil penetration) leads to an E-and A-dependent energy loss, as characterized in detail for thin foils [3]. This energy loss can be simulated quantitatively with software tools, such as SRIM [5], and has been calibrated for the MEFISTO facility [4].

As mentioned above and shown in Fig. B.1 (top right), ions are also neutralized while passing through the rest gas. This mechanism can also generate a neutral beam instead of ions striking a solid surface, with the advantage of negligible energy loss in the process, similar to how ENAs are produced in space (Sec. 1.1). The neutral-atom beam facility at the Los Alamos National Laboratory uses this method to convert 0.45–>10 keV H^+ ions with an intrinsic width in E of ≈ 2 eV into a neutral H beam [6]. However, the efficiency of producing neutral atoms is much lower than with a conversion surface. Magnetic deflection removes remaining ions from the neutral H beam in this facility. The Appendix in the description of the *Van Allen Probes* HOPE instrument presents the Los Alamos Space Plasma Instrument Calibration Facility in detail [7].

B.1.2. *Neutral-atom beams through photo detachment*

An alternative method to the one just described is to produce a neutral-atom beam by turning negative ions into neutral atoms *via* photo detachment with an intense laser beam. Fig. B.1 (bottom right) shows this method schematically, based on descriptions for an H [8] and an O beam [9]. An electrostatic Einzel-lens focusses the negative ion beam exactly in the location where an Ar^+ laser beam (488–512 nm) crosses. The laser beam intensity is higher in the interaction region, the center of the laser resonance chamber, defined by two Brewster windows located between two highly reflective mirrors. Similar to the conversion of positive ions, an electrostatic ion deflector sweeps out any remaining ions. Because the beam does not interact with matter in this conversion method, the energy of the resulting neutral atoms is exactly that of the parent negative ions.

Operation of the ion source at rather high gas pressures (15–25 kPa) improves the negative ion production efficiency. Therefore, differential pumping in two stages is applied to achieve a vacuum better than 10^{-5} Pa in the beamline. The operation of the high-power laser system with the interaction chamber under stable conditions poses additional challenges [9].

B.1.3. *Comparison of different schemes*

As pointed out in Sec. B.1.2, the photo-detachment method produces an atom beam with precisely the energy of the original negative-ion beam, which is a substantial advantage over the conversion through contact with a surface or a thin foil, where the energy loss must be determined and calibrated. However, the production of a stable atom beam over an extended time for instrument calibration is a greater challenge for the laser method. Most importantly, the beam species are limited to elements that produce stable negative ions. Therefore, the photo-detachment method has been demonstrated so far only for H and O. It would be impossible to produce neutral He and Ne beams because these species do not have stable negative ions. The authors are not aware of any operational facility that uses the photo detachment method as of this writing. The one we described (Fig. B.1, bottom right) [8, 9] has been decommissioned.

For an in-depth discussion of ion and ENA calibration of space instruments, we refer to an extensive review [10].

References

1. Ghielmetti, A. G., Balsiger, H., Bänninger, R. *et al.* (1983). Calibration system for satellite and rocket-borne ion mass spectrometers in the energy range from 5 eV/charge to 100 keV/charge, *Rev. Sci. Instr.*, **54**(4), 425–436.
2. Marti, A., Schletti, R., Wurz, P., & Bochsler, P. (2001). New Calibration Facility for Solar Wind Plasma Instrumentation, *Rev. Sci. Instr.*, **72**(2), 1354–1360.
3. Allegrini, F., Ebert, R. W., & Funsten, H. O. (2016). Carbon foils for space plasma instrumentation, *J. Geophys. Res.*, **121**(5), 3931–3950.
4. Wieser, M., & Wurz, P. (2005). Production of a 10 eV–1000 eV neutral particle beam using surface neutralization, *Meas. Sci. Technol.*, **16**(12), 2511–2526.
5. Ziegler, J. F., Biersack J. P., & Ziegler, M. D. (2008). *SRIM the stopping range of ions in matter*, Lulu Press Co., Morrisville, NC.
6. Funsten, H. O., Allegrini, F., Bochsler, P. *et al.* (2009). The interstellar boundary explorer high energy (IBEX-Hi) neutral atom imager, *Space Sci. Rev.*, **146**(1), 75–103. Doi: 10.1007/s11214-009-9504-y.

7. Funsten, H. O., Skoug, R. M., Guthrie, A. A. *et al.* (2013). Helium, Oxygen, Proton, and Electron (HOPE) Mass Spectrometer for the Radiation Belt Storm Probes Mission, *Space Sci. Rev.*, 179, 423–484.

8. Van Zyl, B., Utterback, N. G., & Amme, R. C. (1976). Generation of a fast atomic-hydrogen beam, *Rev. Sci. Instr.*, **47**(7), 814–819.

9. Stephen, T. M., Van Zyl, B., & Amme, R. C. (1996). Generation of a fast atomic-oxygen beam from O^- ions by resonant cavity radiation, *Rev. Sci. Instr.*, **67**(4), 1478–1482.

10. Wüest, M., Evans, D. S., & von Steiger, R. (2007). Eds. *Calibration of Particle Instruments in Space Physic*, ISSI Sci. Rep. **SR-07**, Int'l Space Sci. Inst., Bern, CH.

Index